AF619046

Power Systems

Electrical power has been the technological foundation of industrial societies for many years. Although the systems designed to provide and apply electrical energy have reached a high degree of maturity, unforeseen problems are constantly encountered, necessitating the design of more efficient and reliable systems based on novel technologies. The book series Power Systems is aimed at providing detailed, accurate and sound technical information about these new developments in electrical power engineering. It includes topics on power generation, storage and transmission as well as electrical machines. The monographs and advanced textbooks in this series address researchers, lecturers, industrial engineers and senior students in electrical engineering.

Power Systems is indexed in Scopus

More information about this series at http://www.springer.com/series/4622

Morteza Nazari-Heris • Somayeh Asadi
Behnam Mohammadi-Ivatloo • Moloud Abdar
Houtan Jebelli • Milad Sadat-Mohammadi
Editors

Application of Machine Learning and Deep Learning Methods to Power System Problems

Editors
Morteza Nazari-Heris
Department of Architectural Engineering
Pennsylvania State University
University Park, PA, USA

Somayeh Asadi
Department of Architectural Engineering
Pennsylvania State University
University Park, PA, USA

Behnam Mohammadi-Ivatloo
Department of Energy Technology
Aalborg University
Aalborg, Denmark

Faculty of Electrical
and Computer Engineering
University of Tabriz
Tabriz, Iran

Moloud Abdar
Deakin University
Geelong, VIC, Australia

Houtan Jebelli
Department of Architectural Engineering
Pennsylvania State University
University Park, PA, USA

Milad Sadat-Mohammadi
Department of Architectural Engineering
Pennsylvania State University
University Park, PA, USA

ISSN 1612-1287 ISSN 1860-4676 (electronic)
Power Systems
ISBN 978-3-030-77695-4 ISBN 978-3-030-77696-1 (eBook)
https://doi.org/10.1007/978-3-030-77696-1

This Springer imprint is published by the registered company Springer Nature Switzerland AG
The registered company address is: Gewerbestrasse 11, 6330 Cham, Switzerland

Preface

Considering the modern power systems and their developments in terms of smart operation, high rate of integration of renewable energy sources and emerging technologies in power systems, the importance of the efficiency and quality indexes of power systems, data prediction and energy liberalization, the operation, plan, and control of power networks should be investigated. Additionally, new issues have been observed in the power industry considering the growing rate of load demands and increasing competition in this industry. Accordingly, the application of various machine learning and deep learning methods such as artificial neural networks (ANNs), expert systems, fuzzy systems, evolutionary-based methods, deep neural network (DNN), convolutional neural network (CNN), and long short-term memory (LSTM) has been introduced as effective methods to handle the decision-making process of modeling power systems. The use of machine learning and deep learning, which are data analysis techniques for building analytical models for a variety of subjects (e.g., energy, healthcare, bioinformatics, transportation), is a promising solution to overcome the current challenges of power systems. Machine learning and deep learning, as a part of artificial intelligence family, are very effective methods for facilitating the decision-making process of power systems operation, planning, and control by learning from the raw data, identifying patterns, and making decisions with minimum human intervention.

The advancement and development of power systems, as well as considerable challenges such as the uncertain nature of renewable energy sources such as power output of wind turbines and photovoltaic cells, load demand, and electrical energy market, requires a high-performance approach for appropriate decision-making on the operation, planning, and control of such systems. At the same time, the importance of data clustering and security of power systems highlights the need for a high-performance method to handle the operation, planning, and control of such systems. Accordingly, the use of different machine learning and deep learning methods as effective techniques with acceptable performance (high accuracy) and reliability in dealing with current issues in power systems in terms of management and operation approaches of the system and forecasting the system parameters is discussed in this book. *Application of Machine Learning and Deep Learning Methods to Power System Problems* aims to evaluate the application of machine learning/deep learning to issues and challenges of power systems considering recent developments and advances in planning, operation, and control of such systems in both collecting a comprehensive review based on the literature and applying machine learning to power system planning, operation, and control. The topics covered in this book are as follows:

- Power System Challenges and Issues
- Machine Learning Methods in Energy Engineering
- Overview of the Application of Machine Learning/Deep Learning for Controlling and Planning Power Systems
- Application of Machine Learning/Deep Learning Methods for Clustering in Power Systems
- Application of Machine Learning/Deep Learning Methods for Forecasting Power System Parameters

University Park, PA, USA Morteza Nazari-Heris
University Park, PA, USA Somayeh Asadi
Aalborg, Denmark Behnam Mohammadi-Ivatloo
Geelong, VIC, Austrlia Moloud Abdar
University Park, PA, USA Houtan Jebelli
University Park, PA, USA Milad Sadat-Mohammadi

Contents

Chapter 1
Power System Challenges and Issues

Ali Sharifzadeh, Mohammad Taghi Ameli, and Sasan Azad

1.1 Introduction

In conventional power systems, large power plants have provided balancing in the network parameters and its exchanges. Among different system requirements, a priority after a basic balancing of power and energy is to ensure that power flows and dynamics are within bounds and stable (for the angle, voltage, and frequency) in normal and after events (faults, failures). However, conventional planning and control models, which are well known and standardized, will be challenged by all new concepts and future structures. Thus, the role of modeling and calculations related to power system analysis with new concepts remains of central importance[1].

On the other hand, large blackouts, although infrequent, are costly to society with estimates of direct costs up to billions of dollars [2, 3]. Some other indirect costs such as the failures of communications [4], natural gas network [5, 6], water supply, transportation, and etc. are also causing the challenge. For existing and future networks with extensive energy and money exchanges, blackout is no longer acceptable [2]. Therefore, it is necessary to survey power system security and all network parameters related to stability fast and accurate to be online and real time. In addition, other issues such as greenhouse gas emissions, smart contracts, energy management, and energy storage, each impose new requirements on the power grid. Lack of proper consideration of new concepts and issues and lack of correct prediction in this case make it difficult to manage the network.

All previous tips and the details will be mentioned later, such as renewables, extensive load changes, load forecasting, resiliency, energy transactions, and

A. Sharifzadeh · M. T. Ameli (✉) · S. Azad
Department of Electrical Engineering, Shahid Beheshti University, Tehran, Iran
e-mail: m_ameli@sbu.ac.ir; sa_azad@sbu.ac.ir

M. Nazari-Heris et al. (eds.), *Application of Machine Learning and Deep Learning Methods to Power System Problems*, Power Systems,
https://doi.org/10.1007/978-3-030-77696-1_1

uncertainties, may challenge a power system in the present and future [7]. Also, huge data and increasing interdependence of information technology and energy management infuse much of the current solutions about power systems in the future. Therefore, new challenges required new and intelligent solutions. Older methods will not work properly with new concepts, and new methods will need it.

The most attractive art for power system analysis is machine learning and its various advanced methods, eliminating the need for complex calculations and models. These stately methods, which are fast, robust, and adaptive, can overcome the drawbacks of traditional solutions for several power systems problems [8]. It eliminates the need for system time allocation to perform continuous calculations.

The remainder of this chapter is organized as follows: Sect. 1.1 Introduction, presents a brief of the work, and Sect. 1.2 presents existing and future system challenges and classified according to the types. An overview of machine learning challenges and issues is provided in Sect. 1.3. Finally, Sect. 1.4 presents a conclusion for the chapter.

1.2 Present and Future Challenges

Construction and operation management of a power network always has challenges within itself or various general policies. Addressing issues lead to planning to mitigate the consequences. These challenges may be endogenous or exogenous. The main issues were how to establish connections between different local networks and increase equipment efficiency. In recent years, many issues that can reduce network security or its reliability can make challenges. For example, increasing the penetration rate of distributed generations and renewables and increasing uncertainties on the demand side, such as the number of electric vehicles, have led to new problems. Also, environmental constraints or restructuring to increase all participants' profitability can lead to some new problems. In the future, issues such as online monitoring of the network reduce transmission and distribution of technical or financial costs, as well as online. However, increasing the use of distributed generators (DG) with uncertainty output will remain the major issues.

In the coming years, local energy transactions will be highly regarded. For example, an electric vehicle can be used to supply some of a building's energy consumption or charged by local generator. This issue will be given a lot of attention due to the importance of local exchanges. Due to the decentralization of these exchanges, new platforms will be developed [9]. This subject may not be a challenge, but the multiplicity of these transactions may lead to issues such as legal disputes, local technical or simulation problems, management, and decision-making interferences in distribution networks or microgrids [10]. Figure 1.1 shows the general challenges of future networks at a glance. In the following, the main challenges and issues will be discussed.

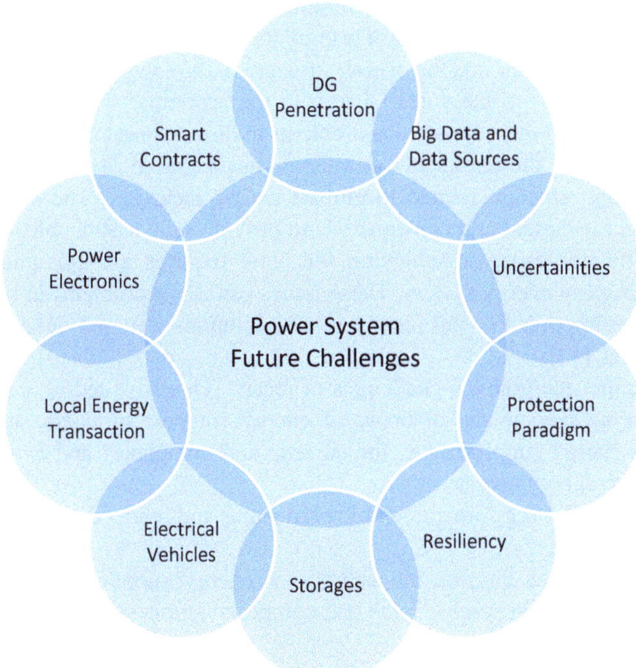

Fig. 1.1 Future challenges to power systems

1.2.1 Greenhouse Gases Emission

One of the most important issues of energy management is the reduction of greenhouse gases and their consequences. One of the sources of greenhouse gas emissions is fossil power plants and industrial consumers that emit large amounts of carbon dioxide and gases. From 1990 to 2013, global energy demand increased by 55%. The European Union, after China (3036 Mtoe), the United States (2188 Mtoe), and Asia (excluding China) (1655 Mtoe), with an estimated gross domestic product (GDP) of 1626 Mtoe, is the fourth most energy-consuming region in the world [11]. In the field of energy production, despite efforts to decarbonize, the share of fossil fuels in world energy has changed slightly over the years [12].

The Lisbon Treaty [13], adopted in 2009, brought about fundamental changes in the main EU laws in the field of energy and European networks of the EU member states. In other words, the treaty sets out four main objectives of EU energy policy.

- Achieve the proper performance of the energy market.
- Ensure the security of energy production in the union.
- Provide solutions to improve energy efficiency and energy savings and create new and renewable forms of energy.
- Enable better connections between different networks.

In general, the European Union's solutions do not affect the right of a member state to determine the conditions of use of its energy resources, choose between different energy sources, and the overall structure of the energy source. However, the Lisbon Treaty refers to measures tailored to the economic situation and technical issues. Especially if there are serious problems in different sectors, especially energy [13], in this regard, the European Commission has begun to develop a strategy for a flexible energy structure related to climate policy measures. The stated goal for families and businesses and consumers is to provide safe, sustainable, competitive, and cost-effective energy. Achieving this goal requires a major change of the planned European energy system. These issues can cause problems in both production and consumption. For this reason, related solutions were provided to overcome the challenges [14].

In particular, the political challenges of recent years have led to a great deal of attention being paid to the diversity of energy sources, suppliers, and safe and sustainable energy supply routes for citizens and companies and to expectations for affordable energy access.

EU policies of energy are guided by three mains and reflect the objectives set out in the Lisbon Treaty.

Cost-effectiveness ensures the activity of energy suppliers in a competitive environment that ensures reasonable and competitive prices for homes, businesses, and industries.

- Security: improving reliability and continuous energy supply
- Sustainability: sustainable energy consumption, with reduced greenhouse gas emissions, pollution, and dependence on fossil fuels

To pursue energy and climate change goals in a long-term strategy, the European Union has agreed on specific targets for 2020–2030 and has drawn up a road map for 2050. The goal of greenhouse gas emissions seems to emerge as a major factor in shaping the EU's domestic/foreign energy policies and climate change.

Greenhouse gas emissions in Europe are expected to decline as follows [15, 16]:

- By 2020: 20% below the 1990 level
- By 2030: 40% reduction
- By 2050: Decrease of at least 80%

Pursuing greenhouse gas reduction programs requires the use of new technologies and concepts in the management and structures of power grids. Given the above and looking at the future of Europe plans, energy by 2040 will rely significantly on renewable resources and new technologies. Some of these new concepts can create uncertainties in the network.

1.2.2 Distributed Generation and Renewables

With the development of industry and the increase of domestic and industrial consumption, the penetration rate of distributed generators and renewables will increase. Therefore, the use of large power plants and fossil fuels will be less. This can reduce the rotational inertia in the network and can lead to instabilities. To this end, changes will occur in the management and operation of the network. This rotational inertia, network requirement, will be partially compensated by electronic power capabilities, and network stability will be improved [17, 18] (Fig. 1.2).

Due to the fact that renewables create an uncertainty in power generation, the use of equipment that can store energy and inject into the network at the appropriate time has been considered. Future grids with high penetration of renewables use microgrids to enable widespread utilization of renewable energy sources for increasing efficiency and reliability [19].

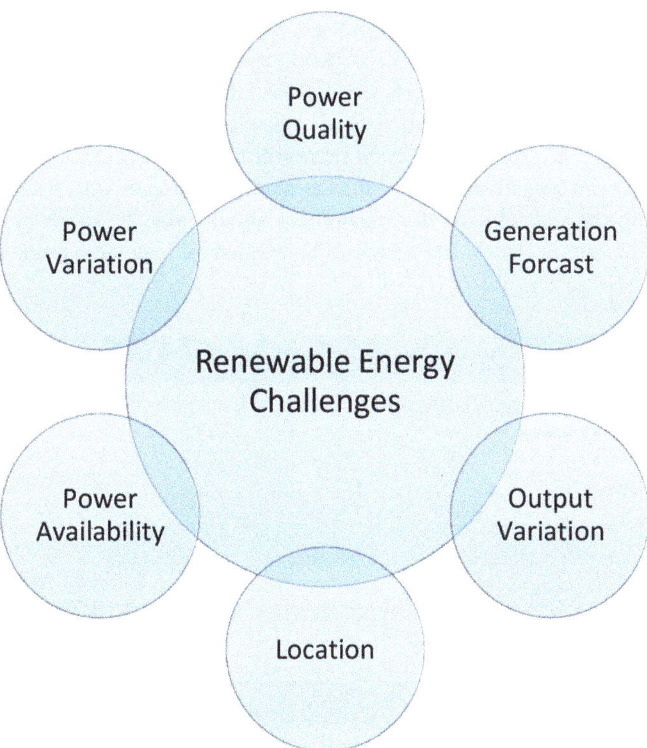

Fig. 1.2 Challenges of renewable energies

1.2.3 Energy Storages Integrity

One of the equipment groups that is widely used is energy storage. With the integrity of storage equipment, operators can perform peak shaving and transient damping better with more flexibility [20, 21]. This equipment usage in the network can cause problems in power flow calculations, and it is necessary for considerations in this regard. Developments in batteries, superconductors, compressed-air energy storage (CAES), power electronic, and etc. have led to the development of storage equipment and are expected to be widely used in transmission and distribution networks in the future (Fig. 1.3).

Storage solutions and technologies available in medium and large capacities create new concepts in general utilization of power networks. In the same way, with the expansion of storage equipment in the network, it is necessary to examine the interactions and effects [22].

The allocation of energy storage in distribution systems is widely considered worldwide. This problem poses two major challenges: storage location and its capacity. If wrong decisions are made in both cases or one of them, the distribution system's performance may be compromised. The placement and selection of storage capacities are done according to the network requirements or depth and optimized. Traditional and intelligent methods are also used in this field [23, 24]. This will also be used extensively in smart home energy management systems [25].

Storages may increase the level of energy dissipation, decrease the voltage profile's status, and negatively affect the network's technical operating conditions. Determining the capacity of this equipment should also be done in a way that optimizes economic issues. In other words, it can store a suitable amount when the

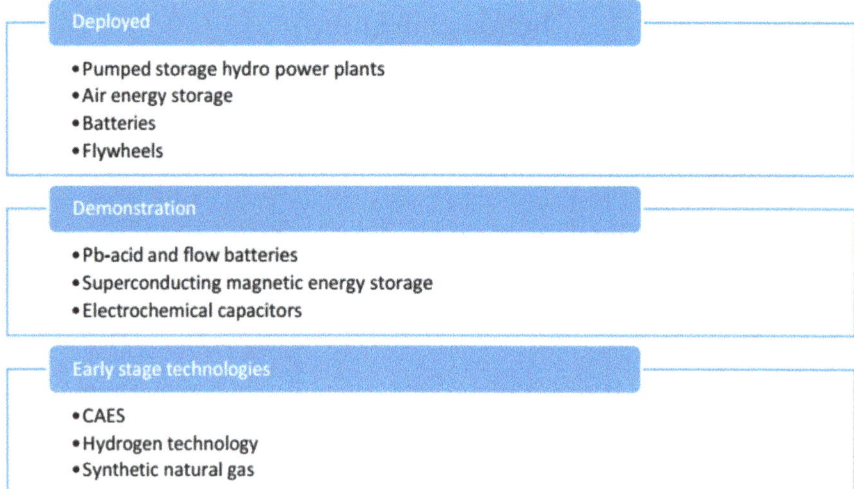

Fig. 1.3 Energy storage technologies

Company Benefits
- Line capacity additional defferal
- Peak load shaving
- Substitute requirement for spinning reserve
- Reduce need for conrengency reserve
- Ability to diversify generation resources
- Grid Voltage and frequency regulation
- Meeting off-grid mandates and demands
- Grid stability

Users Benefits
- Imprtoval power quality and reliability
- Less process intruption and maintenance costs
- Peak load reduction and reduced demand charges
- Ability to use higher onsite renewables
- Overal lowered costs

Renewable Energy (RE) Generation Benefits
- Improved capacity utilization
- Voltage and frequency regulation
- Load levelling and better ability to forcast

Fig. 1.4 Benefits of power storages

energy price is low, and when it is expensive or in case of emergency, it can inject the stored energy into the grid.

A multi-energy system provides more system operation flexibility. This concept will improve reliability and efficiency. Energy storages can integrate various energy sources such as solar, wind, gas, heat, and combined heat and power units at different levels [26]. Additionally, coordinated scheduling of flexible loads and energy storage systems can play an important role in the optimal scheduling of microgrids and lowering the costs [19, 27].

The development of storage equipment and its integration into the power grid will increase maintenance costs because this equipment mainly has expensive parts and impose additional costs. For this purpose, it will be necessary to develop maintenance programs and network operation plans based on these activities (Fig. 1.4).

1.2.4 Electric Vehicles

The spread of electric vehicles, like a double-edged sword, can be used to improve the condition of the network or cause problems if mismanaged. Frequent use of electric vehicles and plugging to the network creates an unpredictable situation. However, managing the charge or discharge of these vehicles and applying new control methods in smart grids can improve the situation.

In future networks, there may be developments that will lead to major changes in this field. New researches cover topics such as wireless charging [28, 29], vehicle to grid (V2G) block-chain management [30], privacy, and security, and smart charge optimization is considered. In the future, with the development of electric vehicles and charging stations, a change in the management of distribution and power networks will be required.

As mentioned above, the advantage of electric vehicle application is that if they are properly managed to charge and discharge. Therefore, network parameters such as voltage and frequency can be controlled. For this reason, the location and capacity of charging stations are of particular importance. Various optimization or intelligent methods have been provided in the placement of this equipment and interact with the network. One of the most important issues addressed in new studies is car charging stations allocation [31]. Point placements are based on optimization methods, but other factors such as feasibility of installation, economic issues, etc. are also involved in this issue. However, all components within the network are currently being used to improve its situation.

1.2.5 Decentralization and Smart Contract

In two-way interactions of consumers and networks owners, new financial structures and different settlement methods are formed. Therefore, by increasing the profitability of all components, the various capabilities that consumers can create will be used properly. The volume of financial transactions and its optimization with different conditions will develop the use of block-chain technology, digital currencies, and smart contract structures [32, 33]. This requires the creation of various legal laws as well as new financial structures. Creating communication platforms and a secure database will also be required although focusing on financial structures may led to some challenges for power network security. But creating flexible rules can drive interactions to improve network parameters and conditions. Smart contracts will create a new perspective on the electricity market. By observing the options of these contracts by the parties, the agreed outputs will be applied, and by deviating from it, some presuppositions or penalties will be activated. By properly designing the various dimensions of this type of contract, the network situation in relation to the market can be controlled appropriately and optimized. Developments and increasing interactions between network components may affect network security and stability. Rapid changes in load values, as well as planning to increase the profitability of components, can reduce operating safety margins. Therefore, software related to controlling, monitoring, and predicting future status should be developed to meet operators' new requirements. To this end, new security features and new methods without modeling will be developed and used. System reliability will also be

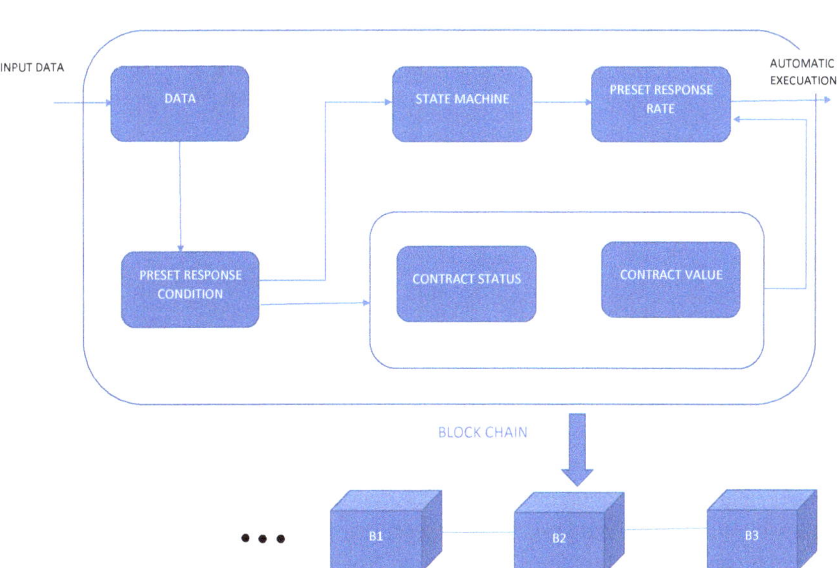

Fig. 1.5 Smart contract configuration

affected, and correction plans will be required. Currently, network status can be extracted using calculations and simulations. Different network modes and emergencies must also be modeled to predict or evaluate the situation (Fig. 1.5).

1.2.6 Reliability and Security Problems

The availability of all components of a power grid and increasing its reliability is one of the main concepts in multi-energy systems. Contribution in different exchange forms of energy such as electricity, heat, and natural gas plays main rules for future consumers and industries. Multi-energy networks can provide more opportunities for players to increase their profit [34]. As the number of devices, market players, and energy forms in the network increases, the network's reliability and security will be affected. These effects may be due to changes in voltage or frequency or other network parameters. Players outside the power grid will also have a big impact on its performance. For this reason, network operators consider it desirable to increase stability in network parameters (Fig. 1.6).

Power system transient stability assessment is important to determine its reliability and continuity of performance. For this purpose, Phasor Measurement Units (PMU) and Remote Terminal Units (RTU) may be widely use in the network. The data collected from these units, after analysis, is used for specific purposes, such as determining the stability of the network or its orientation. This is Supervisory

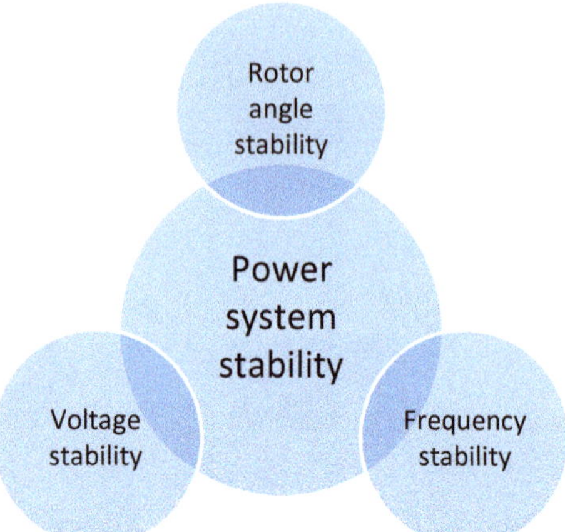

Fig. 1.6 Classification of power system stability

Control and Data Acquisition (SCADA) system responsibility. The amount of processing time of the network status assessment and the accuracy of these calculations can cause some problems. Increasing processing times or decreasing results accuracy can lead to incorrect control decisions or grid blackouts. Also, the network data gathering through PMUs and high-speed communication media is important in the network's proper operation. Using reliable telecommunication infrastructures and capturing network parameters at all levels of voltage, huge volumes of data transmitting pose a great challenge to network designers and operators.

For this reason, operation and optimization of various power network issues and problems led to the new fast and smart solutions. New methods are used to assess the stability and security of the network in studies or in practice. These methods, which have recently used image processing and deep learning processes [35], can provide the desired output with appropriate accuracy. A pattern of network status is provided that identifies and classifies insecure or safe points. These patterns are updated at any time using the information received from PMUs [36]. The problem is the accuracy of the outputs, which, if the network is misdiagnosed as insecure or vice versa, will lead to erroneous decisions and lead to irreparable accidents. For this reason, it is necessary to check the results accuracy so as not to challenge the network. There are several methods for evaluating the accuracy of intelligent methods that have been addressed in recent researches. However, increasing the accuracy and reducing the computation time is very important in evaluating network security. On the other hand, generating train and test data for neural networks is one of the new methods challenges. In recent years, various methods of data generation have been proposed in this field, which are considered in research [37].

Voltage regulation is an important aspect of reliability. Power grid operating regulations require all installations to maintain voltage within a suitable and stable range. This is traditionally achieved using capacitors, tap changers, and distributed generators. All of these methods may be relatively slow. But modern inverters with new technology can enable fast voltage regulation in the network.

1.2.7 Load Forecasting Problems

Extensive use of renewable energy and various loads with high uncertainty in power grids, and the importance of profitability, disrupt short-term and long-term load forecasting. In other words, one of the main challenges in future networks is load management and generation planning based on high uncertainty load forecasts and stochastic energy generation. Therefore, the use of neural networks and deep learning methods in this field is considered important in recent research [38–40] (Fig. 1.7).

1.2.8 Big Data

As a significant application of energy, future networks are complicated interconnected power grid that involves high penetration distributed generations, highly variable loads, smart meters, real-time controllers, and data processors. It continuously generates data with large volume, high velocity, and diverse variety. We need the high volume of data and information exchanged for monitoring and controlling power networks [41].

To this end, the applications of big data and neural networks are becoming more widespread. Free model methods were introduced and used in research. These

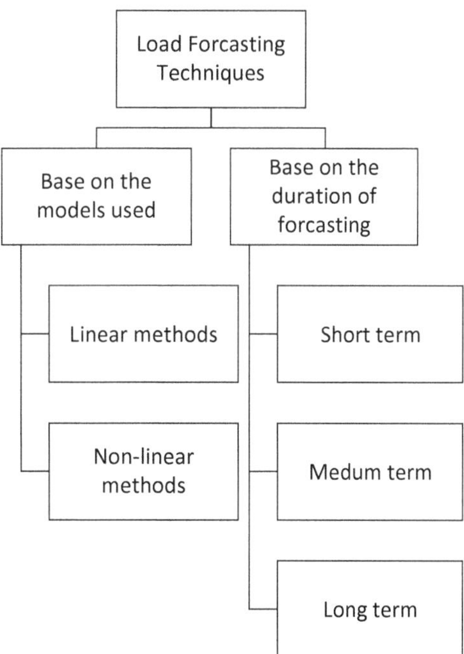

Fig. 1.7 Load forecasting techniques

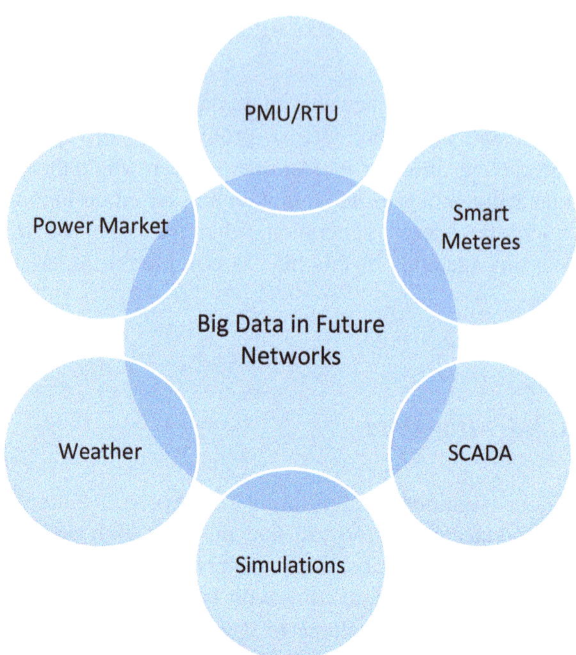

Fig. 1.8 Big data in future networks

methods led to the development of control center applications and improved the system. Relying on telecommunications and large amounts of data can compromise the communication cybersecurity of the system. Creating and increasing telecommunication cybersecurity is one of the most important issues that is addressed. This subject can cause local or major problems (Fig. 1.8).

1.3 Machine Learning (ML) Application and Challenges

Power network management and solving various problems require the use of different methods of complex modeling and calculations. Sometimes, the volume of these calculations may be large enough to interrupt the decision. To solve problems, model-free methods were proposed [42]. Also, artificial neural networks and machine learning methods were used to reduce the volume of calculations. However, these methods did not eliminate much of the need for simulation because true data is required for different learning methods or results testing, which must be either available in the control center's database or generated by simulations. Therefore, big data applications and data mining are also considered. In recent years, focusing on increasing accuracy and reducing processing time, reliable methods based on machine learning models have been proposed. With the development of the proposed models, new processing capabilities and tools have been made available to researchers and engineers. These capabilities are used in data generation, status

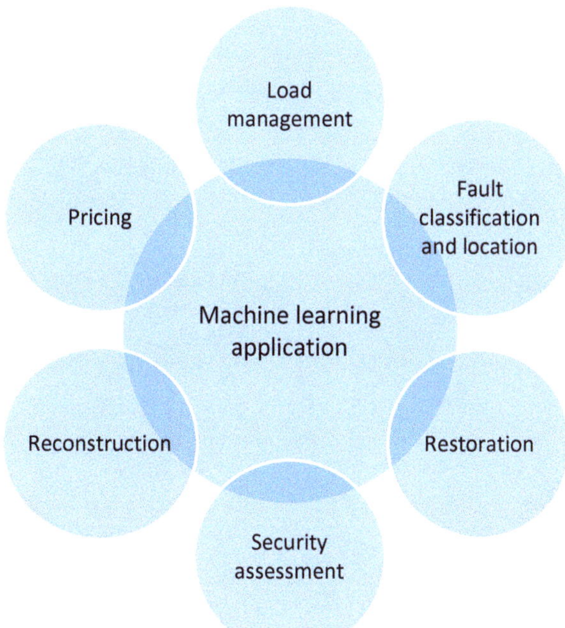

Fig. 1.9 Machine learning applications

assessment, or status forecasting. Figure 1.9 shows a number of applications of machine learning in the field of power engineering.

As shown in Fig. 1.9, there are several applications for machine learning in power grids. In other words, all the challenges in power networks can be solved using different machine learning methods. In recent years, much attention has been paid to develop and improve problem-solving with these methods. For example, in short-term forecasts of load consumption and the presence of large uncertainties, such as renewable generators and electric vehicles, intelligent methods can provide good results in this field. Also, in finding suitable routes to restoring the network after blackouts, intelligent methods can provide suitable patterns based on different grid situations.

In power transactions and electricity markets, machine learning methods can play an effective role in both finding the optimal price and increasing network capabilities. This role will be highlighted when the rate of these transactions is greatly increased and the profitability of network components in interaction with operating conditions is considered. However, given the breadth of new power grid issues and the new complexities that have arisen, the use of these intelligent methods has been considered. However, it needs to be addressed more in the industry. The problem is that many operators are still interested in using traditional methods, and they may prevent the use of smart methods in the operation of power networks.

On the other hand, machine learning and data mining methods led to new challenges in the power system processing. Some of these challenges may be in terms of operating costs, and some may be related to the data usages and applications

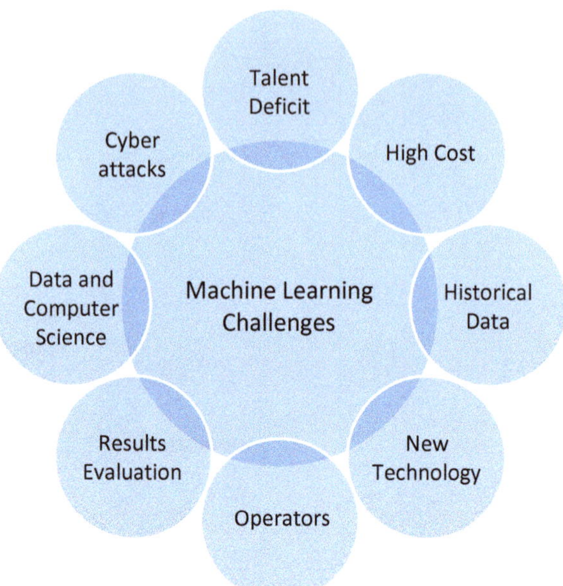

Fig. 1.10 Machine learning challenges

[43]. Figure 1.10 shows some of these challenges. These challenges have not prevented the use of intelligent methods in studies. These challenges may arise due to the requirement for new concepts. These new concepts, including data science, big data, artificial intelligence, and image processing, are used in simulation processes.

For this reason, it may not be easy for older operators to understand. Sometimes, the use of these methods may be opposed by operators. It also requires some structures such as historical databases or result evaluation. Evaluating the results is important because the accuracy of the outputs must be within the appropriate range. Otherwise, decisions may be made with a lot of deviation. Also, if there are deviations in the results, the actual state of the system may be presented incorrectly. Therefore, evaluating the results and its different methods is of great importance.

Engineers and managers who have to work with these structures required data science and computer science training. They should have a wealth of professional talent in this field. On the other hand, creating and using any new structure requires time and money. Solving possible structural problems and improving methods depend on the application of researches in the industries.

The development of machine learning methods and models and dependence on databases and software methods increase hackers and cyberattacks. The relevant protection structures and firewalls can control these attacks.

Different methods of machine learning are rapidly penetrating the power grid operation. This rapid development will lead to a new revolution in various industries and problem-solving strategies.

1.4 Conclusion

As discussed in this chapter, the future power grids will face various challenges and issues. Some of these challenges are related to new equipment, some are related to new concepts, and some are related to new simulation and calculation methods. Also, the large volume of big data and the breadth of calculations will greatly impact operations and decisions. Examining the impact of using new equipment such as storage, renewables, and predicting future network states with many power electronics helps a lot to understand future issues of the power grid.

These challenges will lead to the following main items:

- Renewable and storage penetration
- The impact of consumers on network management
- Big data and large volumes of data exchanges
- New concepts, technologies, and methods

These changes in structure and concepts may also create resistance to the use of new methods. Decisions at the highest levels of management can be helpful in this regard. Using some smart methods can also help provide new solutions. At this stage, new methods have some problems and need more development. However, focusing on new concepts and developing intelligent algorithms such as machine learning effectively reduce the challenges of power grids. In this chapter, by examining the future network, some concepts and methods were evaluated, and general pre-enlightenment challenges were obtained.

References

1. J.P. Lopes, N. Hatziargyriou, J. Mutale, P. Djapic, N. Jenkins, Integrating distributed generation into electric power systems: a review of drivers, challenges and opportunities. Electr. Pow. Syst. Res. **77**(9), 1189–1203 (2007)
2. H. Haes Alhelou, M.E. Hamedani-Golshan, T.C. Njenda, P. Siano, A survey on power system blackout and cascading events: research motivations and challenges. Energies **12**(4), 682 (2019)
3. B.A. Carreras, D.E. Newman, I. Dobson, North American blackout time series statistics and implications for blackout risk. IEEE Trans. Power Syst. **31**(6), 4406–4414 (2016)
4. P.-Y. Kong, Optimal configuration of interdependence between communication network and power grid. IEEE Trans. Ind. Inform. **15**(7), 4054–4065 (2019)
5. H. Ameli, M. Qadrdan, G. Strbac, Coordinated operation strategies for natural gas and power systems in presence of gas-related flexibilities. Energy Syst. Integration **1**(1), 3–13 (2019)
6. V. Shabazbegian, H. Ameli, M.T. Ameli, G. Strbac, Stochastic optimization model for coordinated operation of natural gas and electricity networks. Comput. Chem. Eng. **142**, 107060 (2020)
7. C. Syranidou, J. Linssen, D. Stolten, M. Robinius, Integration of large-scale variable renewable energy sources into the future european power system: on the curtailment challenge. Energies **13**(20), 5490 (2020)
8. S.M. Miraftabzadeh, F. Foiadelli, M. Longo, M. Pasetti, A survey of machine learning applications for power system analytics, in *2019 IEEE International Conference on Environment*

and Electrical Engineering and 2019 IEEE Industrial and Commercial Power Systems Europe (EEEIC/I&CPS Europe), (IEEE, 2019), pp. 1–5
9. D. Vangulick, B. Cornélusse, D. Ernst, Blockchain for peer-to-peer energy exchanges: design and recommendations, in *2018 Power Systems Computation Conference (PSCC)*, (IEEE, 2018), pp. 1–7
10. W. Tushar, C. Yuen, H. Mohsenian-Rad, T. Saha, H.V. Poor, K.L. Wood, Transforming energy networks via peer-to-peer energy trading: The potential of game-theoretic approaches. IEEE Signal Process. Mag. **35**(4), 90–111 (2018)
11. F. Profumo, E. Bompard, Fulli, G, Electricity security: models and methods for supporting the policy decision making in the european union (2016)
12. R.G. Newell, Y. Qian, D. Raimi, *Global Energy Outlook 2015*, National Bureau of Economic Research, 0898-2937 (2016)
13. E. Union, *Treaty of Lisbon: Amending the Treaty on European Union and the Treaty Establishing the European Community* (Office for Official Publications of the European Community, 2007)
14. E.U. Package, A framework strategy for a resilient energy union with a forward-looking climate change policy, in *Communication from the Commission to the European Parliament, the Council, the European Economic and Social Committee, the Committee of the Regions and the European Investment Bank, COM,* vol. 80 (2015)
15. A. Froggatt, A. Hadfield, *Deconstructing the European Energy Union: Governance and 2030 Goals*, EPG Working Paper: EPG 1507 (2015)
16. P. Capros et al., *EU Energy, Transport and GHG Emissions Trends to 2050-Reference Scenario 2013* (European Commission, 2013)
17. K.S. Ratnam, K. Palanisamy, G. Yang, Future low-inertia power systems: requirements, issues, and solutions-a review. Renew. Sustain. Energy Rev. **124**, 109773 (2020)
18. L. Mehigan, D. Al Kez, S. Collins, A. Foley, B. Ó'Gallachóir, P. Deane, Renewables in the European power system and the impact on system rotational inertia. Energy **203**, 117776 (2020)
19. M.A. Mirzaei et al., A novel hybrid two-stage framework for flexible bidding strategy of reconfigurable micro-grid in day-ahead and real-time markets. Int. J. Electr. Power Energy Syst. **123**, 106293 (2020)
20. H.E. Murdock et al., Renewables 2019 Global Status Report, 2019.
21. K.M. Muttaqi, M.R. Islam, D. Sutanto, Future power distribution grids: integration of renewable energy, energy storage, electric vehicles, superconductor, and magnetic bus. IEEE Trans. Appl. Superconductivity **29**(2), 1–5 (2019)
22. J.E. Bistline, D.T. Young, Emissions impacts of future battery storage deployment on regional power systems. Appl. Energy **264**, 114678 (2020)
23. M. Baza, M. Nabil, M. Ismail, M. Mahmoud, E. Serpedin, M.A. Rahman, Blockchain-based charging coordination mechanism for smart grid energy storage units, in *2019 IEEE International Conference on Blockchain (Blockchain)*, (IEEE, 2019), pp. 504–509
24. T. Fu, C. Wang, N. Cheng, Deep learning based joint optimization of renewable energy storage and routing in vehicular energy network. IEEE Internet Things J. (2020)
25. H. Jang, T. Lee, S.M. Kim, J. Lee, S. Park, Energy storage system management method based on deep learning for energy-efficient smart home, in *2020 IEEE International Conference on Consumer Electronics (ICCE)*, (IEEE, 2020), pp. 1–2
26. N. Nasiri et al., A bi-level market-clearing for coordinated regional-local multi-carrier systems in presence of energy storage technologies. Sustain. Cities Soc. **63**, 102439 (2020)
27. M.-N. Heris et al., Evaluation of hydrogen storage technology in risk-constrained stochastic scheduling of multi-carrier energy systems considering power, gas and heating network constraints. Int. J. Hydrogen Energy **45**(55), 30129–30141 (2020)
28. J. Chen, C.W. Yu, W. Ouyang, Efficient wireless charging pad deployment in wireless rechargeable sensor networks. IEEE Access **8**, 39056–39077 (2020)
29. O. Nezamuddin, E.C. dos Santos, Vehicle-to-vehicle in-route wireless charging system, in *2020 IEEE Transportation Electrification Conference & Expo (ITEC)*, (IEEE, 2020), pp. 371–376

30. V. Hassija, V. Chamola, S. Garg, N.G.K. Dara, G. Kaddoum, D.N.K. Jayakody, A blockchain-based framework for lightweight data sharing and energy trading in V2G network. IEEE Trans. Vehicular Technol. (2020)
31. Y. Motoaki, Location-allocation of electric vehicle fast chargers—research and practice. World Electr. Vehicle J. **10**(1), 12 (2019)
32. D. Han, C. Zhang, J. Ping, Z. Yan, Smart contract architecture for decentralized energy trading and management based on blockchains. Energy, 117417 (2020)
33. S. Yu, S. Yang, Y. Li, J. Geng, Distributed energy transaction mechanism design based on smart contract, in *2018 China International Conference on Electricity Distribution (CICED)*, (IEEE, 2018), pp. 2790–2793
34. M.Z. Oskouei, M.A. Mirzaei, B. Mohammadi-Ivatloo, M. Shafiee, M. Marzband, A. Anvari-Moghaddam, A hybrid robust-stochastic approach to evaluate the profit of a multi-energy retailer in tri-layer energy markets. Energy **214**, 118948 (2020)
35. J.-M.H. Arteaga, F. Hancharou, F. Thams, S. Chatzivasileiadis, Deep learning for power system security assessment, in *2019 IEEE Milan PowerTech*, (IEEE, 2019), pp. 1–6
36. P.K. Jaiswal, S. Das, B.K. Panigrahi, PMU based data driven approach for online dynamic security assessment in power systems, in *2019 20th International Conference on Intelligent System Application to Power Systems (ISAP)*, (IEEE, 2019), pp. 1–7
37. F. Thams, A. Venzke, R. Eriksson, S. Chatzivasileiadis, Efficient database generation for data-driven security assessment of power systems. IEEE Trans. Power Syst. **35**(1), 30–41 (2019)
38. A. Al Mamun, M. Sohel, N. Mohammad, M.S.H. Sunny, D.R. Dipta, E. Hossain, A comprehensive review of the load forecasting techniques using single and hybrid predictive models. IEEE Access **8**, 134911–134939 (2020)
39. M.A. Hammad, B. Jereb, B. Rosi, D. Dragan, Methods and models for electric load forecasting: a comprehensive review. Logist. Sustain. Transp. **11**(1), 51–76 (2020)
40. N.M.M. Bendaoud, N. Farah, Using deep learning for short-term load forecasting. Neural Comput. Appl. **32**(18), 15029–15041 (2020)
41. M. Ghorbanian, S.H. Dolatabadi, P. Siano, Big data issues in smart grids: a survey. IEEE Syst. J. **13**(4), 4158–4168 (2019)
42. X. Li, J. Wen, Review of building energy modeling for control and operation. Renew. Sustain. Energy Rev. **37**, 517–537 (2014)
43. M. Shafique et al., Adaptive and energy-efficient architectures for machine learning: challenges, opportunities, and research roadmap, in *2017 IEEE Computer Society Annual Symposium on VLSI (ISVLSI)*, (IEEE, 2017), pp. 627–632

Chapter 2
Introduction and Literature Review of Power System Challenges and Issues

Ali Ardeshiri, Amir Lotfi, Reza Behkam, Arash Moradzadeh, and Ashkan Barzkar

2.1 Introduction

Since the beginning of electrical power system in 1880s, when lamps were used for lighthouse and street lighting purposes and the commercial use of electricity started [1], it has been developed into a great industry and economy. Having a fundamental role in modern era lifestyle, the consumption of electrical power has risen sharply in the twenty-first century, and as a response do demand growth, electricity generation has increased accordingly [2]. Worldwide electrical power demand increases 2.1% per year until 2040, while total energy demand growth is half of electricity. This matter raises the electricity share in total energy demand from 19% in 2018 to 24% in 2040. The electricity consumption is set to grow due to the electrification of transport and heat, increasing incomes of households, and rising demand of air-conditioning, digital, and electronic devices [3, 4].

Nowadays, the increasing energy demand, development of smart grids, and the combination of different types of energy systems have led to complexity of power systems. On the other hand, ever-expanding energy consumption, development of industry and technology systems, high penetration of solar and wind energies have made electricity networks operate in more complex and uncertain conditions.

A. Ardeshiri (✉) · A. Lotfi · A. Barzkar
Electrical Engineering Department, Sharif University of Technology, Tehran, Iran
e-mail: ali.ardeshirilajimi@mail.concordia.ca; Barzkarashkan@ee.sharif.edu

R. Behkam
Electrical Engineering Department, Amirkabir University of Technology, Tehran, Iran
e-mail: reza.behkam@aut.ac.ir

A. Moradzadeh
Faculty of Electrical and Computer Engineering, University of Tabriz, Tabriz, Iran
e-mail: arash.moradzadeh@tabrizu.ac.ir

M. Nazari-Heris et al. (eds.), *Application of Machine Learning and Deep Learning Methods to Power System Problems*, Power Systems,
https://doi.org/10.1007/978-3-030-77696-1_2

Therefore, analysis of traditional power systems requires physical modeling and extensive numerical computation. To analyze behavior of these systems, advanced metering and monitoring systems are utilized which generate huge amount of data. Machine learning, deep learning, and variety of regression, classification, and clustering algorithms are powerful tools to use in these issues. These procedures can be utilized to solve the power system problems and challenges such as planning, operation, fault detection and protection, power system analysis and control, and cybersecurity.

Globally, millions of people around the world have no access or limited access to electricity. This problem is not limited to developing countries. Even developed countries need to deliver electricity to remote areas which are far from the grid, such as islands, deserts, and places surrounded by mountains. Available energy is vital to reach development targets [5, 6]. Moreover, electricity plays an important role in the interconnection of food, energy, and water, which is known as FEW Nexus. Thus, the availability and resilience of electricity is of paramount importance [7, 8].

The main goal of power system is to supply the demand as much as possible. However, this matter is not as simple as it appears. To address the challenges of power system, three categories of challenges are considered in this chapter which are issues regarding power system planning, operation, and control. Planning is the activity and decision-making associated with the development of plans regarding the construction, design, and expansion of power system elements which meets the electricity demand in the future, considering the present situation [9], by predicting the condition and making some assumption of the future [10]. Power system planning can be mainly about reducing the investment costs, pollution, and power outage or increasing reliability, resiliency, and security [11]. Planning phase vary from 1 to 20 years. Once the power system elements are built, these elements need to be operated to supply the demand while following environmental and economic objectives and satisfying technical constraints. At the end, in power system control, a short time before delivering the power to the loads, considering all events, for example, contingencies and outages, happening in generation, transmission, or distribution, the main focus is to maintain the robustness and security of the power grid, while the economy and environment may not be as important as it was in planning and operation [12].

In the following subsections, the challenges and issues regarding planning, operation, and control are explained in details separately.

2.2 Power System Planning

In the other part, which is known as planning studies, the energy supply required by consumers is not provided properly because parts of the power system have lost their adequacy and are not able to provide the desired loads. The purpose of these studies is to determine the type, installation location, number, and time of installation of new

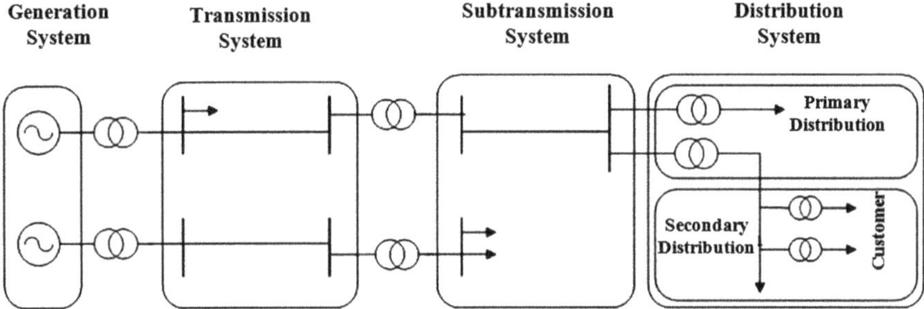

Fig. 2.1 The basic structure of the power system

equipment so that the required network adequacy is retrieved with the least cost [13, 14].

As stated, the purpose of planning studies in power systems is the recovery of the adequacy of equipment to meet the load required by consumers in an optimal way. The inability of power plants to produce the required power, the lack of suitable capacity of lines to transfer the required power, and the lack of capacity or the number of network substations are the inadequacies of the power system. Therefore, it can be concluded that determining the time and capacity required to expand generation, transmission, and distribution to properly supply the load required by consumers to electricity at the lowest possible cost is the goal of the planning issue [15].

Due to the complexity and dimensions of the problem, expansion planning studies are usually considered in three separate sections [16]:

1. Generation Expansion Planning (GEP).
2. Network Expansion Planning (NEP).
3. Substation Expansion Planning (SEP).

In Generation Expansion Planning, the goal is to determine the capacity of generation resources, their installation location, and installation time at the lowest cost. In this purpose, constraints are also considered in such a way that by considering the failures and units leaving the network, the total production capacity in the system is able to supply the maximum amount of consumer load, and their non-supply does not exceed a certain amount [13, 17].

Line expansion planning can also be considered in three sections: transmission lines, sub-transmission, and distribution. Assuming that load of substations and generation centers in the line expansion planning is clear and the generation sector and substations have the necessary adequacy, the type, capacity, location, and time of need for lines are considered [16, 18].

In a power system, as shown in Fig. 2.1, power is generated in energy-generating sources and transmitted through transmission lines to transmission substations. Each transmission substation feeds a number of sub-transmission substations through the sub-transmission lines. Finally, 20 kV feeders deliver power from the sub-transmission substations to the distribution substations, and through this, the consumers are fed. In order to feed consumers adequately in normal conditions or in the event of an accident, the total load of the substations must be calculated taking into account the allowable capacity of the equipment installed in the substation. As the load grows, the capacity of the existing substation equipment may not meet the load consumption. To solve this problem, existing substations must be expanded or solved by installing new substations on the network.

In general, the amount of network load downstream and the position of the network affect the capacity of each substation. With the implementation of SEP in a network, it is determined what equipment of the mentioned substations should be increased in capacity when and to what extent or in what time and place how many new posts should be built with what capacity that the network will regain its lost adequacy [16, 18].

As mentioned, the purpose of implementing SEP is to determine a combination of expanding existing substations or constructing new substations to provide safe and desirable supply to loads at the lowest cost.

In a given geographical area, consumers are scattered throughout the study area. Using transmission lines, sub-transmission, and distribution, electrical energy must be delivered to all consumers in the desired geographical area. After generating electricity in power plants, it is transferred to the sub-transmission substations through transmission lines, transmission substations, and sub-transmission lines. The sub-transmission substations are closer to the consumers and transfer the power to several distribution substations through 20 kV feeders after reducing the voltage level. Transmission substations with a capacity of several hundred MVA, sub-transmission substations with a capacity of several tens of MVA, and distribution substations with a capacity of about a fraction of MVA to a few MVA are constructed. The geographical distribution of consumers and the amount of their load determine the number and density of required substations. The manner of establishing the sub-transmission substations is effective on the construction site of the transmission substations, and the manner of establishing the distribution substations is also effective on the construction site of the sub-transmission substations. Given this dependency, distribution planning for distribution, sub-transmission, and transmission substations should be done simultaneously. But because the problem of simultaneous development planning of these posts is very complex due to the existence of many variables and constraints, in order to achieve a practical answer in a certain time, the designers of the power system divide the problem of planning the development of posts into three separate parts. These are the expansion of distribution substations, expansion of sub-transmission substations, and expansion of transmission substations, each of which is solved separately [16].

Thus, first, according to the location of low-pressure loads, the arrangement of low-pressure lines and the capacity and location of distribution substations are determined. Then these substations are considered as load points, and their capacity is also considered as the value of these load points, and according to this, the arrangement of medium pressure lines and the location and capacity of the sub-transmission substations are determined optimally. Downstream (distribution) and upstream (transmission) networks are not considered to simplify the task at this stage. After obtaining the location of the two capacities of the sub-transmission substations, these substations are considered as load points with the amount of load, and according to that, the arrangement of the above distribution lines, the location, and capacity of the transfer substations are determined.

As mentioned, in order to reduce the complexity and dimensions of the problem in each of the stages of separate expansion, the arrangement of downstream or upstream networks is not considered. However, this can take the answers to the problem of developing the whole system away from the optimal value. For example, suppose that in planning the expansion of a sub-transmission network including transmission substations, sub-transmission lines, and sub-transmission substations, the problem of simultaneous expansion is divided into two subthemes of the expansion of sub-transmission substations and the development of transmission substations. In this case, in planning the sub-transmission substations, the designer tries to reduce the length of the feeders and reduce costs, the substations should be built as close as possible to the load centers, and this will cause these substations have more distance from the transmission substations, which in addition to increasing the cost of connection to the transmission system will lead to other technical problems. This is shown schematically in Fig. 2.2.

Network expansion planning is based on projected loads, while projected loads are not based solely on network performance in the future and will be affected by factors based on network performance in the past and influential factors in the future. In this regard, examining the effect of uncertainty on effective parameters will have a significant impact on the planning response [19].

Load uncertainty studies are one of the important inputs in estimating the time of equipment addition to maintain network adequacy. As mentioned, the main causes of uncertainties are the emergence of new loads, rising prices, changing laws, and weather conditions (temperature and wind, etc.) [20]. In addition, in recent years, there has been a wave of changes in power networks that have led to uncertainties. These include restructuring of distribution networks, privatization, free market or competitive market, alternative energy sources, distributed generation, and new protection and communication technologies in the field of energy [21].

In most of the proposed models for network development planning for simplification, the effect of uncertainty on the problem parameters has been neglected [22].

Expansion planning for large networks will not have a real answer without considering these factors because as the network grows, both the probability of uncertainty and the amount of uncertainty in the parameters of the problem increase.

Due to recent developments in the field of the logic of power systems operation, in the face of increasing system capacity, innovative options have been proposed. One of the options that have been considered by many researchers recently is distributed generation (DG) [23].

DGs represent production units that are connected to the grid near low-load capacities. The use of DGs improves the voltage profile and reduces losses, which in turn reduces the cost of operating the network. Also, with the installation of DGs in the network, investment costs are reduced because the installation of DGs can delay the construction of new equipment or strengthen existing equipment [24].

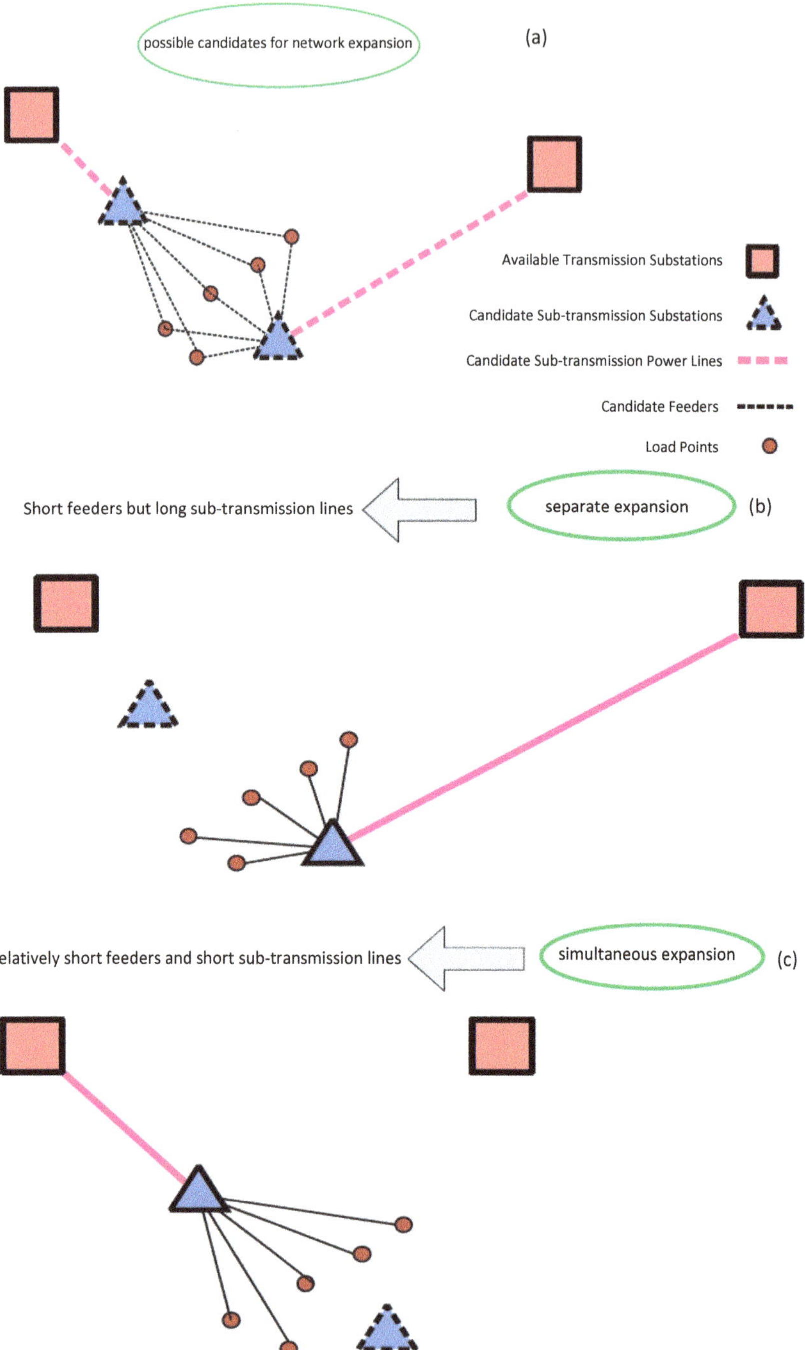

Fig. 2.2 Comparison of separate and simultaneous expansion schematically

In addition, DGs will play an important role in increasing the quality and reliability of customer service. Given the potential benefits of DGs and the declining price of DG resources, these resources are expected to play an important role in the future of power systems [25].

2.3 Power System Operation Challenges

The power system operation deals with several challenges all the time. Some of these challenges, i.e., **disturbance/unexpected events** and **cyberattack,** could happen every current time, but there is challenge, i.e., **climate change**, which will effect gradually on power system operation, and its result will be noticeable after several decades. These challenges and their impacts on the power system operation are summarized in the following.

2.3.1 Climate Change Impact on Power System Operation

Many various aspects (especially generation aspects) of energy sector is susceptible to the influences of climate changes and environmental conditions. Some of these interactions are summarized in below.

- **Solar.** Cloud cover will decline in low- to midlatitude regions [26]. While due to rising temperature, decreasing efficiency will counterbalance the increase in solar resources. Based on prediction of regional studies, solar generation changes will be less than ±10% by the end of the century [27–29].
- **Wind.** According to the several different studies, it is predicted that around Europe and North America, average wind speeds would remain within three different percent, ±15% [30], ±20% [31, 32], and ±30% [33] of current values. Moreover, it is evaluated that there will be no considerable change in wind resources over China [34] or Southern Africa [35].
- **Wave.** Due to changes in wind pattern and sea-level rise, wave resources could be affected. Studies indicated that wave generations have no changes in the Persian Gulf [36] and Menorca [37]. In addition, wave generation changes in the UK will be less than 3% [38].
- **Hydropower.** Global studies were done all over the world, and according to results, their conclusions differ about the impact of rising temperatures and changing precipitation patterns. In some studies, it is concluded that climate changes will have little influences on total global resources potential [39, 40], while the conclusion of other study indicated that due to climate changes, global hydropower volume will decline up to 6.1% in the 2080s [41].
- **Bioenergy.** Rising temperature and changing precipitation patterns affected the areas that appreciate to grow bioenergy and average yields. Yields will increase as

much as +28% at high latitude and will decline as much as −16% at low latitude in the 2050s [42]. Moreover, it is predicted to shift the suitable land northward for several crops [43], switchgrass in the USA [44], and miscanthus globally [45].

- **Thermal power stations.** As a result of reduction in the thermal efficiencies due to rising temperatures, power plant output will reduce about 0.4–0.7% per degree [46–48]. The consequence of the decline in water resources for cooling of power stations is reduction in the load, and it could be more serious which causes shutdown of power stations.
- **Transmission lines.** Rising temperature will decrease the overhead lines' transmission capacity. For an instance in the USA, it is expected reduction in transmission capacity of overhead lines will be approximately 5.8% at the peak summer demand times [49].

2.3.2 *Disturbance/Unexpected Events Impact on Power System Operation*

Unexpected events have considerable influences on power operation. Large or small disturbances happen due to several reasons such as human fault, impact of extreme weather events [50], and failure of power equipment due to production/installation error or equipment aging [51]. As a result of improvement in the life span of power equipment due to utilizing superconducting technologies, it is possible to reach more reliability of power system operation and reduce the failure possibility of power applications because of the aging factor [52, 53]. Moreover, some of other type of disturbances, their possible causes, and their effect are listed in Table 2.1 [54]:

Numerous studies have so far addressed the issues of protection and fault detection in transmission lines and distribution networks in the power systems using deep learning and machine learning techniques.

Power systems mainly due to various environmental, mechanical, and electrical factors suffer from two categories (11 types) of short circuit faults called asymmetric faults (AG, BG, CG, ABG, BCG, ACG, AB, BC, and AC) and symmetric faults (ABC and ABCG). In [55], a hybrid machine learning application based on morphological median filter (MMF) and decision tree (DT) is utilized to identify and classify the types of symmetric and asymmetric faults in the transmission line. In other similar studies [56, 57], the DT method has been employed to identify the type and location of symmetric and asymmetric faults related to the transmission line. In a valuable study [58], deep learning applications called convolutional neural network (CNN) and deep reinforcement learning have been used to identify the location of single-phase to ground short-circuit faults in the transmission line. In this paper, comparing and evaluating the results introduces the deep reinforcement learning procedure as the superior method.

Identification and classification of tested faults in the IEEE 123 bus benchmark distribution system using deep learning applications called the graph-CNN is done in

Table 2.1 Disturbance and unexpected events and their features

Disturbance type	Description	Possible causes	Symptoms/ effects	Potential solutions
Power outage	Total interruption of electrical supply: – *Momentary outage* lasts 03.5 cycles, 3 s – *Temporary outage* last from 3 s to 1 min – *long-term outage* last longer than 1 min	– Accidents, acts of nature, etc. which require the proper operation of utility equipment (fuses, reclosers, etc.) – Internal short circuits requiring the proper operation of a customer's breakers and fuses.	– System shutdown – Loss of computer/controller memory – Hardware damage – Product loss or damage	Uninterruptible power supply (UPS)
Transient (surge)	A sub-cycle disturbance in the AC waveform, evidenced by a sharp, brief discontinuity of the waveform	Surges are caused by storms (lightning); operation of utility fuses, reclosers, and breakers; and turning on or off large equipment and capacitor switching (customer and utility).	– Computer lockup, processing errors and data loss – Burned circuit boards, electrical insulation damage, and equipment damage	– Surge protectors – Uninterruptible – Power supplies with built-in surge suppression – Isolation transformers – Constant voltage transformer – Line reactors
Sag/swell	Any short-term (0.5 cycles, 1 min) decrease (sag) or increase (swell) in voltage. Sags account for up to 87% of all power disturbances (according to a Bell Labs study).	– Major equipment shutdown and/or restart – Short circuits – Utility equipment failure or utility switching	– Memory loss and data errors – Equipment shutdown – Flickering lights – Motors stalling or stopping and decreased motor life	– Uninterruptible power supply – Constant voltage transformers – Voltage regulators – Power electronic sag correctors
Noise	An unwanted high-frequency electrical signal that alters the normal voltage pattern (sine wave)	– Interference from radio or TV transmission – Operation of electronic equipment	– Lockup of sensitive equipment – Data loss and processing errors – Distorted audio and video reception	– Uninterruptible power supply – Isolation transformers – Power line filters
Harmonic distortion	The alteration of the normal voltage pattern (sine wave) due	Electronic ballasts and other nonlinear loads like switch-	– Overheating of electrical equipment and	– Harmonic filters – Isolation

(continued)

Table 2.1 (continued)

Disturbance type	Description	Possible causes	Symptoms/ effects	Potential solutions
	to equipment generating frequencies other than the standard 60 cycles per second	mode power supplies and variable frequency drives	wiring – Decreased motor performance – Improper operation of breakers, relays, or fuses	transformers – Improved wiring and grounding – Isolated loads – Line reactors
Undervoltage/ overvoltage	Any long-term change lasting more than a minute below or above normal voltage	Overloaded wiring or equipment. Large load swings or improper transformer settings. Undersized wiring and faulty or poor electrical connections	– Dim or bright lights – Equipment shutdown – Overheating of motors or lights – Reduced efficiency or life of electrical equipment	– Uninterruptible power supply – Constant voltage transformers – Verify electrical connections and wiring – Relocate equipment – Reduced voltage motor starters and voltage ride-through equipment

[59]. The method proposed in this paper converts the signals received from the system into two-dimensional images and then considers them as the CNN input. In [60], trip fault is identified in the China Southern power grid using the Support Vector Machine (SVM) and Long Short-Term Memory (LSTM) techniques. The performance evaluation of the proposed models in this paper introduces the LSTM method as the superior solution. The localization of single-phase to ground short-circuit faults in the distribution network is done by one of the deep learning applications called Stacked Auto-Encoder in [61]. In another valuable study [62], the identification of symmetric and asymmetric faults in the IEEE 39-bus distribution system is performed using the LSTM method. In this paper, the performance of the proposed method is compared with other machine learning techniques called SVM, DT, random forest (RF), and k-nearest neighbor (k-NN).

In addition to the abovementioned damages related to transmission lines and distribution networks, equipment installed in the power systems such as power transformers, electrical machines, and power electronics equipment are also damaged by various factors. Timely detection and elimination of abnormalities in this equipment can prevent serious damage to the power system. In this regard, the applications of machine learning and deep learning are mainly utilized.

The detection of mechanical faults related to the power transformer windings in [63] has been done using the SVM technique. In [64], the CNN procedure is selected as one of the deep learning applications for locating high impedance and low impedance short-circuit faults in transformer windings. Minor short circuit fault localization in transformer windings has been performed in [65] by using regression version of SVM. In this study, early detection of fault location prevents serious damage to the transformer windings and power system security.

In [66], the high-resistance grounding faults detection in a high-voltage direct current (HVDC) transmission lines with considering double-end unsynchronized has been performed by employing Hilbert-Huang (HH) transform and one-dimensional CNN techniques. An optimized adaptive neuro-fuzzy inference system (ANFIS) neural network and HH transform have been introduced in [67] for localization of internal and external faults in a voltage-sourced converter-HVDC (VSC-HVDC). Diagnosing three common transients including GF, lightning fault, and lightning disturbance on a VSC-HVDC transmission lines has been done by employing one of the deep learning applications called deep belief network in [68].

2.3.3 Cyberattack Impact on Power System Operation

Nowadays, in our modern digital society with complex and modern infrastructure, the power systems are one of the most crucial components which play a role as a backbone for its economic activities and securities. Modern power systems are considerably dependent on communication systems for their operation, as a result, susceptive to cyberattacks. Consequently, every country is interested to boost the security of their power system operation against cyber risks and threats which could cause cascading effects, power outage, and personal data breaches [69–71].

These cyberattack challenges affect sufficiency of TSOs and DSOs (Transmission and Distribution System Operators) to insure flexibility and resilience, dependability, stableness, security of the supply, and power quality for electricity's consumers. Power systems functions against risks of cyberattacks include several aspects, i.e., generation, transmission, and distribution stability and reliability, communication between systems and equipment, information on the operating circumstance of generation, transmission, distribution equipment, black start capability, and application performance and ability to recover [72].

So far, many studies have investigated and detected the types of cyberattacks in the power systems. Introducing and reviewing the challenges in cyber-physical systems, especially the security issues of these systems, have been discussed and evaluated in some valuable review papers [73–78]. The cyber-physical system generalities have been described in detail in [73]. In [74], the reliability effects of cyber-physical layers in the power system has been evaluated. In [75], after evaluating the security of the cyber-physical system, the existing challenges in this regard has been reviewed, and appropriate solutions are introduced. The effects of cyber-physical attacks in the smart grid has been investigated in [76]. In a valuable review paper [77], the limitations, issues, and future trends of cyber-physical systems security have been surveyed. Projection, prediction, and forecasting of various types of cyberattacks in the power system are introduced and evaluated in [78]. A review of the literature and evaluation of studies conducted in recent years shows that many solutions have been introduced to cyberattacks detection in the power systems so that machine learning and deep learning methods have found a special place in this field [79–83].

In the following, susceptibility associated with three important power system applications, i.e., state estimation, automatic generation control, and voltage control, against cyberattack are summarized.

1. Attacks on State Estimation.

   State estimation plays an important role in energy management application (control centers) and many operational decisions [84, 85]. Control centers run an elaborate algorithm to process measurement data from different locations such as power injections, line flows, voltage measurements, and the status information of breakers and switches. Based on the process result, the system states, i.e., voltage magnitude and angle, were estimated. According to the mentioned estimation, decisions such as ramping of generators, opening or closing of lines, and changing of transformer tapes are made. The attacks to state estimation can be done in several ways like modifying the sensor data at the meter level, a cyber-interference at the communication layer, and deceiving the operator with wrong configuration of the topology. The operator can figure out the presence of unreasonable data and omit them for estimation target if the attack is not arranged. Although the estimation cannot be conceivable, in the case of system, it becomes unobservable as a result of discarding too many measurements [86]. Totally, the purpose behind leading attacks on state estimation of power system can cause large-scale blackouts, for an instance, the case of Ukrainian attack of 2015. Table 2.1 shows the disturbance type, the description, possible causes, symptoms, effects, and potential solutions of these disturbances.
2. Attacks on Automatic Generation Control.

   In an interconnected power system, due to automatic generation control (AGC), the power output is regulated in which power system frequency changes in defined limits, and the power exchange between neighboring zones remain within acceptable scheduled values [85, 87]. The AGC operates such as a closed-loop feedback control system and results in minimization of human intervening. Control signals duty from the AGC should be in the range of several seconds. Consequently, it is not possible that data validation algorithms are utilized, so attackers are able to manipulate the measurements without detailed mathematics [88]. As a result of the attack on the AGC, a loud generation mismatch will be occurring which causes center of inertia motion in the generation frequencies. If the frequency falls sufficient, underfrequency relays will react and isolate large regions from the main grid. If maintenance of load generation balance in the isolated regions will not happen, the insolation regions will go on the cumulative manner and cause a large-scale load shedding or blackout [86].

3. Attack on Voltage Control.

   In the long distribution network, stability of the voltage is vital component. The increase in loading and voltage drop across the lines can cause a cascading effect and finally lead to collapse. Load Tap Changing (LTC) transformer are utilized to enhance voltage stability of a distribution system. According to the value of the voltage sensed by LTC transformer, the position of the tapes changes. The voltage value at the end of line is sent through communication channel to the control center where the LTC transformer is placed. As a result, the system is vulnerable to attacks. In the two following ways, the sensor data can be played [86]:

(a) The sensors are comprised to report values lower than the actual ones under normal operation situations. Consequently, the actuators are excited to increase the tap setting, and system operates inefficiency at a higher voltage.
(b) The voltage drops actually happen but the value data tampered in way that the base case values are read. According to the read data, it is concluded that everything is fine, but the voltage drops could increase as much as result in voltage collapse.

2.4 Power System Control

One of the main functions of electricity is to convert energy from other sources to electrical power and then deliver the energy to the consumers in near or far distances. Usually, electricity is used in other converted types of energy, for example, mechanical energy, light, and heat. One of the main advantages of electricity is that it can be controlled relatively easier than other forms of energy with high reliability and efficiency. A professionally controlled power system must have the following conditions [1, 89]:

1. The power system must meet the fluctuating demand considering both active and reactive power. Dissimilar to other forms of energy, electrical power cannot be easily and efficiently stored in adequate amounts yet. Thus, sufficient reserve including both spinning and non-spinning reserve must be considered and controlled.
2. The predefined power system standards must be maintained at all times, known as "Power System Quality," which are frequency consistency, voltage consistence, and high reliability [1].

In order to meet the aforementioned necessities, several complex devices are needed to be integrated which is depicted in Fig. 2.3.

In a generator, the controlling system includes prime mover (governor) and excitation controllers. Prime mover controller deals with speed standards and controls input energy variables such as boiler temperature, flows, and pressure.

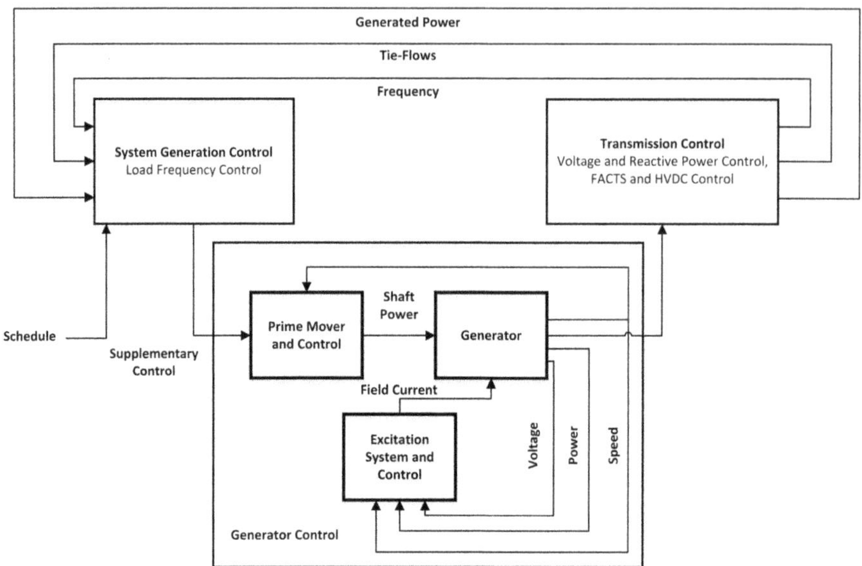

Fig. 2.3 Power system control

Excitation control regulates and maintains the voltage of the generator and injects reactive power [1].

System generation control sets the output power of the generators by calculating the network losses and loads and then balancing the generator output power against them. This helps the system to achieve the desired power exchange with neighboring networks (tie flows) [1].

The control of transmission consists of voltage and power control elements, which are switched reactors and capacitors, synchronous condensers, phase-shifter transformers, Static Var Compensator (SVC), tap-changing transformers, and HVDC transmission controllers [1, 89].

The aforementioned system concerns about the perfect performance of power system by controlling the frequency and voltages of the grid while keeping the other variables within a certain limit. In addition, these controllers have a huge effect on the dynamic performance of the grid, which demonstrates the tolerance of the grid against disturbances and faults [1, 90].

The objective of power systems control may be different, depending on the system operating condition. For normal conditions, the aim of the controlling system is to keep the voltage and frequency in the predetermined range and run the grid with efficiency. However, in abnormal conditions like contingencies or blackouts, the primary goal of the controlling system is to restore the grid and try to go back to the normal condition. In this case, efficiency and cost are the secondary concerns. These abnormal conditions are caused by single disturbances, leading to the collapse of the grid, which was known to be a secure system earlier. Human error, equipment and element malfunction, extreme climate events (such as heavy snow or rain, thunder,

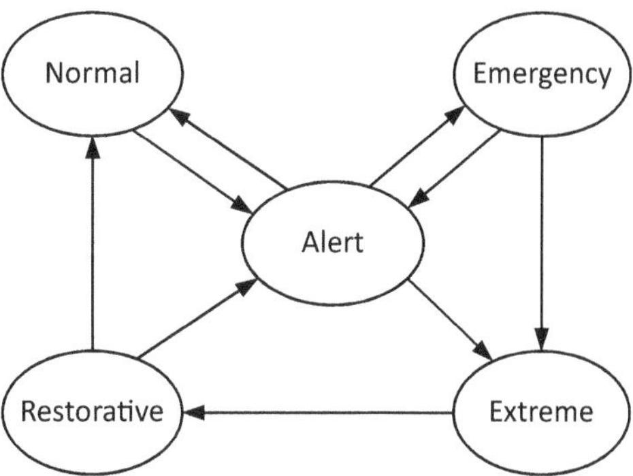

Fig. 2.4 Power system conditions

or tornado), and poor and insecure power network design may result in cascading breakdown and outages [1, 90].

In this chapter, five system conditions are introduced which are Normal, Alert, Emergency, Extreme, and Restorative. These condition are depicted in Fig. 2.4 [1, 90].

The system parameters are within the normal limit, and no element is overloaded. The grid performs efficiently with security and is able to handle a single contingency with no constraint violation [1, 90].

In alert condition, the security of the grid reduces although the system parameters are still within the standard range. However, the parameters are prone to go below or above the normal limits due to possible upcoming conditions such as thunder or storms. An equipment overload leads the grid to emergency condition; however, if the disturbance is harsh, it will place the grid directly in extreme condition. In order to bring the grid back to the normal condition, preventive actions such as increased generation or generation shifting can be considered [1].

The system will move from alert to emergency condition if enough harsh disturbances happen in alert mode. In emergency condition, the equipment are overloaded, and voltage of buses are out of the specified range; however, the system can be returned to alert condition by commencing emergency control, which are load curtailment, excitation control, fault clearing, generation runback and tripping, HVDC modulation, and fast valving [1].

If the aforementioned actions are not carried out on time or ineffectively, the grid will enter the extreme condition, in which cascading outages and partial grid shutdown will happen. In this condition, system separation and load shedding are the solution to prevent the blackout of the entire grid [1].

After extreme condition, a set of measures are taken to reconnect all the equipment and loads to the grid and restore the system. The system might enter normal or alert condition based on the taken actions and grid condition [1].

Obtaining a framework for control strategies and operator manual for each condition for power system is necessary to handle each condition fast and efficiently. Power system control will help the operator to restore the grid back to its normal condition. If the disturbance is not large, power system controls can deal with the matter; nonetheless, if the disturbance is huge, the operator needs to take measures, such as equipment switching or generator dispatch to bring the grid back to normal condition [1].

2.4.1 Power Grid Criteria for Control and Stability

A reliable power grid must be intact and solid against a different number of disturbances. So it is vital that the decision makers design the grid and operate it so that system can tolerate the disturbances and still be able to operate within the predetermined boundaries and supple the demand with minimum load loss, especially without cascading outages [1, 91].

Determining design, operating, and controlling criteria is of paramount importance in avoiding system disturbances after harsh contingencies. The implementation of criteria will guarantee that in the worst-case scenario, for all regular happening contingencies, the system will enter from normal condition to alert, instead of entering the other critical conditions, which are emergency and extreme conditions. After the grid enters the alert condition, the operators will commence taking steps to return the grid back to normal condition [1].

2.4.2 Regular Contingencies

These types of contingencies have common and have high probability of happening. These contingencies include:

1. Three-phase fault on any transmission line, generator, transformer, or bus bar, including fault clearing and reclosing.
2. Phase to ground fault on any transmission line, generator, transformer, circuit breaker, or bus bar due to circuit breaker's malfunction, signal channel, or relay.
3. Phase to ground fault on any phase, on any tower, which is cleared quickly and in normal time.
4. Any equipment loss without fault.
5. Simultaneous loss of DC bipolar generator poles.

Following the aforementioned faults, the criteria necessitates that the grid be sustained and loading and voltages of equipment be kept within the standard range. The standards are applied to the following situations [1].

1. All live and active equipment.
2. An equipment out of service, including a transmission circuit, generator, or a transformer.

2.4.3 Extreme Contingencies

These contingencies increase the risk of exceeding the voltage or loading of equipment. However, the probability of these contingencies happening is low. The extreme contingencies are as follows:

1. Loss of a whole generation station capacity.
2. Loss of all transmission lines in a right of way, connecting the generation, substation, and switching station to each other.
3. A three-phase fault on any transmission line, generator, transformer, or bus bar, without quick fault clearing due to malfunction in reclosing equipment, causing angle swing in generators.
4. An unexpected adding or removal of a large load, or malfunction of a special protection system, which is generation or load rejection, to name a few [1].

2.4.4 Frequency Control

Frequency in power grid is a fluctuating variable, which demonstrates the balance of load and generation [92]. In fact, frequency deviation is caused by the imbalance of load and generation, which will cause equipment damaging, protection devices triggering, transmission lines overloading, and deficiency in power system operation, reliability, and security [90]. Frequency is controlled and kept within the predetermined and standard range by the system operator. In the UK, two frequency ranges are introduced, which are operational range (49.8–50.2 Hz) and statutory range (49.5–50.5 Hz). Table 2.2 shows the case description of frequency deviations in UK [92, 93]. As a result of integration of renewable energy generators, frequency control by controlling conventional power plants becomes more difficult.

The frequency of power grid is proportional to the generator's rotor speed, so frequency control can be done by controlling the speed of generator turbine. This can be done by implementing governing control systems that monitor the rotor speed and follow the load change and redirect the frequency back to the standard value [90, 94].

Depending on the range of frequency deviation, different control loops are needed to restore the grid frequency stability, which are presented in Fig. 2.5.

Table 2.2 Frequency range and the description of required action to restore the frequency

Frequency range (Hz)	Description
49.8–50.2	Frequency within the normal range and maximum frequency deviation happens with a loss of generation or load up to ±300 MW
49.5–50.5	Maximum frequency deviation for loss of generation or load more than 300 MW and less than 1320 MW
Less than 49.2	Maximum frequency deviation for loss of generation more than 1320 MW and less than 1800 MW. Frequency must be returned to 49.5 Hz in 1 min

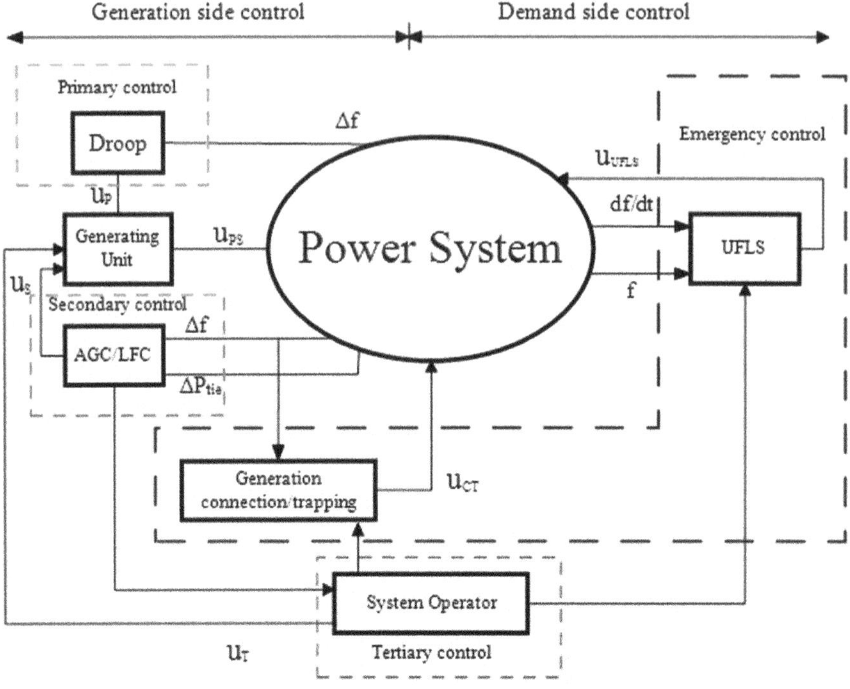

Fig. 2.5 Frequency control system [90]

When the load and generation become unbalanced, at the first step for very small frequency deviation, inertial response is activated, where the kinetic energy is stored in the rotor of the generators and lasts for a few seconds [95]. If the frequency deviation goes beyond a certain limit, primary control responds for small frequency deviation in normal condition. Secondary control, known as Load Frequency Control (LFC), comes into action considering the available reserve power when larger frequency deviations happen in non-normal conditions [90, 94]. If the secondary control loop is not enough, a tertiary control is implemented manually, and the frequency is brought back to the standard value by unit commitment and generation dispatch [95].

Two of the main objectives of LFC is to maintain grid frequency and control power interchanges with neighboring areas [90].

Nowadays, frequency control has become more important due to the changing of structure, complexity, and size of power grid. In addition, LFC has led to reducing power system economy and reliability pressure by controlling grid frequency and tie-line flows closer to predetermined values. This helps power grid with energy exchanges and electricity trading.

2.4.5 Voltage and Angle Control

The three integral parameters of power grid are nodal voltage magnitudes, nodal voltage angles, and grid frequency. Also, these parameters define the condition in which the grid situates in a particular time. Voltage and angle stability can be grouped into small and large disturbance stability. This stability points out to the damping of power swings in any part of the grid and voltage deviation beyond a certain predetermined value. Voltage and angle stability can be maintained sharply by using specific control devices which are implemented in power grid and acquire the dynamic response of the grid. These devices are FACTS, AVR, and PSS [90].

The generators operate at a fixed voltage by implementing AVR. AVR controls the excitation current of the generator. The direct current of the generator generates flux for the rotor. PSS is a controller that is located within the turbine-governing system and creates a complementary control loop to AVR system in a generator. A general implementation of PSS and AVR is shown in Fig. 2.6 [90].

ΔV_{PSS} is zero in the steady state so that it does not cause any distortion in the voltage regulation process. However, the generator speed is not constant in the transient state as the rotor swing cause ΔV to change. The voltage change is reduced when PSS generates a damping signal ΔV_{PSS} that is in phase with generator speed change Δw [90].

Voltage control consists of multiple control loops on several system levels. AVR loop maintains the generator terminal voltages besides on lower system levels and reacts in less than a second. Meanwhile, the secondary voltage control that set the voltage reference value is activated in tens of seconds or minutes. Secondary voltage control is used to coordinate setting of set points of AVR and several reactive power sources in a power grid to maintain the stability. Moreover, voltage stability can be increased by implementing higher control levels which respond in several minutes and are called as tertiary voltage control. Tertiary voltage control mainly considers the grid economic optimization and tries to minimize the cost of generation [90, 95].

Many challenges exist in voltage and frequency control in power grid. A number of these challenges are explained below [95, 96]:

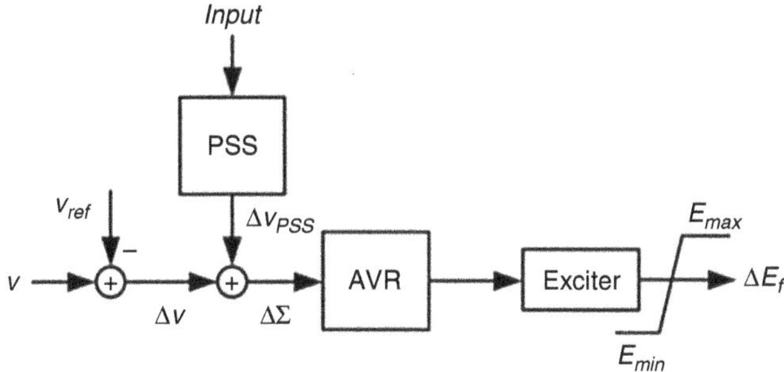

Fig. 2.6 Voltage control system [90]

1. Calculating and specifying AVR set points and input signals:
   For voltage control, AVR set points are not determined globally for the best performance of the power grid. Also, supplementary signals are on local contemplation in order to reach better stability. The need exists to create a thorough and global strategy for voltage stability and control for dynamic and static conditions.
2. Placement and control of FACTS:
   One of the biggest concerns of power system voltage control is to figure out the best locations and sites for reactive power control systems to support the grid during normal conditions and contingencies.
3. Setting the standard limits for grid parameters:
   Different components and equipment of power grid must be held within a certain standard limit. Having in mind that different standard limits have been introduced, these standard limits must be determined through analysis tools which lead to the best operation condition.
4. Novel security idea:
   The minimum number of equipment and system components that can be lost without cascading failure must be determined. This is separate from N-1 security method. This novel security idea introduces the number of outages and contingencies need to cause a blackout.
5. Fast reactive power calculation in contingencies:
   Traditional linear contingency analysis overlooks reactive power and voltage challenges. The issue of fast computation of contingencies' impact on voltage exists.
6. New control necessities:
   Due to the high penetration of renewable energy sources and nonsynchronous generators which are based on converters in order to be able to connect to the grid, these generators do not provide high inertia during frequency deviations. Thus, new control systems are needed to respond faster with higher precision to the disturbances.

The power system is a nonlinear system that its performance is affected by a wide variety of equipment and devices, which have various response time and characteristics. The features and characteristics of every major equipment and device of the power grid have an impact on the system stability. Thus, in order to examine and learn power system stability and control, a good grasp of these characteristics is vital. Another challenge of power system control is the complexity of physical side of different categories of power system, their assessment methods, and action to increase the performance of power system stability [1].

References

1. P. Kundur, *Power System Stability and Control* (McGraw-Hill, New York, 1993)
2. M. Fotuhi-Firuzabad, A. Safdarian, M. Moeini-Aghtaie, R. Ghorani, M. Rastegar, H. Farzin, Upcoming challenges of future electric power systems: sustainability and resiliency. Sci. Iranica **23**, 1565 (2016)
3. IEA, World Energy Outlook (2019)
4. A. Moradzadeh, O. Sadeghian, K. Pourhossein, B. Mohammadi-Ivatloo, A. Anvari-Moghaddam, Improving residential load disaggregation for sustainable development of energy via principal component analysis. Sustainability (Switzerland) **12**(8), 3158 (2020). https://doi.org/10.3390/SU12083158
5. C. Canizares, J. Nathwani, D. Kammen, Electricity for all: issues, challenges, and solutions for energy-disadvantaged communities, in *Proceedings of the IEEE*, vol. 107 (2019)
6. A. Moradzadeh, A. Mansour-Saatloo, B. Mohammadi-Ivatloo, A. Anvari-Moghaddam, Performance evaluation of two machine learning techniques in heating and cooling loads forecasting of residential buildings. Appl. Sci. (Switzerland) **10**(11), 3829 (2020). https://doi.org/10.3390/app10113829
7. A. Lotfi, B. Mohammadi-Ivatloo, S. Asadi, Introduction to FEW Nexus, in *Food-Energy-Water Nexus Resilience and Sustainable Development*, ed. by S. Asadi, B. Mohammadi-Ivatloo, (Springer, Switzerland, 2020)
8. O. Sadeghian, A. Moradzadeh, B. Mohammadi-Ivatloo, M. Abapour, F.P.G. Marquez, Generation units maintenance in combined heat and power integrated systems using the mixed integer quadratic programming approach. Energies **13**(11), 2840 (2020). https://doi.org/10.3390/en13112840
9. A. Demir, N. Hadžijahić, Power system planning: part I—basic principles, in *Advanced Technologies, Systems, and Applications II*, ed. by M. Hadžikadić, S. Avdaković, vol. 28, (Springer, New York, 2018)
10. A.M. Al-Shaalan, Essential aspects of power system planning in developing countries. J. King Saud Univ. Eng. Sci. **23**, 27–32 (2011)
11. A. Lotfi, S.H. Hosseini, Composite distributed generation and transmission expansion planning considering security. World Acad. Sci. Eng. Technol. Int. J. Energy Power Eng. **11** (2017)
12. A.J. Conejo, L. Baringo, Power Systems. In Power System Operations, pp. 1–15, (Springer, Cham, 2018)
13. X. Wang, J.R. McDonald, *Modern Power System Planning* (McGraw-Hill, New York, 1994)
14. A. Moradzadeh, K. Khaffafi, Comparison and evaluation of the performance of various types of neural networks for planning issues related to optimal management of charging and discharging electric cars in intelligent power grids. Emerg. Sci. J. **1**(4), 201–207 (2017). https://doi.org/10.28991/ijse-01123
15. A. Moradzadeh, B. Mohammadi-Ivatloo, M. Abapour, A. Anvari-Moghaddam, S. Gholami Farkoush, S.B. Rhee, A practical solution based on convolutional neural network for non-intrusive load monitoring. J. Ambient Intell. Humaniz. Comput. (2021). https://doi.org/10.1007/s12652-020-02720-6
16. H. Seifi, M.S. Sepasian, *Electric Power System Planning: Issues, Algorithms and Solutions* (Springer, New York, 2011)
17. R. Hemmati, R.A. Hooshmand, A. Khodabakhshian, Comprehensive review of generation and transmission expansion planning. IET Gener. Transm. Distrib. **7**(9), 955–964 (2013)
18. G. Latorre, R.D. Crus, J.M. Areiza, A. Villegas, Classification of publications and models on transmission expansion planning. IEEE Trans. Power Syst. **18**(2), 938–946 (2003)
19. R.S. Najafi, H. Khatami, Optimal and robust distribution system planning to forecasting uncertainty. Tabriz J. Electr. Eng. **46**(2), 323–332 (2016)
20. A. Moradzadeh, K. Pourhossein, Early detection of turn-to-turn faults in power transformer winding: an experimental study, in *Proceedings 2019 International Aegean Conference on Electrical Machines and Power Electronics, ACEMP 2019 and 2019 International Conference*

on Optimization of Electrical and Electronic Equipment, OPTIM 2019 (2019), pp. 199–204, https://doi.org/10.1109/ACEMP-OPTIM44294.2019.9007169
21. A. Mansour-Saatloo, A. Moradzadeh, B. Mohammadi-Ivatloo, A. Ahmadian, A. Elkamel, Machine learning based PEVs load extraction and analysis. Electronics (Switzerland) **9**(7), 1–15 (2020). https://doi.org/10.3390/electronics9071150
22. S.N. Ravandanegh, N. Jahanyari, A. Amini, N. Taghizadeghan, Smart distribution grid multi-stage expansion planning under load forecasting uncertainty. IET Gener. Transm. Distrib. **10** (5), 1136–1144 (2016)
23. P. Prakash, D.K. Khatod, Optimal sizing and siting techniques for distributed generation in distribution systems: A review. Renew. Sust. Energ. Rev. **57**, 111–130 (2016)
24. A.R. Jordehi, Allocation of distributed generation units in electric power systems: A review. Renew. Sust. Energ. Rev. **56**, 893–905 (2016)
25. J.P. Lopes, N. Hatziargyriou, J. Mutale, P. Djapic, N. Jenkins, Integrating distributed generation into electric power systems: A review of drivers, challenges and opportunities. Electr. Power Syst. Res. **77**(9), 1189–1203 (2007)
26. A. Patt, S. Pfenninger, J. Lilliestam, Vulnerability of solar energy infrastructure and output to climate change. Clim. Change **121**, 93–102 (2013). https://doi.org/10.1007/s10584-013-0887-0
27. J.A. Crook, L.A. Jones, M. Forster, R. Crook, Climate change impacts on future photovoltaic and concentrated solar power energy output. Energy Environ. Sci. **4**, 3101–3109 (2011). https://doi.org/10.1039/c1ee01495a
28. M. Gaetani, T. Huld, E. Vignati, F. Monforti-ferrario, A. Dosio, F. Raes, The near future availability of photovoltaic energy in Europe and Africa in climate-aerosol modeling experiments. Renew. Sust. Energ. Rev. **38**, 706–716 (2014). https://doi.org/10.1016/j.rser.2014.07.041
29. I.S. Panagea, I.K. Tsanis, A.G. Koutroulis, M.G. Grillakis, Climate change impact on photovoltaic energy output : the case of Greece. Adv Meteorol **2014**, 63–86 (2014)
30. S.C. Pryor, R.J. Barthelmie, Climate change impacts on wind energy: a review. Renewable and sustainable energy reviews **14**, 430–437 (2010). https://doi.org/10.1016/j.rser.2009.07.028
31. I. Tobin et al., Assessing climate change impacts on European wind energy from ENSEMBLES high-resolution climate projections. Clim. Change **128**, 99–112 (2015). https://doi.org/10.1007/s10584-014-1291-0
32. R. Davy, N. Gnatiuk, L. Pettersson, L. Bobylev, Climate change impacts on wind energy potential in the European domain with a focus on the Black Sea. Renew. Sustain. Energy Rev. **2016**, 1–8 (2017). https://doi.org/10.1016/j.rser.2017.05.253
33. C.S. Santos, D. Carvalho, A. Rocha, M. Gómez-Gesteira, Potential impacts of climate change on European wind energy resource under the CMIP5 future climate projections. Renew. Energy **101**(2017), 29–40 (2020). https://doi.org/10.1016/j.renene.2016.08.036
34. L. Chen, S.C. Pryor, D. Li, Assessing the performance of Intergovernmental Panel on Climate Change AR5 climate models in simulating and projecting wind speeds over China. Journal of Geophysical Research: Atmospheres **117**, 1–15 (2012). https://doi.org/10.1029/2012JD017533
35. C. Fant, C.A. Schlosser, K. Strzepek, The impact of climate change on wind and solar resources in southern Africa. Appl. Energy (2015). https://doi.org/10.1016/j.apenergy.2015.03.042
36. B. Kamranzad, A. Etemad-shahidi, V. Chegini, Climate change impact on wave energy in the Persian Gulf. (2015). https://doi.org/10.1007/s10236-015-0833-y
37. J.P. Sierra, M. Casas-prat, E. Campins, Impact of climate change on wave energy resource : the case of Menorca (Spain). Renew. Energy **101**, 275–285 (2017). https://doi.org/10.1016/j.renene.2016.08.060
38. D.E. Reeve, Y. Chen, S. Pan, V. Magar, D.J. Simmonds, A. Zacharioudaki, An investigation of the impacts of climate change on wave energy generation : The Wave Hub, Cornwall, UK. Renew. Energy **36**(9), 2404–2413 (2011). https://doi.org/10.1016/j.renene.2011.02.020
39. B. Hamududu, A. Killingtveit, E. Engineering, Assessing climate change impacts on global hydropower. Energies **5**(2), 305–322 (2012). https://doi.org/10.3390/en5020305

40. S.W.D. Turner, J. Yi, S. Galelli, Science of the total environment examining global electricity supply vulnerability to climate change using a high-fidelity hydropower dam model. Sci. Total Environ. (2017). https://doi.org/10.1016/j.scitotenv.2017.03.022
41. M.T.H. Van Vliet, D. Wiberg, S. Leduc, K. Riahi, Power-generation system vulnerability and adaptation to changes in climate and water resources. Nature Climate Change **6**(4), 375–380 (2016). https://doi.org/10.1038/NCLIMATE2903
42. H. Haberl et al., Global bioenergy potentials from agricultural land in 2050: sensitivity to climate change, diets and yields. Biomass and bioenergy **35**(12), 4753–4769 (2011). https://doi.org/10.1016/j.biombioe.2011.04.035
43. G. Tuck, M.J. Glendining, P. Smith, J.I. House, M. Wattenbach, The potential distribution of bioenergy crops in Europe under present and future climate. Biomass Bioenergy **30**, 183–197 (2006). https://doi.org/10.1016/j.biombioe.2005.11.019
44. J.N. Barney, J.M. Ditomaso, Bioclimatic predictions of habitat suitability for the biofuel switchgrass in North America under current and future climate scenarios. Biomass Bioenergy **34**(1), 124–133 (2010). https://doi.org/10.1016/j.biombioe.2009.10.009
45. H.A. Hager, S.E. Sinasac, Z. Gedalof, J.A. Newman, Predicting potential global distributions of two miscanthus grasses : implications for horticulture, biofuel production, and biological invasions. PLoS One **9**(6), e100032 (2014). https://doi.org/10.1371/journal.pone.0100032
46. C. Chuang, D. Sue, Performance effects of combined cycle power plant with variable condenser pressure and loading. Energy **30**, 1793–1801 (2005). https://doi.org/10.1016/j.energy.2004.10.003
47. A. Durmayaz, O.S. Sogut, Influence of cooling water temperature on the efficiency of a pressurized-water reactor nuclear-power plant. International Journal of Energy Research, **2005**, 799–810 (2006). https://doi.org/10.1002/er.1186
48. K. Linnerud, T.K. Mideksa, G.S. Eskeland, The impact of climate change on nuclear power supply. Energy J. **32**, 149–168 (2011)
49. M. Bartos et al., Environ. Res. Lett. **11** (2016)
50. R. Contreras-Lisperguer, K. De-Cuba, The potential impact of climate change on the energy sector in the Caribbean region. Organization of American States, Washington DC (2008)
51. W. Li, E. Vaahedi, P. Choudhury, Power system equipment aging. IEEE Power Energy Mag **4** (3), 52–58 (2006). https://doi.org/10.1109/MPAE.2006.1632454
52. A. Moradnouri, A. Ardeshiri, M. Vakilian, A. Hekmati, M. Fardmanesh, Survey on high-temperature superconducting transformer windings design. J. Superconductivity Novel Magnet. **33**, 2581–2599 (2020). https://doi.org/10.1007/s10948-020-05539-6
53. S.S. Kalsi, *Application of High-Temperature Superconductors to Electric Power Equipment* (IEEE Press, Wiley, 2011)
54. We Energies, Disturbance types and solutions [Online], https://www.we-energies.com/safety/power-quality/disturbance-types
55. R. Godse, S. Bhat, Mathematical morphology-based feature-extraction technique for detection and classification of faults on power transmission line. IEEE Access **8**, 38459–38471 (2020). https://doi.org/10.1109/ACCESS.2020.2975431
56. M.M. Taheri, H. Seyedi, B. Mohammadi-ivatloo, DT-based relaying scheme for fault classification in transmission lines using MODP. IET Generation Transm. Distrib. **11**(11), 2796–2804 (2017). https://doi.org/10.1049/iet-gtd.2016.1821
57. M. Mohammad Taheri, H. Seyedi, M. Nojavan, M. Khoshbouy, B. Mohammadi Ivatloo, High-speed decision tree based series-compensated transmission lines protection using differential phase angle of superimposed current. IEEE Trans. Power Deliv. **33**(6), 3130–3138 (2018). https://doi.org/10.1109/TPWRD.2018.2861841
58. H. Teimourzadeh, A. Moradzadeh, M. Shoaran, B. Mohammadi-Ivatloo, R. Razzaghi, High impedance single-phase faults diagnosis in transmission lines via deep reinforcement learning of transfer functions. IEEE Access (2021). https://doi.org/10.1109/ACCESS.2021.3051411

59. K. Chen, J. Hu, Y. Zhang, Z. Yu, J. He, Fault location in power distribution systems via deep graph convolutional networks. IEEE J. Sel. Areas Commun. **38**(1), 119–131 (2020). https://doi.org/10.1109/JSAC.2019.2951964
60. S. Zhang, Y. Wang, M. Liu, Z. Bao, Data-based line trip fault prediction in power systems using LSTM networks and SVM. IEEE Access **6**, 7675–7686 (2018). https://doi.org/10.1109/ACCESS.2017.2785763
61. G. Luo, Y. Tan, M. Li, M. Cheng, Y. Liu, J. He, Stacked auto-encoder-based fault location in distribution network. IEEE Access **8**, 28043–28053 (2020). https://doi.org/10.1109/ACCESS.2020.2971582
62. B. Li, J. Wu, L. Hao, M. Shao, R. Zhang, W. Zhao, Anti-jitter and refined power system transient stability assessment based on long-short term memory network. IEEE Access **8**, 35231–35244 (2020). https://doi.org/10.1109/ACCESS.2020.2974915
63. J. Liu, Z. Zhao, C. Tang, C. Yao, C. Li, S. Islam, Classifying transformer winding deformation fault types and degrees using FRA based on support vector machine. IEEE Access **7**, 112494–112504 (2019). https://doi.org/10.1109/access.2019.2932497
64. A. Moradzadeh, K. Pourhossein, Short circuit location in transformer winding using deep learning of its frequency responses, in *Proceedings 2019 International Aegean Conference on Electrical Machines and Power Electronics, ACEMP 2019 and 2019 International Conference on Optimization of Electrical and Electronic Equipment, OPTIM 2019* (2019), pp. 268–273, https://doi.org/10.1109/ACEMP-OPTIM44294.2019.9007176
65. A. Moradzadeh, K. Pourhossein, Application of support vector machines to locate minor short circuits in transformer windings, in *2019 54th International Universities Power Engineering Conference (UPEC)*, (2019), pp. 1–6
66. S. Lan, M.-J. Chen, D.-Y. Chen, A novel HVDC double-terminal non-synchronous fault location method based on convolutional neural network. IEEE Trans. Power Deliv. **34**(3), 848–857 (2019). https://doi.org/10.1109/TPWRD.2019.2901594
67. R. Rohani, A. Koochaki, A hybrid method based on optimized neuro-fuzzy system and effective features for fault location in VSC-HVDC systems. IEEE Access **8**, 70861–70869 (2020). https://doi.org/10.1109/ACCESS.2020.2986919
68. G. Luo, J. Hei, C. Yao, J. He, M. Li, An end-to-end transient recognition method for VSC-HVDC based on deep belief network. J. Mod. Power Syst. Clean Energy **8**(6), 1070–1079 (2020). https://doi.org/10.35833/MPCE.2020.000190
69. SGTF_EG2, *2nd Interim Report Recommendations for the European Commission on Implementation of a Network Code on Cybersecurity* (2018)
70. The European Economic and Social Committee and the Committee of the Regions Cybersecurity strategy of the E. U. European Commission. Joint communication to the European parliament, the council, *An open, safe and secure cyberspace* (2013)
71. ANL_GSS_15/4, *Analysis of critical infrastructure dependencies and interdependencies, Argonne-risk and infrastructure science center*, (2015)
72. A. Dagoumas, Assessing the impact of cybersecurity attacks on power systems. Energies (2019). https://doi.org/10.3390/en12040725
73. A. Humayed, J. Lin, F. Li, B. Luo, Cyber-physical systems security - a survey. IEEE Internet Things J. **4**(6), 1802–1831 (2017). https://doi.org/10.1109/JIOT.2017.2703172
74. B. Jimada-Ojuolape, J. Teh, Surveys on the reliability impacts of power system cyber–physical layers. Sustain. Cities Soc. **62**, 102384 (2020). https://doi.org/10.1016/j.scs.2020.102384
75. Y. Ashibani, Q.H. Mahmoud, Cyber physical systems security: Analysis, challenges and solutions. Comput. Secur. **68**, 81–97 (2017). https://doi.org/10.1016/j.cose.2017.04.005
76. H. He, J. Yan, Cyber-physical attacks and defences in the smart grid: A survey. IET Cyber-Phys. Syst. Theory Appl. **1**(1), 13–27 (2016). https://doi.org/10.1049/iet-cps.2016.0019
77. J.P.A. Yaacoub, O. Salman, H.N. Noura, N. Kaaniche, A. Chehab, M. Malli, Cyber-physical systems security: Limitations, issues and future trends. Microprocess. Microsyst. **77**, 103201 (2020). https://doi.org/10.1016/j.micpro.2020.103201

78. M. Husak, J. Komarkova, E. Bou-Harb, P. Celeda, Survey of attack projection, prediction, and forecasting in cyber security. IEEE Commun. Surv. Tutorials **21**(1), 640–660 (2019). https://doi.org/10.1109/COMST.2018.2871866
79. Y. Wang, M.M. Amin, J. Fu, H.B. Moussa, A novel data analytical approach for false data injection cyber-physical attack mitigation in smart grids. IEEE Access **5**, 26022–26033 (2017). https://doi.org/10.1109/ACCESS.2017.2769099
80. H. Karimipour, A. Dehghantanha, R.M. Parizi, K.-K.R. Choo, H. Leung, A deep and scalable unsupervised machine learning system for cyber-attack detection in large-scale smart grids. IEEE Access **7**, 80778–80788 (2019). https://doi.org/10.1109/ACCESS.2019.2920326
81. J.J.Q. Yu, Y. Hou, V.O.K. Li, Online false data injection attack detection with wavelet transform and deep neural networks. IEEE Trans. Ind. Informat. **14**(7), 3271–3280 (2018). https://doi.org/10.1109/TII.2018.2825243
82. A. Al-Abassi, H. Karimipour, A. Dehghantanha, R.M. Parizi, An ensemble deep learning-based cyber-attack detection in industrial control system. IEEE Access **8**, 83965–83973 (2020). https://doi.org/10.1109/ACCESS.2020.2992249
83. S. Soltan, P. Mittal, H.V. Poor, Line failure detection after a cyber-physical attack on the grid using Bayesian regression. IEEE Trans. Power Syst. **34**(5), 3758–3768 (2019). https://doi.org/10.1109/TPWRS.2019.2910396
84. F.C. Schweppe, J. Wildes, Power system static-state estimation, part i: Exact model. IEEE Trans. Power Apparatus Syst. **59**(1), 120–125 (1970)
85. A.J. Wood, B.F. Wollenberg, *Power Generation Operation and Control* (Wiley, New York, 2003)
86. K. Chatterjee, V. Padmini, S.A. Khaparde, Review of cyber attacks on power system operations, in *IEEE Region 10 Symposium, Conference Paper*, (2017)
87. D. P. Kothari and I. J. Padmini, Power System Engineering, New Delhi: Tata McGraw Hill Education, 2008
88. P.M. Esfahani, M. Vrakopoulou, K. Margellos, J. Lygeros, G. Andersson, Cyber Attack in a Two-Area Power System : Impact Identification using Reachability, In Proceedings of the 2010 American control conference, pp. 962–967. IEEE (2010)
89. B.F. Wollenberg, Power system operation and control, in *Power System Stability and Control*, 3rd edn., (CRC Press, 2017). https://doi.org/10.4324/b12113
90. H. Bevrani, *Robust Power System Frequency Control (Power Electronics and Power Systems)* (Springer, New York, 2009)
91. A. Moradzadeh, K. Pourhossein, B. Mohammadi-Ivatloo, F. Mohammadi, Locating inter-turn faults in transformer windings using isometric feature mapping of frequency response traces. IEEE Trans. Ind. Informat., **17**, 1–1 (2020). https://doi.org/10.1109/tii.2020.3016966
92. Z.A. Obaid, L.M. Cipcigan, L. Abrahim, M.T. Muhssin, Frequency control of future power systems: Reviewing and evaluating challenges and new control methods. J. Mod. Power Syst. Clean Energy **7**(1), 9–25 (2019). https://doi.org/10.1007/s40565-018-0441-1
93. F. Teng, Y. Mu, H. Jia, J. Wu, P. Zeng, G. Strbac, Challenges of primary frequency control and benefits of primary frequency response support from electric vehicles. Energy Procedia **88**, 985–990 (2016). https://doi.org/10.1016/j.egypro.2016.06.123
94. M.J. Bryant, R. Ghanbari, M. Jalili, P. Sokolowski, L. Meegahapola, Frequency Control Challenges in Power Systems with High Renewable Power Generation: An Australian Perspective, RMIT University (2019)
95. H.T. Nguyen, G. Yang, A.H. Nielsen, P.H. Jensen, Challenges and research opportunities of frequency control in low inertia systems, in *E3S Web of Conferences*, vol. 115, (2019). https://doi.org/10.1051/e3sconf/201911502001
96. P.W. Sauer, Reactive power and voltage control issues in electric power systems, in *Applied Mathematics for Restructured Electric Power Systems. Power Electronics and Power Systems*, ed. by J. H. Chow, F. F. Wu, J. Momoh, (Springer, Boston, 2005)

Chapter 3
Machine Learning and Power System Planning: Opportunities and Challenges

Mohammad Hosein Asgharinejad Keisami, Sasan Azad, Reza Mohammadi Chabanloo, Morteza Nazari-Heris, and Somayeh Asadi

Nomenclature

b, b_{i}, b_{C}, b_{f}, b_{o}	Corresponding biases
C_t	Cell state at time t
$\widetilde{C}_t$	Vector of new candidate values
$f(x)$	The plane that fits closest to the data
H_t	Output at timestamp t
n	Total number of features
s_t	State at timestamp t
W_{i}, W_{f}	Weight matrixes corresponding to the input and forget gates
W^T	Weights
x_i	Input label at time stamp i
x_t	Input at timestamp t
y_i	Output label at time stamp i
$y^{-}_{j,t}$	The predicted value of feature j at time t
$y_{j,\ t}$	Actual value of feature j at time t
ρ	Correlation coefficient
σ	Standard deviation

M. H. Asgharinejad Keisami · S. Azad (✉) · R. Mohammadi Chabanloo
Department of Electrical Engineering, Shahid Beheshti University, Tehran, Iran
e-mail: m_asgharinejad@sbu.ac.ir; sa_azad@sbu.ac.ir; re_mohammadi@sbu.ac.ir

M. Nazari-Heris · S. Asadi
Department of Architectural Engineering, Pennsylvania State University, University Park, PA, USA
e-mail: mun369@psu.edu; m.nazariheris.2015@ieee.org; sxa51@psu.edu; asadi@engr.psu.edu

M. Nazari-Heris et al. (eds.), *Application of Machine Learning and Deep Learning Methods to Power System Problems*, Power Systems,
https://doi.org/10.1007/978-3-030-77696-1_3

3.1 Introduction

The application of machine learning (ML) as a modern method to solve engineering problems has been the center of attention in most research topics, especially power systems. The underlying need for using such methods can be traced to the rise of intermittency and uncertainty in the power systems. This increased unpredictability of power consumption, of course, can be attributed to a variety of different factors. These factors include but are not limited to the unpredictable nature of consumer behavior, the economic development of countries and increased gross domestic product (GDP), the increasing penetration of modern technologies in developing societies such as the Internet, and the subsequent increase in demand. The exponential growth of using renewables as a clean, free alternative to fossil fuel resources, climate change, and shifting weather patterns is also a major contributing factor. Such significant changes pose a formidable challenge to the power engineering community regarding how to plan and operate such power systems with acceptable reliability and how to predict and be prepared for the future. Here is where, in particular, the ML methods shine bright. ML allows us to infer knowledge from data, and by implementing different algorithms, such as artificial neural networks, the unpredictable events can be predicted by an acceptable degree [1]. This is an essential feature of ML in various fields, from science to engineering and humanities. ML methods are widely used in power engineering in recent years because of their effectiveness and the immense variety of ML approaches, each with its benefits and costs.

It is noteworthy that terms like machine learning, artificial intelligence, and deep learning are sometimes incorrectly used interchangeably in the literatus. Figure 3.1

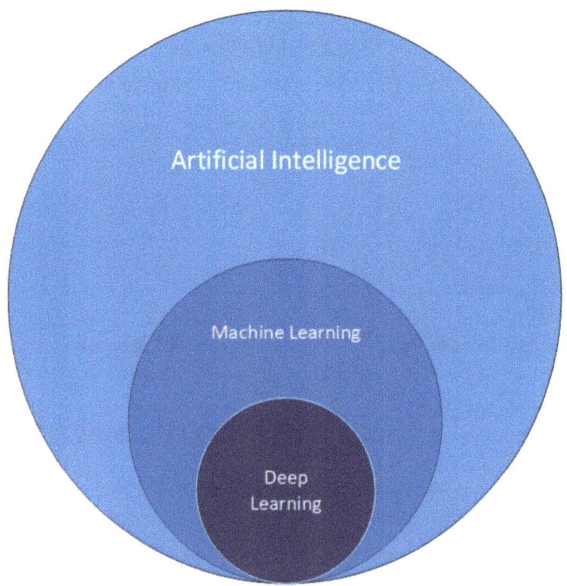

Fig. 3.1 The Van diagram of artificial intelligence, machine learning, and deep learning

demonstrates the relation between artificial intelligence, machine learning, and deep learning to clear such misconceptions.

In this chapter, in particular, the authors will inspect the use of ML in power systems planning. First, the most used methods of ML and deep learning will be briefly introduced. The authors will then show and examine the papers and studies done on applying such networks and techniques in solving the power planning problems. The examples of applications of ML in load forecasting and optimization problems will be studied, and in the final section, the concluding remarks are provided.

3.1.1 Machine Learning Methods

ML approaches vary in their architecture and methods, but in general, they can be categorized as three different main categories as follows [2]:

1. Supervised machine learning.
2. Unsupervised ML.
3. Reinforcement machine learning.

In short, the supervised learning methods require a labeled data set for training in order to create a link between the input and the expected output. Examples of these approaches include neural networks, decision tree networks, support vector machine learning, linear regression, logistic regression, and nearest neighbor.

On the other side, unsupervised learning approaches do not require a labeled data set to train and attempt to identify existing patterns. Examples for this approach include a self-organizing map (SOM), adaptive resonance theory (ART), *K*-means, principal component analysis (PCA), and support vector machine (SVM) [3].

In the reinforcement learning approaches, artificial intelligent agents interact with the system and observe the results. The feedback provided by the interaction between the dynamic system and the agents guides them into reaching the desired output. This mechanism is often simulated as a concept of a cumulative reward. Examples of these approaches are Q-learning, Monte Carlo, state-action-reward-state-action (SARSA), deep Q-learning, and deep belief networks.

Of course, other approaches implement two or more of the mentioned methods to increase the performance or the accuracy of the result. These methods, known as ensemble methods, can also be found in the literature and are worth mentioning. Figure 3.2 shows the basic overview of ML methods and their classifications [4].

Different algorithms are used in the field of machine learning, such as artificial neural networks [4], generalized neural networks [5], and fuzzy logic models [6]. These algorithms are not necessarily a subset of machine learning but computer systems that are popular and used in machine learning methods.

One of the most popular methods of ML in the power system is supervised methods. These methods usually implement the different architectures of ANN. ANNs are computer models inspired by natural neural networks such as the human brain and can be used for recognition purposes based on prior knowledge

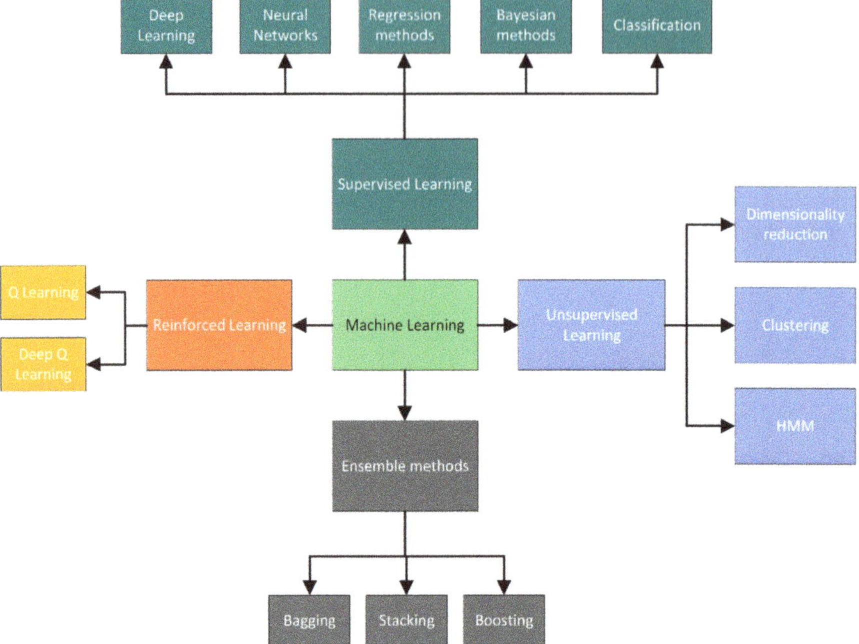

Fig. 3.2 Classification of different ML methods

and training networks. These models are useful and adoptive mathematical solutions to various problems that can estimate the relationships between data. The underlying architecture of ANN is made of layers of mathematical neurons connected through links. These links have an attributed weight, which determines the strength of the connection between neurons. As the network undergoes training, the weight attributed to these links is adjusted and weakened or strengthened. Models that use ANN with multiple hidden layers of neurons are often called deep networks. These models usually have connections between consecutive layers but do not have connections within the same layer [7].

There are a variety of models used in power systems planning. These ML models include support vector machine (SVM), long short-term memory (LSTM), and other regression models. These models have found a favorite place among researchers who use these tools in power system planning problems.

3.2 Literature Review

Machine learning methods are used in various applications and are related directly or indirectly to the problem of power system planning. For example, in renewable energy planning, machine learning methods are heavily used in Wind Power Forecasting applications and wind power estimations [8–10].

They are also used in Photo Voltaic (PV) power forecasting, where ML methods can be hugely beneficial for predicting and countering intermittency. Examples of these applications are introduced in [11–14].

The authors in [4] have surveyed recent publications on using ML models in smart grid applications. The reviewed papers cover a variety of methods and models used. They are divided into four main categories: electric load and price forecasting, ML application in fault and failure analysis, ML application in demand-side management, ML application in cyberspace security, and other miscellaneous applications.

For example, authors in [15] have proposed a framework for planning and operating distributed energy resources by decentralizing optimal power flow and learning control policies for each DER using machine learning.

In [16], based on Deep Neural Networks, a scenario generation model for a system with high penetration of renewable resources has been proposed. Authors have used data-driven approaches to capture renewable energy production patterns.

In the subject of clustering consumer load data, authors of [17] have used a Deep Learning-based Yearly Load Profile (YLP) feature extraction method from consumer load data. The feature extraction methods can be used to cluster consumers and compress metering data and is a useful tool for power system planning.

Authors in [18] have proposed a deep learning approach that combines the EMD method with the long short-term memory network. They have used it for electricity demand estimation in a day-ahead market.

To schedule solutions of multiple electric vehicle charging stations in a system in the presence of PV and Energy Storage System (ESS), authors have implemented a deep reinforcement method in [19].

Different types of loads must be considered when approaching a load planning problem. Moreover, different consumers have different consumption patterns. For this reason, a deep regression and stump tree-based ensembles models (DRTSEM) have been introduced by authors in [20] that are used for load planning. The model considers various parameters such as weight decay and leaning rate and the total number of hidden layers to achieve accurate prediction.

Compared to other applications of ML in engineering problems, load forecasting is a mature area. ML tools have played a critical role in solving the load forecasting problems and moving toward more accurate predictions using weather data and hierarchical forecasts of various zones and regions. The ML-based short-term forecast models have done a great job predicting the demand and countering the intermittent and uncertain nature of renewables in networks with high penetration of renewable power generations. Popular ML models used in the area of load forecasting include long short-term memory (LSTM), recurrent neural network (RNN), and random forest models [4]. However, one of the areas with excellent ML potential is the planning frameworks of the energy systems. In the authors' opinion in [4], the smart grid planning and operation problem can greatly benefit from using ML methods.

These studies further emphasize that the ML methods offer an excellent solution to various planning problems, and such methods encourage the researchers to implement ML tools on more problems. ML methods have been the center of attention for their application in energy management and smart cities [21] and smart dispatch systems that manage the generation by implementing artificial intelligence [22]. Some ML methods, such as reinforcement learning, are also favored by researchers in building energy control and power system planning. Using a game theory-based multi-agent method, the authors propose a demand-side management scheme that aims to control, operate, and plan the heating, ventilation, and air-conditioning systems (HVAC) and minimize the social cost in [23].

Using ML methods in an online regime is a subject of various studies, such as voltage stability observation methods [24]; however, most ML applications are made in an offline state in the planning framework. These applications include short-term [25], medium-term, and long-term load forecasting. The main challenge in load forecasting is to improve prediction accuracy when faced with big data gathered from multiple sources or various uncertainties involving weather forecast and the intermittent nature of renewable energies [26].

In [27], the authors have studied a generation expansion plan in a multi-period timescale with the objective function of minimizing the expansion costs and pollution.

The competitive aspect of such a plan in the presence of environmental policies and penalties is considered in [28]. In alignment with the subject of ML methods, the authors have considered the uncertainties of electricity price and wind speed, where ML methods are used to forecast the load using rough neural networks. The presence of environmental restrictions has made finding optimal generation expansion planning in a competitive environment challenging.

The authors in [29] believe that the total emissions can be reduced to 45% of the original amount and maintain an adequate income for generation companies participating in nearby electricity markets by conforming to pollution control policies. The study uses an ANN architecture named a multilayer perceptron. The perceptron consists of two layers, a hidden layer with ten rough neurons and a sigmoid activation function and one output layer with a linear activation function and one rough neuron. This paper's assumed neurons are based on a rough structure or RNN [27]. The basic principle of RNN is to apply the rough set theory to the neural networks in one or more links such as designing, training, or learning links. The overall purpose of these networks is to improve the performance of traditional neural networks. Neuron models based on the rough set theory are named rough neurons, and neural networks made from rough neurons are called rough neural networks. The rough neural networks accelerate the training speed of models and increase neural networks' performance but do not pose too much change in the structure of neural networks in the process [28].

3.3 Load Forecasting

Load forecasting is one of the most important studies done for power system planning and often the first stage of future expansion plans [29]. It is evident that to propose a plan for further developing a given network, the planners must understand the behavior and future electricity consumption trends. Without this understanding, proposed expansion plans will either underestimate the required energy and not provide the demand or overestimate, resulting in unnecessary investment costs.

Load forecasting is often categorized into four different categories: long-term forecasting (years), medium-term forecasting (month to a year), short-term forecasting (a day to a week), and very short-term forecasting (minutes to hours) [30].

Load forecasting can be used for both system operation and power system planning. In the case of power system operation problems, the system operators usually require short-term to very-short term studies. In power system planning, such as generation expansion planning, it is the long-term load forecasting study. Various factors affect the future of load demands. They range from the upcoming trends in technology to social welfare and economic issues and policies. This uncertainty has made load forecasting the right subject for ML methods that specialize in predicting the future and dealing with uncertainties.

In [31], authors have proposed a hybrid AI and DL method for load forecasting. The results of the load forecasting problem are implemented to inform a maintenance plan.

Authors use deep learning framework and convolutional neural network (CNN) and long short-term memory (LSTM) in [32] for short-term electric load forecasting and individual residential customers and compare the effectiveness of the proposed method with other conventional methods used in the literature.

Two methods named input attention mechanism (IAM) and hidden connection mechanism (HCM) are applied to the problem of short-term load forecasting in [33] to enhance the accuracy and efficiency of RNN-based load forecasting models.

For customer short-term load forecasting, an LSTM framework has been implemented in [34], in which, according to the authors, this method outperformed similar state-of-the-art methods in speed and accuracy.

The daily load is nonlinear and dynamic and therefore hard to predict. For this reason, authors in [35] have used a deep learning method named bespoke-gated recurrent neural network for daily peak load forecasting.

The following section examines a method introduced in [36] for the long-term load forecasting done with supervised ML. There are multiple regression models used in load forecasting, and each one has its advantage and disadvantages. The authors in [36] have used four different ML models for load forecasting, and they have showcased the results. We will introduce the fundamentals of two different models in the following section: subsets of supervised machine learning.

3.3.1 SVM (Support Vector Machine)

One of the popular algorithms used for classification and regression problems is the SVM. Vapnik and AT&T Bell Laboratories developed SVM as one of the most robust and widely used algorithms in the field. The SVM, in general, has two unique properties. It can maximize the margin of separation between two classes, and it supports nonlinear functions by utilizing different kernels [37]. There are similarities between SVM and ANN. Simultaneously, the ANN aims at estimation error minimization of training data; the SVM, on the other hand, is trying to follow the structural risk minimization principle to improve generalization error [37]. The SVM creates an optimal hyperplane classifier. By maximizing the margin of separation, the classifiers aim to classify the data without the error for minimizing the empirical risk. The empirical risk is defined as the average loss of an estimator for a finite set of data. It also minimizes the expected risk, described as the hypothesis value of loss function. The empirical risk minimizer is then used to define the SVM performance's theoretical boundaries by minimizing the training data's error. Using risk minimization is to measure the estimator's performance and find the estimator that minimizes risk over distribution [37].

First, assume that we have a training set $\{(x_1, y_1), (x_1, y_2)...(x_i, y_i)\}$ $x_i \in R^n$. The dependent variables are denoted $y_i \in R$ that are associated with each input, and n is the number of features. The main objective of SVM is to map a plane $f(x) : R^n \to R$ $f(x)$ such that it fits closest to the data and is formulated as follows:

$$f(x) = W^T \cdot X + b \tag{3.1}$$

Here, b is defined as the bias.

$f(x)$ is the function that has the most deviations σ from targets y_i. It should be noted that deviations of the resulting hyperplane that are smaller than σ are not considered. It is required that the transformation function $\phi : R^n \to R^s (s > n)$ should be applied to all of the points in the input space. This is done so that we capture the nonlinear nature of the data.

$$f(s) = W^T \cdot \phi(X) + b \tag{3.2}$$

The input space is therefore mapped to the linear feature space via the transformation and with higher dimensionality. We can write the SVR solution problem as a convex optimization problem. After the calculation, a pair of Lagrange multipliers are associated with each training point. After training the SVM and finding the optimal hyperplanes, support vectors are defined as points outside σ band and distinctly have at least one zero Lagrange multipliers [38].

3.3.2 Long Short-Term Memory Network (LSTM)

Recurrent neural networks' main idea arises from the fact that traditional ANN is not able to make a direct connection between the previous data and the following information and correcting the error. The idea of backpropagation can solve this shortcoming of the traditional ANN. These networks, commonly referred to as

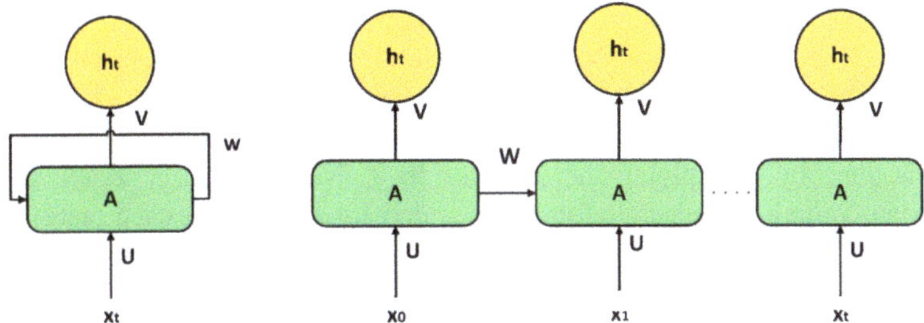

Fig. 3.3 The architecture of the RNN unit [36]

recurrent neural networks (RNN), use backpropagation to compare the error from the network's input and output until the error is below a certain threshold [38]. One instance of using such networks is in text prediction software, where the network is trained to predict the following words based on the typed words so far.

The backpropagation is done through network loops. The output of previous timestamps is used as input for the current timestamps. A simple structure of RNN can be seen in Fig. 3.3. In Fig. 3.3, x_t is input at timestamp t, s_t represents the state at timestamp t, and h_t is output at timestamp t. s_t is the current state as calculated based on the x_t and the previous hidden state s_{t-1} [36, 38].

RNN performs very well in situations where dependencies in the data are short term. However, as the gap between dependencies increases, the performance of the RNN networks decreases drastically. This problem was addressed by introducing LSTM networks by Hochreiter and Schmidhuber (1997); these networks include memory cells and gates responsible for regulating information flow across the network.

Each unit of LSTM consists of a memory part and three different types of gates responsible for handling information flow. The mechanism of an LSTM unit can be described in the following steps. First, the LSTM decides what information should be forgotten and thus removed from the cell state. This operation is performed through a sigmoid layer named the "forget gate." It produces an output number between 0 and 1 based on the h_{t-1} and x_t in the cell state C_{t-1}. Number 1 represents keeping the information, and number 0 represents forgetting this information, while any number between represents the amount of information percentage-wise that should be overlooked [39]. Please note that C_t is the cell state at time t, and W_{i} and W_{f} are weight matrixes corresponding to the input and forget gates. b_{i}, b_{C}, b_{f}, and b_{o} are all corresponding biases to the equations.

The following step is deciding which new information is going to be stored in the cell state. First, a sigmoid layer named "input gate" generates a number determining what values should be updated. Then a tanh layer creates new candidate values $\widetilde{C}_t$ in the form of a vector that is suitable to be added to the state.

$$i_t = \sigma(W_{\mathrm{i}} \cdot [h_{t-1}, x_t] + b_{\mathrm{i}}) \tag{3.3}$$

$$\widetilde{C}_t = \tanh\left(W_C \cdot [h_{t-1}, x_t] + b_C\right) \tag{3.4}$$

The old cell C_{t-1} is then updated to the new cell state C_t [38].

The next step is to forget the information that was determined to be unnecessary. This is done by multiplying the old state by f_t. Next, the C_t is added, and new candidate values are generated and are scaled for how much each state value is updated.

$$f_t = \sigma(W_f \cdot [x_t . h_{t-1}] + b_f) \tag{3.5}$$

$$C_t = f_t \odot C_{t-1} \oplus i_t \odot \widetilde{C}_t \tag{3.6}$$

The last step is the output step that is based on the cell state. The output gate, a sigmoid layer, decides the cells' parts that will be included in the output. The *tanh* operator is then acting on the cell state, which generates values between −1 and 1. These results are then multiplied by the sigmoid gate's output, and thus, the desired outputs are generated [38].

$$o_t = \sigma(W_o[h_{t-1}, x_t] + b_o) \tag{3.7}$$

$$h_t = o_t \otimes \tanh(C_t) \tag{3.8}$$

Figure 3.4. describes the general structure of a typical LSTM unit with different gates and inputs and outputs.

The authors in [36] then collect data on electricity consumption, and they use clustering techniques to categorize them. Specifically, the method of clustering used in the reference is *K*-means clustering. The main goal of clustering is categorizing and classifying data based on similarities in one or more properties of objects.

Using data clustering methods, system planners can find trends and patterns between consumers and data groups. This trend characterization can be implemented

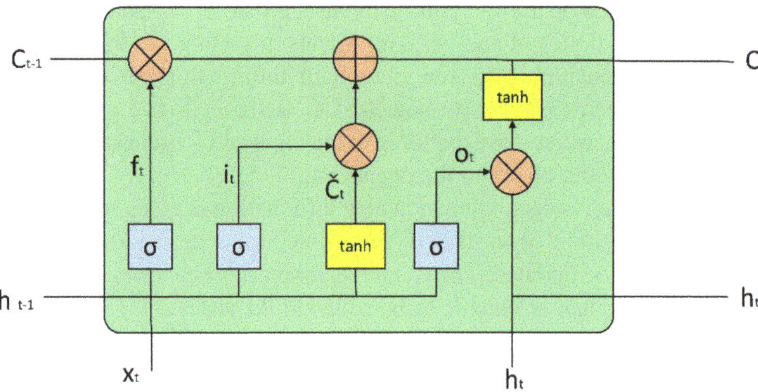

Fig. 3.4 Structure of an LSTM unit [36]

to understand the behavior of electricity consumption and load demand. By identifying similar groups, ML solutions can provide a better understanding of energy planning problems. As previously mentioned, the authors use ML algorithms to analyze the load characteristics and predict the future. The ML algorithms' primary basis, mainly the SVM and LTSM algorithm, was discussed in the previous section.

In [36], the authors have implemented four regression models, LSTM, SVM, ANN, and RNN, to perform a load demand estimation for the union territory Chandigarh in India.

To compare different ML methods, the authors in [36] have used three different evaluation measures. The first root means squared error (RMSE), one of the most used and common measures, is defined as the standard deviation of the differences between actual and predicted values.

This measure can be formulated as follows:

$$\mathrm{RMSE} = \sqrt{\frac{1}{2n}\sum_{t=1}^{2}\sum_{j=1}^{n}\left(y_{j,t} - y^{-}_{j,t}\right)^{2}} \tag{3.9}$$

$y^{-}_{j,t}$ is defined as the predicted value, and the $y_{j,\ t}$ is the actual value at time t.

The second correlation coefficient approximates the strength of the relationship between real-time observations and the predicted values and is defined as follows:

$$\rho = \mathrm{Correl}(y, y^{-}) = \frac{\mathrm{Covariance}(y, y^{-})}{\sigma_y \cdot \sigma_{y^-}} \tag{3.10}$$

Finally, the amount of deviation of the forecasted values from the actual values can be expressed as the mean absolute percentage error (% Error), which can be mathematically described as:

$$\%\mathrm{Error} = \frac{1}{2n}\sum_{t=1}^{2}\sum_{j=1}^{n}\left|\frac{\left(y^{-}_{j,t} - y_{j,t}\right)}{y_{j,t}}\right| \tag{3.11}$$

3.3.3 *Results of Load Forecasting Using Machine Learning*

According to [36], after comparing the prediction of ML methods and actual data for a given cluster of data, the LSTM network model seems to have made a more accurate prediction than SVM, ANN, and RNN models. For example, using the LSTM model, the average error is between 7 and 10% and considered acceptable. LSTM networks' advantages over SVM can be summarized as follows: LSTM can access current input timesteps. Simultaneously, for the SVM to be trained, several successive inputs are required from a properly selected time window. By including real-time demand observation, the LSTM method can support active learning; however, the SVM regression model only supports static learning.

3.4 ML in Power Systems Optimization Problems

The application of ML methods in power system planning is not limited to load forecasting. The long-term planning problems often require an optimization method to minimize the investment and operation cost or find the most suitable location capacity or other variables in the scheme of a planning problem for the planning horizon. By elongating the planning problems in years, technology advancement's inherent nature introduces uncertainties to the problem. The two most common optimization methods used for planning problems under uncertainties include stochastic programming (SP) and robust optimization (RO). The SP aims to optimize the expected value's objective functions, where the expectation is calculated regarding the probability distribution of the random variables. These variables represent the uncertainty in the problem. On the other hand, RO methods have a min-max approach to the uncertainty, and in contrast to the SP, they do not require defining the specific probability distribution functions.

In [40], the authors aim to solve long-term electric peak load and demand forecasting using optimization methods and ML tools. These studies are a critical part of generation expansion planning studies, where having a better understanding of the yearly peak load is directly responsible for better planning solutions. Therefore, the authors first implement an SVM method for long-term load forecasting problems and particle swarm optimization (PSO) method as a heuristic optimization algorithm.

Moreover, the authors in [40] try to minimize the forecasting error using a hybrid forecasting method, which uses three different ML models. These models using as a hybrid model are auto-regressive integrated moving average (ARIMA), artificial neural network (ANN), and the proposed support vector regression technique. The introduced method prioritizes each forecasting method based on the resulted errors over the existing data. The yearly peak load and total energy demand of Iran National Electric Energy system are the aimed results.

Therefore, RO establishes a set of possible realizations such as ranges of variation based on the known information about the uncertain parameters and then optimizes toward feasible solutions for all of the uncertain parameters within the uncertainty set [40]. These uncertainties open an opportunity for the ML models to be introduced to the planning problem. The authors in [40] have proposed a method for the problem of long-term energy planning. To show that there is a discrepancy among the associated stochastic solutions and other robust solutions published in the literature, they first devise an SP method considering different sources of uncertainty and multiple different PDFs to emphasize and manifest their potentials and their weaknesses in the long-term investment models in the energy planning problem. The authors then proposed a distributionally robust optimization method combined with ML that allows them to "numerically tractable, recourse-based robust formulation of the EnergyScope model," with less sensitivity to the chosen type of PDF. The authors of [40] proposed that DRO solutions vary less regarding the underlying distribution, and they claimed that it produces a more robust decision. Compared to the other DRO uses, a large dimension of underlying uncertainty has been implemented in their work that poses a significant computational challenge. Therefore, ML tools have been used to rank and decide which uncertainty parameter is the most significant and should be included in the ambiguity set definition.

Combining the ML algorithms and the DRO approach enables the authors to select the critical variables more systematically. The ML algorithm used is the Extreme Gradient Boosting (XGBoost) method, one of the models based on the regression tree model and therefore considered a supervised machine learning. In short, the XGboost algorithms sequentially makes decision trees according to the margin of error left by the predictive variables of the previous tree. This is repeated until a maximum threshold of adjustment is reached for all trees' combined performance, and the algorithm is stabilized [39]. As a result, they have been able to use their DRO method and, at the same time, cut the computational burden of considering a large number of uncertainties with the help of machine learning. Using ML algorithms to rank the variables in the problem and reduce the computational power required and time needed in an optimization problem in power system planning with multiple sources of uncertainty is one of the applications of ML in the power system planning problem.

3.5 Conclusion

In this chapter, first, the authors discussed the necessity of ML in modern-day engineering problems by focusing on different types of ML and four different categories of ML methods. In the next section, overall applications of ML in power systems engineering were investigated with a focus on power system planning. It seemed that the problem of load forecasting had been an excellent topic to introduce the application of ML in power system planning. After manifesting the structure of two supervised ML methods (i.e., SVM and LSTM), the authors reviewed the load forecasting problem's results with different ML methods according to referenced sources. In the end, the authors briefly mentioned other applications of ML tools in the power planning problem, such as the long-term optimization problem.

In conclusion, the ML methods are excellent tools in power system engineers' hands to tackle various problems. In that regard, by correctly using and implementing ML, researchers can solve problems infused with uncertainty using artificial intelligence's immense potential. As some researchers in the field have pointed out, artificial intelligence systems in power engineering and distribution and planning have great potential and require more research and resources [20]. The field of artificial intelligence and, therefore, ML has been exponentially progressing for decades. The achievements often result in unexpected researchers; hence, this field's only expected trajectory seems to be upward.

References

1. Y. Chen. Bridging Machine Learning to Power System Operation and Control (2020). https://doi.org/10.13140/RG.2.2.25463.98720. https://www.researchgate.net/publication/339988648_Bridging_Machine_Learning_to_Power_System_Operation_and_Control
2. C.M. Bishop, *Pattern Recognition and Machine Learning* (Springer, New York, 2006)
3. X. Kong, X. Liu, R. Shi, K.Y. Lee, Wind speed prediction using reduced support vector machines with feature selection. Neurocomputing **169**, 449–456 (2015)
4. M.S. Ibrahim, W. Dong, Q. Yang, Machine learning driven smart electric power systems: current trends and new perspectives. Appl. Energy **272**, 115237 (2020)
5. A. Mansour-Saatloo, A. Moradzadeh, B. Mohammadi-Ivatloo, A. Ahmadian, A. Elkamel, Machine learning based PEVs load extraction and analysis. Electronics **9**(7), 1150 (2020)
6. H. Yang, Z. Jiang, L. Haiyan, A hybrid wind speed forecasting system based on a 'decomposition and ensemble' strategy and fuzzy time series. Energies **10**(9), 1422 (2017)
7. E. Mocanu, P.H. Nguyen, M. Gibescu, Deep learning for power system data analysis, in *Big Data Application in Power Systems*, (Elsevier, Amsterdam, 2018), pp. 125–158
8. X. He et al., Research on a novel combination system on the basis of deep learning and swarm intelligence optimization algorithm for wind speed forecasting. IEEE Access **8**, 51482–51499 (2020)
9. J. Yan et al., Forecasting the high penetration of wind power on multiple scales using multi-to-multi mapping. IEEE Trans. Power Syst. **33**(3), 3276–3284 (2017)
10. T. Hu et al., Distribution-free probability density forecast through deep neural networks. IEEE Trans. Neural Netw. Learn. Syst. **31**(2), 612–625 (2019)
11. H. Zhou et al., Short-term photovoltaic power forecasting based on long short term memory neural network and attention mechanism. IEEE Access **7**, 78063–78074 (2019)
12. Ray, Biplob, et al. A New Data Driven Long-Term Solar Yield Analysis Model of Photovoltaic Power Plants. IEEE Access 8 (2020): 136223–136233. https://ieeexplore.ieee.org/document/9149581
13. Z. Zhen et al., Deep learning based surface irradiance mapping model for solar PV power forecasting using sky image. IEEE Trans. Ind. Appl. (2020)
14. G.W. Chang, L. Heng-Jiu, Integrating gray data preprocessor and deep belief network for day-ahead PV power output forecast. IEEE Trans. Sustain. Energy **11**(1), 185–194 (2018)
15. R. Dobbe et al., Toward distributed energy services: decentralizing optimal power flow with machine learning. IEEE Trans. on Smart Grid **11**(2), 1296–1306 (2019)
16. Y. Chen et al., Model-free renewable scenario generation using generative adversarial networks. IEEE Trans. Power Syst. **33**(3), 3265–3275 (2018)
17. S. Ryu et al., Convolutional autoencoder based feature extraction and clustering for customer load analysis. IEEE Trans. Power Syst. **35**(2), 1048–1060 (2019)
18. J. Bedi, D. Toshniwal, Empirical mode decomposition based deep learning for electricity demand forecasting. IEEE Access **6**, 49144–49156 (2018)
19. M.J. Shin, D.-H. Choi, J. Kim, Cooperative management for PV/ESS-enabled electric vehicle charging stations: a multiagent deep reinforcement learning approach. IEEE Trans. Ind. Inform. **16**(5), 3493–3503 (2019)
20. T. Ahmad, D. Zhang, Novel deep regression and stump tree-based ensemble models for real-time load demand planning and management. IEEE Access **8**, 48030–48048 (2020)

21. M. Zekić-Sušac, S. Mitrović, A. Has, Machine learning based system for managing energy efficiency of public sector as an approach towards smart cities. Int. J. Inf. Manag. **50**, 102074 (2020)
22. L. Yin et al., A review of machine learning for new generation smart dispatch in power systems. Eng. Appl. Artif. Intell. **88**, 103372 (2020)
23. J. Hao, Multi-agent reinforcement learning embedded game for the optimization of building energy control and power system planning. arXiv preprint arXiv:1901.07333 (2019). https://arxiv.org/abs/1901.07333
24. V. Malbasa et al., Voltage stability prediction using active machine learning. IEEE Trans. Smart Grid **8**(6), 3117–3124 (2017)
25. U. Munawar, Z. Wang, A framework of using machine learning approaches for short-term solar power forecasting. J. Electr. Eng. Technol. **15**(2), 561–569 (2020)
26. D. Zhang, X. Han, C. Deng, Review on the research and practice of deep learning and reinforcement learning in smart grids. CSEE J. Power Energy Syst. **4**(3), 362–370 (2018)
27. R.A. Mehrabadi, M.P. Moghaddam, M.K. Sheikh-El-Eslami, Generation expansion planning in multi electricity markets considering environmental impacts. J. Clean. Prod. **243**, 118611 (2020)
28. H. Liao et al., An overview on rough neural networks. Neural Comput. & Applic. **27**(7), 1805–1816 (2016)
29. S. Madadi, M. Nazari-Heris, B. Mohammadi-Ivatloo, S. Tohidi, Implementation of genetic-algorithm-based forecasting model to power system problems, in *Handbook of Research on Predictive Modeling and Optimization Methods in Science and Engineering*, (IGI Global, Philadelphia, 2018), pp. 140–155
30. I. Koprinska, M. Rana, V.G. Agelidis, Correlation and instance based feature selection for electricity load forecasting. Knowl.-Based Syst. **82**, 29–40 (2015)
31. S. Motepe, A.N. Hasan, R. Stopforth, Improving load forecasting process for a power distribution network using hybrid AI and deep learning algorithms. IEEE Access **7**, 82584–82598 (2019)
32. M. Alhussein, K. Aurangzeb, S.I. Haider, Hybrid CNN-LSTM model for short-term individual household load forecasting. IEEE Access **8**, 180544–180557 (2020)
33. M. Zhang, Z. Yu, X. Zhenghua, Short-term load forecasting using recurrent neural networks with input attention mechanism and hidden connection mechanism. IEEE Access **8**, 186514–186529 (2020)
34. W. Kong et al., Short-term residential load forecasting based on LSTM recurrent neural network. IEEE Trans. Smart Grid **10**(1), 841–851 (2017)
35. Z. Yu et al., Deep learning for daily peak load forecasting–a novel gated recurrent neural network combining dynamic time warping. IEEE Access **7**, 17184–17194 (2019)
36. J. Bedi, D. Toshniwal, Deep learning framework to forecast electricity demand. Appl. Energy **238**, 1312–1326 (2019)
37. Rastgoufard, Samin. Applications of Artificial Intelligence in Power Systems. (2018). https://scholarworks.uno.edu/td/2487/
38. C. Olah, Understanding lstm networks (2015), https://colah.github.io/posts/2015-08-Understanding-LSTMs/
39. E. Guevara et al., A machine learning and distributionally robust optimization framework for strategic energy planning under uncertainty. Appl. Energy **271**, 115005 (2020)
40. M.-R. Kazemzadeh, A. Amjadian, T. Amraee, A hybrid data mining driven algorithm for long term electric peak load and energy demand forecasting. Energy **204**, 117948 (2020)

Chapter 4
Introduction to Machine Learning Methods in Energy Engineering

Arash Moradzadeh, Behnam Mohammadi-Ivatloo, Kazem Pourhossein, Morteza Nazari-Heris, and Somayeh Asadi

4.1 Introduction

Increasing energy consumption and the penetration of renewable energy sources in traditional power systems have complicated power systems. Hence, the traditional power system has moved toward smartening in order to increase sustainability as well as supply energy demand [1, 2]. These smartening are performed through information infrastructures such as the Internet, which is also sometimes exposed to cyberattacks. On the other hand, the intelligent power system and the use of

A. Moradzadeh
Faculty of Electrical and Computer Engineering, University of Tabriz, Tabriz, Iran
e-mail: arash.moradzadeh@tabrizu.ac.ir

B. Mohammadi-Ivatloo (✉)
Faculty of Electrical and Computer Engineering, University of Tabriz, Tabriz, Iran

Department of Energy Technology, Aalborg University, Aalborg, Denmark
e-mail: bmohammadi@tabrizu.ac.ir; bmo@et.aau.dk

K. Pourhossein
Department of Electrical Engineering, Tabriz Branch, Islamic Azad University, Tabriz, Iran
e-mail: k.pourhossein@iaut.ac.ir

M. Nazari-Heris
Faculty of Electrical and Computer Engineering, University of Tabriz, Tabriz, Iran

Department of Architectural Engineering, Pennsylvania State University, University Park, PA, USA
e-mail: mun369@psu.edu; m.nazariheris.2015@ieee.org

S. Asadi
Department of Architectural Engineering, Pennsylvania State University, University Park, PA, USA
e-mail: sxa51@psu.edu; asadi@engr.psu.edu

M. Nazari-Heris et al. (eds.), *Application of Machine Learning and Deep Learning Methods to Power System Problems*, Power Systems,
https://doi.org/10.1007/978-3-030-77696-1_4

advanced monitoring systems have caused the users of power systems to face big data [3, 4]. One beneficial way to deal with big data of the power systems for proper operation and planning methodologies is the application of machine learning and deep learning methods. Data mining has been introduced as a powerful tool for analyzing huge volumes of data. This branch of data science is named based on searching for valuable information and features in a large database and identifying its patterns. Data mining has easily answered questions in science and technology that have been impossible or time consuming to resolve manually. Using statistical techniques for data assessment, feature extraction, pattern recognition, and finding relationships between data is the main idea of data mining [5, 6]. Data mining can be used in all data-based issues in science and technology. Today, this science is used dramatically in many research and industrial fields such as medicine, engineering, and even security issues related to face recognition and fingerprinting. Accordingly, in this chapter, the applications of data mining in issues related to energy engineering are discussed [7]. The fundamental of machine learning methods is automated learning from past experiences and examples without explicit programming. Various machine learning approaches have been introduced to apply to engineering problems such as supervised and unsupervised learning methods, reinforcement learning, etc. The high performance of machine learning methods in data process, classification, and prediction have made these methods popular in several engineering areas. Such methodologies have been widely applied in computer vision, speech recognition, object recognition, and content-based retrieval from multimedia [6, 8]. In the energy industry, machine learning methods have been widely used for dealing with exploration and processes of oil and gas, prediction of solar radiation, obtaining optimal operation of reactors, prediction of wind power output, forecasting of fault in power systems, and prediction of energy demands [7].

The remainder of this chapter is as follows: Section 4.2 has focused on data mining and its applications in energy engineering. Then the machine learning approaches in energy engineering is studied in Sect. 4.3, and deep learning approaches in energy engineering is investigated in Sect. 4.4. The evaluation metrics of regression and classification applications is analyzed in Sect. 4.5, and the chapter is concluded in Sect. 4.6.

4.2 Data Mining and Its Applications in Energy Engineering

The wide range of issues related to energy and power systems and especially the big data related to each of the functions of these systems has led to data mining that has high applications in power and energy systems. Data mining has a variety of techniques and algorithms, each of which is used in some way to solve problems related to energy engineering [3, 9].

Today, generating energy from renewable energy sources, transmission, and injecting it into the power grid and its distribution is fraught with many problems that have posed many challenges. On the other hand, providing long-term and short-term planning for traditional and renewable power plants to meet the demand for load and supply the required and sustainable energy is another challenging issue that is widely discussed today [10, 11]. The literature mainly deals with issues related to energy production in the power system. Today, the most widely covered issues in energy engineering are energy management and conservation in consumers' side. So that extensive studies and projects are done annually to manage and increase energy efficiency in residential and industrial applications [12–14].

In recent decades, solving such problems related to power systems has been done in various ways. A review of the literature in recent years shows the significant use of data mining applications such as artificial neural network (ANN) algorithms, machine learning, and deep learning. Due to their high capabilities in data processing, each of these algorithms has been able to solve problems related to power and energy systems. Data mining applications in energy and power systems can be classified into a variety of forecasts, detections, and analyses. Forecasting programs include load forecasting [15–17], load disaggregation [18–20], forecasting of renewable energy sources such as wind and solar energies [21–24], plug-in electric vehicles load forecasting [25–28], price forecasting in various electricity markets [29–31], and measuring the potential of different regions for the construction of the various wind and solar power plants. Diagnostic operations by data mining applications can be considered as detecting various types of faults and damages related to equipment in energy and power systems [32–36], as well as its various parts such as transmission lines and distribution networks [37, 38]. In addition, today, the cyber-physical system security and the cyberattacks detections on power and energy systems are dramatically done by data mining applications [39, 40].

Each of these programs is performed by specific algorithms. All data mining techniques are able to solve the abovementioned problems, but choosing the correct algorithm to do anything depends on the data and the type of problem. For example, load forecasting problems are mainly solved with regression algorithms due to the fact that they are based on continuous-type data. Classification algorithms mainly do detection and localization of faults and damages. Each of the regression and classification applications has suitable algorithms that can be used depending on the problem data type.

As mentioned in the above literature, machine learning and deep learning algorithms are used extensively in power and energy systems issues. The most important problem that power and energy system users mainly face is the lack of complete familiarity with the types and applications of machine learning and deep learning algorithms. Accordingly, in the following sections of this chapter, various techniques of the machine learning and deep learning will be introduced so that the structure, formulation, and application of each of them are described in detail to be used as a guide for users of power and energy systems in the employ of data mining applications.

4.3 Machine Learning Schemes in Energy Engineering

Machine learning is a tool for presenting a model or an estimate of the future based on the learnings of past events. In machine learning applications, computer algorithms predict future behavior based on learning from past experiences [41]. Machine learning algorithms based on the performed training create a mathematical model for the training data to make this prediction. The use of machine learning tools for emerging fields such as smart buildings is one of the research trends that has recently attracted the attention of research communities in several disciplines, including computer science and electrical engineering, civil engineering, and architecture. Machine learning has many training algorithms, each of which is used for specific applications of data mining science [42].

This section introduces the various machine learning algorithms used in energy engineering applications.

4.3.1 Machine Learning Methods: Application, Formulation, and Structure

Machine learning techniques are divided into two categories of regression and classification applications. In the continuation of this section, we will get acquainted with the types and applications of each of the machine learning algorithms in energy engineering.

4.3.1.1 Support Vector Machine (SMV)

SVM was introduced in 1995 by Cortes and Vapnik as a nonparametric statistical learning method [43]. The SVM is one of the most widely used supervised machine learning algorithms to solve classification, pattern recognition, and prediction problems. The main purpose of the SVM is to create a linear mapping and transform the nonlinear input area into a high-dimensional space to find a hyperplane. This hyperplane is achieved by maximizing the margin between the separating data [44]. Thus, by achieving this goal, the dataset is broken down into a number of predefined classes and separated in a manner consistent with the training samples. Figure 4.1 shows an example of an SVM structure for classification in a two-dimensional input space [45].

Linear mapping in SVM for the training dataset $Z = \{X_i, Y_i | i = 1, 2, 3, \ldots, n\}$ can be obtained as [44]:

$$\gamma = \omega^T \theta(x) + b \tag{4.1}$$

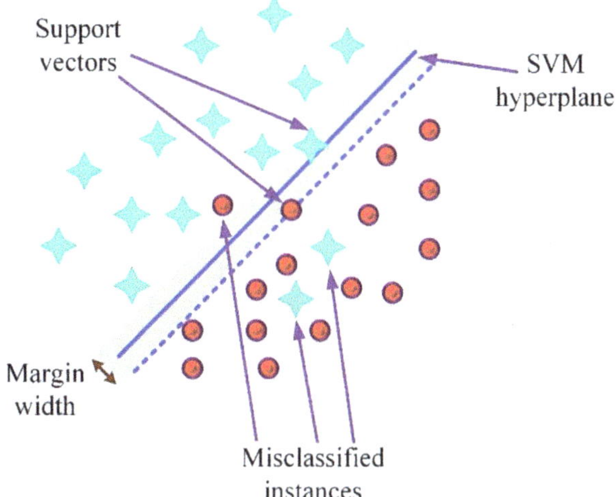

Fig. 4.1 Main structure of the SVM [45]

where ω shows the weight vector, b represents the bias, and $\theta(X)$ demonstrates the agent of a nonlinear mapping function. Calculation of support vectors related to each class are described as:

$$\begin{cases} b + W^T \cdot X_i = +1, & \text{for } d_i = +1 \\ b + W^T \cdot X_i = -1, & \text{for } d_i = -1 \end{cases} \tag{4.2}$$

where d_i is the related class, i.e., $d_i = +1$ corresponding to class A and $d_i = -1$ related to class B.

Solving the following inequality determines the training data for classes A and B, and the last objective function is achieved by Eq. (4.4) [45]:

$$\{d_i(b + w^t x_i) \geq 1 \tag{4.3}$$

$$f(x) = \text{sign}\left(\sum_{i=1}^{N} a_{0,i}\left(x^T x_i\right) + b\right) \tag{4.4}$$

where X and N show the input vectors and the support vector numbers attained in training stage, respectively. The positive parameters $a_{0,\ i}$ are employed to represent the support vectors between the input vectors.

A vector-mapping function as $\varphi(x)$ is employed to transfer inseparable data to a high-dimensional linear space and separate them. This transform allows data to be categorized using the drawn linear hyperplane. Finally, the decision function is used as follows [45]:

$$f(x) = \text{sign}\left(\sum_{i=1}^{N} a_{0,i}(\varphi(x)\varphi(x_i)) + b\right) \tag{4.5}$$

4.3.1.2 Group Method Data Handling (GMDH)

GMDH is a machine learning application and is one of the nonlinear regression modeling algorithms that was introduced in 1968 by Ivakhnenko [46]. Modeling complex systems, performance approximation, feature extraction, and discovering the relationship between input and output data in them are the most important applications of the GMDH [47]. Building a forward-forward neural network based on a quadratic transfer function through supervised learning is the main objective of the GMDH. Figure 4.2 shows the structure of the GMDH network. The mapping between the input and output variables in this structure is performed by the nonlinear function of the Volterra series according to Eq. (4.6). Volterra series analysis is performed using Eq. (4.7) [48].

$$\widehat{Y} = a_0 + \sum_{i=1}^{m} a_i X_i + \sum_{i=1}^{m}\sum_{j=1}^{m} a_{ij} X_i X_j + \sum_{i=1}^{m}\sum_{j=1}^{m}\sum_{k=1}^{m} X_j a_{ijk} X_i X_j X_k + \cdots \tag{4.6}$$

$$G(X_i, X_j) = a_0 + a_1 X_i + a_2 X_j + a_3 X_i^2 + a_1 X_j^2 + a_5 X_i X_j \tag{4.7}$$

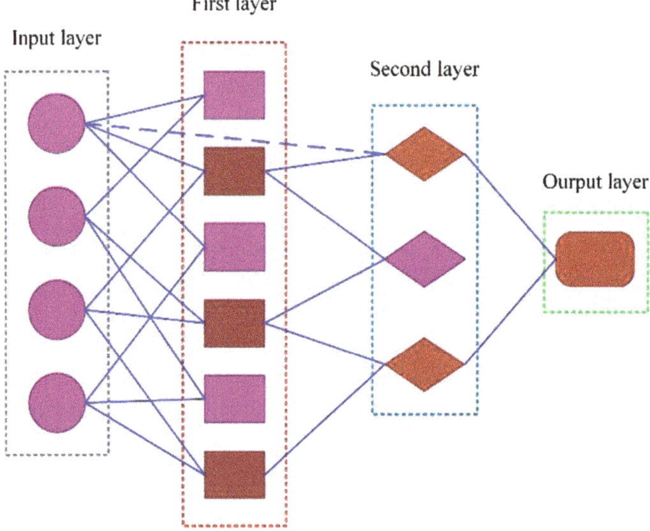

Fig. 4.2 Principle structure of the GMDH

where $\widehat{Y}$ shows the model prediction of output, X represents the inputs, m is the number of inputs, and a demonstrates the weights. Finally, by considering the principle of least squares error [49], the G function is defined to estimate the output parameter y as follows:

$$E = \frac{\sum_{i=1}^{M} (Y_i - G_i O)^2}{M} \tag{4.8}$$

$$y_i = f(X_{i1}, X_{i2}, X_{i3}, \ldots, X_{im}), \quad i = 1, 2, 3, \ldots, m \tag{4.9}$$

4.3.1.3 Support Vector Regression (SVR)

SVR was introduced as a regression version of the SVM in 1995 by Vapnik [50]. The SVR as a supportive regression suggests the e-intensive loss function (e-SVR) performance. Exact linear mapping between input and output variables, nonlinear data modeling without the overfitting problems in the training phase, and high processing speed are the most important advantages of the SVR [15, 51]. Based on mathematical relations, the general model of linear regression is as follows:

$$f(x) = \langle w, x \rangle + b \tag{4.10}$$

Calculating the weight vector based on the reduction of quadratic errors is the basis principle of the traditional regression models, while the e-SVR is based on absolute error optimization. The main objective of the e-SVR is to develop a function in which all errors lie under a certain value e but with the best generalization capacity possible and is applied based on the following relationships [15]:

$$\begin{aligned} &\text{Minimize } \frac{1}{2}\|w\|^2 \\ &\text{Subject to } \begin{cases} y_i - (\langle W, X_i \rangle + b) \leq \varepsilon \\ (\langle W, X_i \rangle + b) - y_i \leq \varepsilon \end{cases} \end{aligned} \tag{4.11}$$

$$\xi_{|\varepsilon|} = \begin{cases} 0 & \text{if } |y_i - \widehat{y}_i| < \varepsilon \\ |y_i - \widehat{y}_i| - \varepsilon & \text{otherwise} \end{cases} \tag{4.12}$$

where ξ_i and ξ_i^* represent the loose variables, y_i represents the measured output, $\widehat{y}_i$ is the predicted output, and e is a user-defined parameter. The points inside and outside the e-intensive contain the values of Eqs. (4.13) and (4.14), respectively [15, 52]:

$$\xi_i = 0 \; and \; \xi_i^* = 0 \tag{4.13}$$

$$(\xi_i > 0 \text{ and } \xi_i^* = 0) \quad \text{or} \quad (\xi_i = 0 \text{ and } \xi_i^* > 0) \tag{4.14}$$

Equation (4.11) is optimized by the Lagrangian coefficient during a dual standard optimization. Finally, after calculating the Lagrangian coefficient, the following optimized expression is obtained:

$$f(X) = \sum_{i=1}^{n} (\alpha_i - \alpha_i^*) \langle X_i, X \rangle + b \tag{4.15}$$

where α_i and α_i^* represent the Lagrange coefficients.

4.3.1.4 General Regression Neural Network (GRNN)

The GRNN was introduced in 1999 by Specht as one of the radial basic function (RBF) networks to establish nonlinear relationships between input and output variables [53, 54]. This network was a suitable alternative to the feedback error propagation training algorithm in the feed-forward neural network with very strong efficiency in regression applications. One of the most important advantages of the GRNN is its ability to present good performance for very small datasets. Despite an input parameter, it still can predict well, whereas the available data from the measurements of an operating system is usually never enough for a backpropagation neural network (BPNN) [25, 55]. In addition, the GRNN is a time series-based regression algorithm that can extract and model a continuous time series relationship between input and output variables. Figure 4.3 shows the interconnected structure of the GRNN.

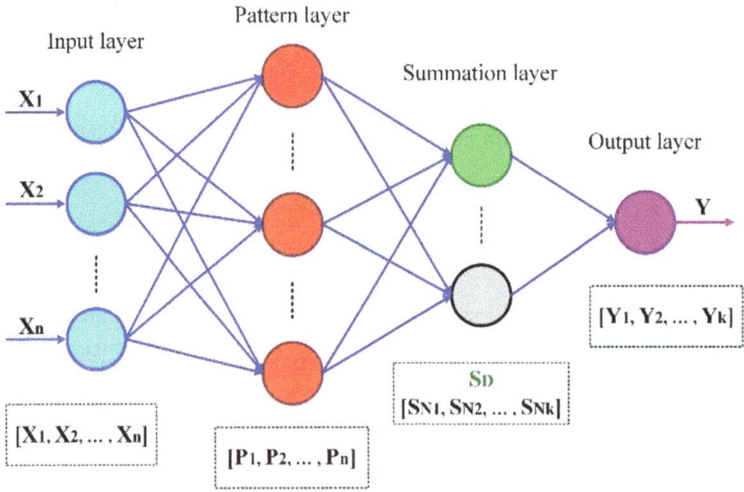

Fig. 4.3 Basic principle of the GRNN

In general, estimating the surface of regression types on the independent variable via GRNN is described in terms of mathematical formulation as follows [55]:

$$\widehat{Y} = E\left(\frac{y}{X}\right) = \frac{\int_{-\infty}^{\infty} yf(X, y)\mathrm{d}y}{\int_{-\infty}^{\infty} f(X, y)\mathrm{d}y} \tag{4.16}$$

where X shows an input vector, and Y is the forecasted output value by the GRNN. The density function should be estimated by the Parzen nonparametric prediction using the dataset of samples. This prediction is based on Gaussian kernel estimator and can be written as [56]:

$$\widehat{f}(X, y) = \frac{1}{n(2\pi)^{\frac{p+1}{2}}\sigma^{p+1}} \times \sum_{i=1}^{n} \exp\left[-\frac{(X - X_i)^2(X - X_i)}{2\sigma^2}\right] \exp\left[-\frac{(X - X_i)^2}{2\sigma^2}\right] \tag{4.17}$$

where σ represents the smoothing parameter, and X_i and Y_i show the data point of random variable x and y, respectively. The n is the sample sizes and p is the dimension number of random variable x.

Finally, the Euclidean distance between the selected input sample and the training dataset, the pattern for estimating the final output, is determined as follows [25]:

$$D_i = (X - X_i)^T(X - X_i) \tag{4.18}$$

4.3.1.5 Decision Tree

Decision tree is one of the most common data mining techniques that is produced through a two-step process of learning and classification. The decision tree uses a tree structure to describe, classify, and generalize the dataset [57]. This method's learning process is done by dividing the input dataset into two sets of training and test. After extracting the features and completing the training process, a decision tree production algorithm generates the decision tree by considering the training data as input [58]. So far, various algorithms have been introduced and used to generate decision trees. C4.5 [59], classification and regression trees [60], and ID3 [61] are widely used examples of these algorithms.

4.3.1.6 *k*-Means

k-means is one of the fastest unsupervised learning algorithms in machine learning applications used for clustering issues. This algorithm divides the input data into separate clusters through its feature vectors. The basic principle of data clustering is eigenvalues of data. So that data with the same eigenvalues are placed in a cluster [62]. The integer k must determine the number of clusters before starting the clustering operation. After determining the number of clusters, the main goal of k-means is to minimize the sum of squares within the cluster (the sum of the functions of the distance from each point of the cluster to the center k) [63, 64]. To do this, k-mean clustering for the input dataset $X_1, X_2, \ldots, X_n$, where each X corresponds to an attribute vector, divides n objects into $k(\leq n)$ and sets $S = \{S_1, S_2, \ldots, S_k\}$. Equation (4.1) describes this behavior in the mathematical model [63]:

$$\underset{S}{\arg\min} \sum_{i=1}^{k} \sum_{x \in S_i} \|X - U_i\|^2 \tag{4.19}$$

where U_i shows a typical feature vector of S_i and is the centroid of S_i.

In the next step, the distance between two data or two objects is calculated by a distance function. So far, various distance functions such as Euclidean and cosine distances have been employed to do this. Each of these distance functions has unique functions, the choice of which depends on the type of problem [62, 63].

4.4 Deep Learning Schemes in Energy Engineering

In the last decade, deep learning has been considered as the most important advancement of science in the field of computer science so that today its impact and the significant role are felt in all scientific fields and industrial applications. Deep learning can be considered an evolution of artificial intelligence based on deep and continuous learnings [65, 66]. Today, the special position of deep learning applications in issues related to energy and power systems is clearly visible. In addition, deep learning has been able to practically compensate for some of the machine learning disabilities. Various deep learning algorithms are implemented in four popular ways such as a convolutional neural network (CNN), sparse autoencoders, restricted Boltzmann machine (RBM), and long short-term memory (LSTM) [66, 67]. This section introduces the structure, formulation, and applications of these algorithms in energy engineering.

4.4.1 Deep Learning Methods: Application, Formulation, and Structure

Deep learning algorithms are divided into two categories of regression and classification applications. In the continuation of this section, we will get acquainted with each of the four types and applications of each of the deep learning algorithms in energy engineering.

4.4.1.1 Convolutional Neural Network (CNN)

CNN was first introduced and used by LeCun in 1990 for handwritten digit identification. Since then, with the efforts of researchers, this algorithm has improved day by day and is used in other applications such as computer vision, image processing, speech recognition, and some predictive applications [66, 68]. The CNN has a default structure like Fig. 4.4. As can be seen, this structure consists of convolution layers, pooling layers, fully connected layers, and, finally, the classification layer. The basic principle in the structure of the CNN is the automatic extraction of hidden features in the input data structure through several layers in a row by the filters used in the convolution layers [69].

Each of these layers has unique applications in the structure of the CNN. Convolution layers, which use multiple kernels as filters in their structure, are responsible for extracting the features of the input data. The pooling layers are employed to aggregate the extracted features from convolution layers in order to prevent overfitting. In the CNN structure, the features pooled by each pooling layer are used as input for the next convolution layer. The last pooling layer collects the final extracted features and converts them into a feature space as input for the fully connected layers. The structure of fully connected layers acts like a feed-forward neural network and is responsible for determining the weight and bias associated with data connections. After weighing and passing the training step in this layer, finally, in the last layer of the CNN structure, a Softmax function is utilized to classify the data and present the final output [70, 71].

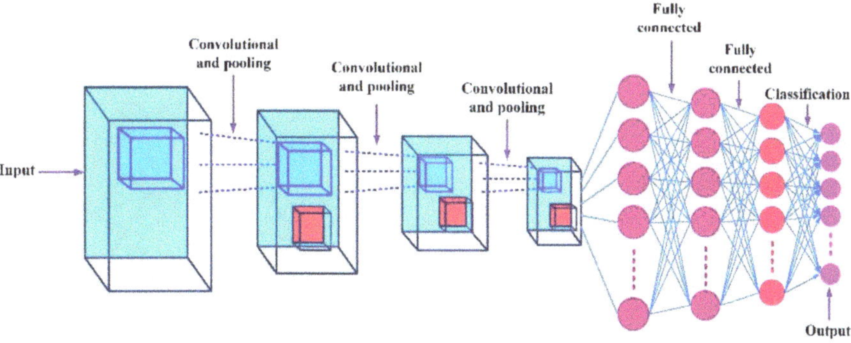

Fig. 4.4 Layer-to-layer structure of the CNN [69]

4.4.1.2 Autoencoders

Autoencoders are one of the deep learning applications used for unsupervised training of data. These networks have a structure similar to feed-forward neural networks and consist of two components, including an encoder and a decoder. As Fig. 4.5 shows, an input layer, several hidden layers, and an output layer of the same size as the input layer form the structure of these networks [72]. Minimizing the mean reconstruction loss is one of the most important issues to consider in the training of autoencoders. In the training phase of these networks, with the proper reconstruction of the input data, the maximum information about the original input data is stored by the neurons in the hidden layers. To activate in the layers of this network, nonlinear activation functions such as ReLU, tanh, and sigmoid are employed. These functions make the network architecture more abstract for learning ideal and hierarchical features and enable the network to identify complex and useful patterns [66, 68]. Autoencoders are divided into three categories including sparse autoencoder (SAE) [73], denoising autoencoder (DAE) [74], and contractive autoencoder (CAE) [75].

4.4.1.3 Recurrent Neural Network (RNN)

RNNs are one of the deep learning algorithms and one of the most interesting applicationsof backpropagation networks. Based on their structure, these networks are used for problems that have consecutive and interconnected inputs. The RNNs are very powerful dynamic systems and perform better than other deep learning algorithms for the series and sequential data [66]. Each input sequence is processed separately in the training phase and held as a state vector in its hidden units. At the end of the training phase, a strong database of the history of all past elements is available. When the outputs of the hidden layer are considered as discrete over time, it is as if they are the output of different neurons from a deep multilayer network

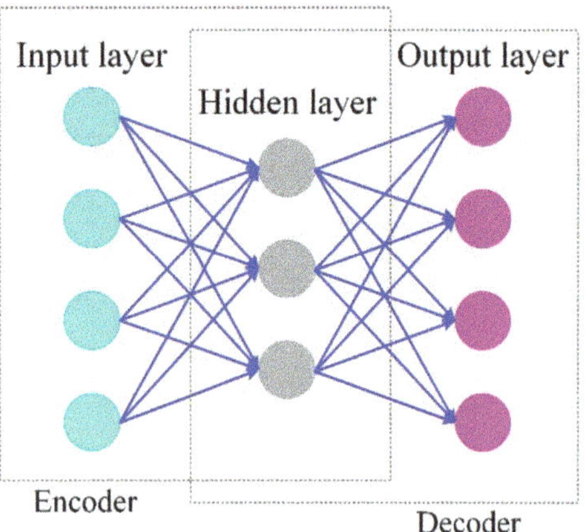

Fig. 4.5 Principle structure of the autoencoders

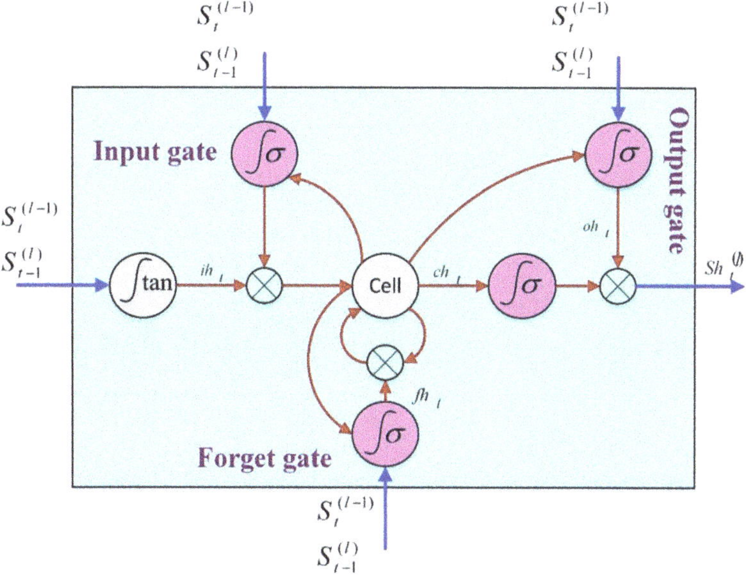

Fig. 4.6 Structure of an LSTM unit

[65]. The main purpose of the RNNs is to learn the long-term dependencies of data and to model them for future predictions. But theoretical and empirical evidence has shown that learning and modeling information for a long time has some problems. Despite all the advantages that these networks have, it has been proven that their training is problematic due to the growth or shrinkage of the backpropagated gradients at each time step. So during different stages, they usually have problems such as explosion or overfitting [76, 77]. To solve this problem and process long-term data, the idea of network reinforcement was used. The first proposal to do so was the long short-term memory (LSTM) networks, introduced in 1997 by Hockeriter et al. [78]. The LSTM has the ability to train for tasks that require knowledge of multiple previous modes by retaining information about the previous status. The network has dramatically reduced the RNN restrictions and eliminated problems with gradients loss and explosions. Figure 4.6 shows the structure of an LSTM unit. It is observed that the signal flow in this structure is done by blocks of the memory cell state, and the adjustment of this structure is done using input, forget, and output gates. Everything stored, read, and written in this cell is controlled by these gates [15, 79]. The mathematical formulation of the LSTM is described in detail in [15].

4.4.1.4 Restricted Boltzmann Machine (RBM)

An RBM is an advanced neural network and one of the deep learning applications. RBM can utilize unsupervised learning algorithms to solve regression problems,

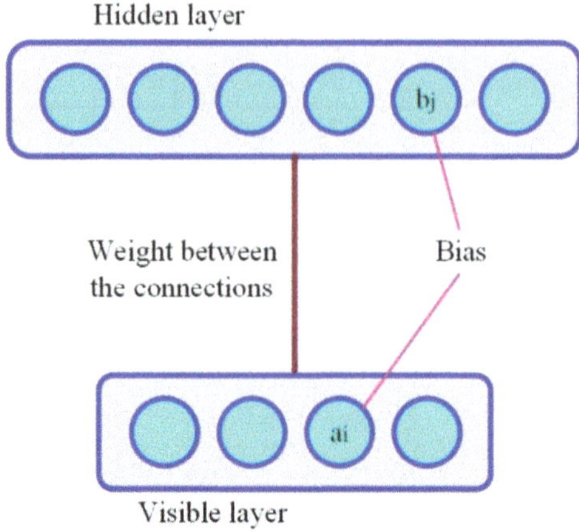

Fig. 4.7 Interconnected structure of the RBM

categorize, predict, and construct nonlinear models of unlabeled data [65]. As shown in Fig. 4.7, the RBM consists of a visible input layer and a single hidden layer that are completely interconnected, but there is no interlayer communication in this diagram. The visible layers receive the input data, and the input-related features are extracted in the hidden layers [65, 72]. The network is trained to perform possible reconstruction of the inputs by increasing the performance of the vector probability in the visible units. In the training phase, by repeatedly updating the weight and network bias connections using an algorithm called contrastive divergence, the probability of entering a specific dataset according to network parameters can be maximized. This helps to extract useful and new features from the inputs in the hidden layer. The RBMs are divided into two types of Deep Belief Network (DBN) and Deep Boltzmann Machine (DBM) [68]. The structure and function of each of these types are different, and each of which are described in detail in [80, 81].

4.5 Evaluating the Results of Regression and Classification Applications

Machine learning and deep learning techniques are used based on the various algorithms introduced in this chapter. Each of these algorithms specifically deals with classification and regression problems. The results of machine learning and deep learning algorithms will not be equal for the same data. But their effectiveness and performance can be achieved by evaluating the results [15, 82]. In comparing and evaluating the results, care should be taken to make comparisons for the same data. So far, many statistical evaluation indicators have been introduced and utilized for this evaluation [83]. It should be noted that each of the classification and regression functions has its own unique performance evaluation indicators. Indicators that evaluate the results of regression applications include the Coefficient of

Variation (CV), Mean Absolute Error (MAE), Mean Absolute Percentage Error (MAPE), Mean Bias Error (MBE), Mean Squared Error (MSE), Root Mean Square Error (RMSE), R – Squared (R^2), Error Rate (δ_i), Relative Absolute Error (RAE), Coefficient of Variation of Root Mean Square Error (CVRMSE), Weighted Mean Absolute percent Error (WMAPE), and Mean Bias Error (MBR) [83–86]. Each of these indicators has unique definitions and functions. CV is defined as the relative standard deviation and a standardized measure of dispersion of a probability distribution. MAE is the measure of errors between paired observations. MBE is used as an indicator to estimate the average bias in the model and to decide on measures to correct the model bias. MSE is used to measure the difference between the mean squares of the predicted and actual values. R^2 is one of the most important statistical indicators in regression applications. This index is used to measure the variance ratio of a dependent variable as described by an independent variable in the model. RAE is used to evaluate the performance of regression models and is expressed in a ratio by comparing an average error with the errors for each predicted value. Each of these indicators is calculated based on the following equations:

$$\mathrm{CV} = \frac{\sqrt{\frac{\sum_{i=1}^{n}(y_i-x_i)^2}{n}}}{\overline{x}} \tag{4.20}$$

$$\mathrm{MAE} = \frac{1}{n}\sum_{i=1}^{n}|y_i - x_i| \tag{4.21}$$

$$\mathrm{MAPE} = \frac{1}{n}\sum_{i=1}^{n}\left|\frac{y_i - x_i}{x_i}\right| \times 100 \tag{4.22}$$

$$\mathrm{MBE} = \frac{\frac{\sum_{i=1}^{n}(y_i-x_i)}{n}}{\overline{x}} \tag{4.23}$$

$$\mathrm{MSE} = \frac{1}{n}\sum_{i=1}^{n}(y_i - x_i)^2 \tag{4.24}$$

$$\mathrm{RMSE} = \sqrt{\frac{\sum_{i=1}^{n}(y_i - x_i)^2}{n}} \tag{4.25}$$

$$R^2 = 1 - \frac{\sum_{i=1}^{n}(y_i - x_i)^2}{\sum_{i=1}^{n}(x_i - \overline{x})^2} \tag{4.26}$$

$$\delta_i = x_i(1 - x_i)(y_i - x_i) \tag{4.27}$$

$$\mathrm{RAE} = \frac{\sum_{i=1}^{n}|y_i - x_i|}{\sum_{i=1}^{n}|x_i - \overline{x}|} \tag{4.28}$$

$$\mathrm{CVRMSE} = \frac{\sqrt{\sum_{i=1}^{n}(x_i - y_i)^2}}{\left(\sum_{i=1}^{n}x_i\right)/n} \tag{4.29}$$

$$\mathrm{WMAPE} = \frac{\sum_{i=1}^{n}\left|\frac{x_i - y_i}{x_i}\right| \times x_i}{\sum_{i=1}^{n}x_i} \tag{4.30}$$

$$\mathrm{MBR} = \frac{1}{n}\sum_{i=1}^{n}(y_i - x_i) \tag{4.31}$$

where y_i and x_i represent the predicted value and target value at ith time point, respectively, $\bar{x}$ shows the average of target values, and n illustrates the number of total data samples.

Statistical evaluation indicators such as accuracy (Acc), error rate (err), precision (p), recall (r), and $F1$-Score are used to evaluate the performance of the results obtained from classification algorithms [87, 88]. The computational formulation of each of these performance evaluation indicators is as follows:

$$\mathrm{Acc} = \frac{\mathrm{tp} + \mathrm{tn}}{\mathrm{tp} + \mathrm{fp} + \mathrm{tn} + \mathrm{fn}} \tag{4.32}$$

$$\mathrm{err} = \frac{\mathrm{fp} + \mathrm{fn}}{\mathrm{tp} + \mathrm{fp} + \mathrm{tn} + \mathrm{fn}} \tag{4.33}$$

$$p = \frac{\mathrm{tp}}{\mathrm{tp} + \mathrm{fp}} \tag{4.34}$$

$$r = \frac{\mathrm{tp}}{\mathrm{tp} + \mathrm{fn}} \tag{4.35}$$

$$F1\text{-Score} = \frac{2 \times p \times r}{p + r} \tag{4.36}$$

where tp and tn are the true positive and true negative, respectively. fp and fn shows the false positive and false negative, respectively.

As mentioned in the literature, data mining applications have found a special position in the various sciences and industries. Accordingly, using machine learning applications to predict the energy price in order to participate in electricity markets can be of great benefit to both consumers and energy producers. In addition, the use of these intelligent algorithms as a tool to estimate the impact of electric vehicles on reducing environmental pollution in metropolitan areas and improving the monitoring of various components of power systems can be considered as future research on this issue.

4.6 Conclusions

This chapter studied the application of machine learning and deep learning methods in energy system problems by concentrating on the literature review and studies of application, formulation, and structure of such methods. Accordingly, the introduction and main areas of studies around machine learning and deep learning methods were provided. Then the basics of various machine learning and deep learning methods were discussed. Literature on machine learning and deep learning methods, training process, and applications of each method were introduced and analyzed. To introduce and highlight the role of such methods in energy systems problems, some methods were selected, and their learning, structure, formulation, mode of operation, and application were discussed. The modeling and formulation of machine learning and deep learning were introduced in detail by focusing on their basics. This chapter can be beneficial for researchers working on the application of machine learning and deep learning approaches to deal with problems of energy systems.

References

1. O. Sadeghian, A. Moradzadeh, B. Mohammadi-Ivatloo, B. Mohammadi-Ivatloo, M. Abapour, F.P.G. Marquez, Generation units maintenance in combined heat and power integrated systems using the mixed integer quadratic programming approach. Energies **13**(11), 2840 (2020). https://doi.org/10.3390/en13112840
2. S. Abapour, M. Nazari-Heris, B. Mohammadi-Ivatloo, M. Tarafdar Hagh, Game theory approaches for the solution of power system problems: a comprehensive review. Arch. Comput. Methods Eng. **27**(1), 81–103 (2020). https://doi.org/10.1007/s11831-018-9299-7
3. S. Pan, T. Morris, U. Adhikari, Developing a hybrid intrusion detection system using data mining for power systems. IEEE Trans. Smart Grid **6**(6), 3104–3113 (2015). https://doi.org/10.1109/TSG.2015.2409775
4. E. Hossain, I. Khan, F. Un-Noor, S.S. Sikander, M.S.H. Sunny, Application of big data and machine learning in smart grid, and associated security concerns: a review. IEEE Access **7**, 13960–13988 (2019). https://doi.org/10.1109/ACCESS.2019.2894819
5. I.H. Witten, E. Frank, M.A. Hall, C.J. Pal, *Data Mining: Practical Machine Learning Tools and Techniques* (Elsevier, New York, 2016)
6. Z. Feng, Y. Zhu, A survey on trajectory data mining: techniques and applications. IEEE Access **4**, 2056–2067 (2016). https://doi.org/10.1109/ACCESS.2016.2553681
7. Y. Zheng, Trajectory data mining: an overview. ACM Trans. Intell. Syst. Technol. **6**(3), 1–41 (2015). https://doi.org/10.1145/2743025
8. A. Moradzadeh, K. Pourhossein, PCA-assisted location of small short circuit in transformer winding, in *2020 28th Iranian Conference on Electrical Engineering (ICEE)*, (2020), pp. 1–6. https://doi.org/10.1109/icee50131.2020.9260815
9. M. Bagheri, R. Esfilar, M.S. Golchi, C.A. Kennedy, A comparative data mining approach for the prediction of energy recovery potential from various municipal solid waste. Renew. Sust. Energ. Rev. **116**, 109423 (2019). https://doi.org/10.1016/j.rser.2019.109423
10. Y. Noorollahi, A. Golshanfard, A. Aligholian, B. Mohammadi-ivatloo, S. Nielsen, A. Hajinezhad, Sustainable energy system planning for an industrial zone by integrating electric vehicles as energy storage. J. Energy Storage **30**, 101553 (2020). https://doi.org/10.1016/j.est.2020.101553

11. M. Ghahramani, M. Nazari-Heris, K. Zare, B. Mohammadi-Ivatloo, *Robust Optimal Planning and Operation of Electrical Energy Systems* (Springer, Cham, 2019)
12. D.A.C. Narciso, F.G. Martins, Application of machine learning tools for energy efficiency in industry: a review. Energy Rep. **6**, 1181–1199 (2020). https://doi.org/10.1016/j.egyr.2020.04.035
13. A. Moradzadeh, A. Mansour-Saatloo, B. Mohammadi-Ivatloo, A. Anvari-Moghaddam, Performance evaluation of two machine learning techniques in heating and cooling loads forecasting of residential buildings. Appl Sci (Switzerland) **10**(11), 3829 (2020). https://doi.org/10.3390/app10113829
14. S. Fathi, R. Srinivasan, A. Fenner, S. Fathi, Machine learning applications in urban building energy performance forecasting: a systematic review. Renew. Sust. Energ. Rev. **133**, 110287 (2020). https://doi.org/10.1016/j.rser.2020.110287
15. A. Moradzadeh, S. Zakeri, M. Shoaran, B. Mohammadi-Ivatloo, F. Mohamamdi, Short-term load forecasting of microgrid via hybrid support vector regression and long short-term memory algorithms. Sustainability (Switzerland) **12**(17), 7076 (2020). https://doi.org/10.3390/su12177076
16. S.S. Roy, P. Samui, I. Nagtode, H. Jain, V. Shivaramakrishnan, B. Mohammadi-ivatloo, Forecasting heating and cooling loads of buildings: A comparative performance analysis. J. Ambient. Intell. Humaniz. Comput. **11**(3), 1253–1264 (2020). https://doi.org/10.1007/s12652-019-01317-y
17. G. Chitalia, M. Pipattanasomporn, V. Garg, S. Rahman, Robust short-term electrical load forecasting framework for commercial buildings using deep recurrent neural networks. Appl. Energy **278**, 115410 (2020). https://doi.org/10.1016/j.apenergy.2020.115410
18. A. Moradzadeh, O. Sadeghian, K. Pourhossein, B. Mohammadi-Ivatloo, A. Anvari-Moghaddam, Improving residential load disaggregation for sustainable development of energy via principal component analysis. Sustainability (Switzerland) **12**(8), 3158 (2020). https://doi.org/10.3390/SU12083158
19. Y.T. Quek, W.L. Woo, T. Logenthiran, Load disaggregation using one-directional convolutional stacked long short-term memory recurrent neural network. IEEE Syst. J. **14**(1), 1395–1404 (2020). https://doi.org/10.1109/JSYST.2019.2919668
20. M. Kaselimi, E. Protopapadakis, A. Voulodimos, N. Doulamis, A. Doulamis, Multi-channel recurrent convolutional neural networks for energy disaggregation. IEEE Access **7**, 81047–81056 (2019). https://doi.org/10.1109/ACCESS.2019.2923742
21. A. Ahmadi, M. Nabipour, B. Mohammadi-Ivatloo, A.M. Amani, S. Rho, M.J. Piran, Long-term wind power forecasting using tree-based learning algorithms. IEEE Access **8**, 151511–151522 (2020). https://doi.org/10.1109/ACCESS.2020.3017442
22. M. AlKandari, I. Ahmad, Solar power generation forecasting using ensemble approach based on deep learning and statistical methods. Appl. Comput. Inform. (2019). https://doi.org/10.1016/j.aci.2019.11.002
23. H. Demolli, A.S. Dokuz, A. Ecemis, M. Gokcek, Wind power forecasting based on daily wind speed data using machine learning algorithms. Energy Convers. Manag. **198**, 111823 (2019). https://doi.org/10.1016/j.enconman.2019.111823
24. C. Voyant et al., Machine learning methods for solar radiation forecasting: A review. Renew. Energy **105**, 569–582 (2017). https://doi.org/10.1016/j.renene.2016.12.095
25. A. Mansour-Saatloo, A. Moradzadeh, B. Mohammadi-Ivatloo, A. Ahmadian, A. Elkamel, Machine learning based PEVs load extraction and analysis. Electronics (Switzerland) **9**(7), 1–15 (2020). https://doi.org/10.3390/electronics9071150
26. A. Ahmadian, M. Sedghi, H. Fgaier, B. Mohammadi-ivatloo, M.A. Golkar, A. Elkamel, PEVs data mining based on factor analysis method for energy storage and DG planning in active distribution network: introducing S2S effect. Energy **175**, 265–277 (2019). https://doi.org/10.1016/j.energy.2019.03.097

27. H. Jahangir et al., Charging demand of plug-in electric vehicles: forecasting travel behavior based on a novel rough artificial neural network approach. J. Clean. Prod. **229**, 1029–1044 (2019). https://doi.org/10.1016/j.jclepro.2019.04.345
28. A. Moradzadeh, K. Khaffafi, Comparison and evaluation of the performance of various types of neural networks for planning issues related to optimal management of charging and discharging electric cars in intelligent power grids. Emerg. Sci. J. **1**(4), 201–207 (2017). https://doi.org/10.28991/ijse-01123
29. J. Nowotarski, R. Weron, Recent advances in electricity price forecasting: a review of probabilistic forecasting. Renew. Sust. Energ. Rev. **81**, 1548–1568 (2018). https://doi.org/10.1016/j.rser.2017.05.234
30. R. Weron, Electricity price forecasting: a review of the state-of-the-art with a look into the future. Int. J. Forecast. **30**(4), 1030–1081 (2014). https://doi.org/10.1016/j.ijforecast.2014.08.008
31. K. Wang, C. Xu, Y. Zhang, S. Guo, A.Y. Zomaya, Robust big data analytics for electricity price forecasting in the smart grid. IEEE Trans. Big Data **5**(1), 34–45 (2017). https://doi.org/10.1109/tbdata.2017.2723563
32. A. Moradzadeh, K. Pourhossein, B. Mohammadi-Ivatloo, F. Mohammadi, Locating inter-turn faults in transformer windings using isometric feature mapping of frequency response traces. IEEE Trans. Ind. Inform. (2020). https://doi.org/10.1109/tii.2020.3016966
33. A. Moradzadeh, K. Pourhossein, Application of support vector machines to locate minor short circuits in transformer windings, in *2019 54th International Universities Power Engineering Conference (UPEC)*, (2019), pp. 1–6
34. H. Momeni, N. Sadoogi, M. Farrokhifar, H.F. Gharibeh, Fault diagnosis in photovoltaic arrays using GBSSL method and proposing a fault correction system. IEEE Trans. Ind. Inform. **16**(8), 5300–5308 (2020). https://doi.org/10.1109/TII.2019.2908992
35. D.N. Coelho, G.A. Barreto, C.M.S. Medeiros, J.D.A. Santos, Performance comparison of classifiers in the detection of short circuit incipient fault in a three-phase induction motor, in *2014 IEEE Symposium on Computational Intelligence for Engineering Solutions (CIES)*, (2014), pp. 42–48. https://doi.org/10.1109/CIES.2014.7011829
36. A. Moradzadeh, K. Pourhossein, Early detection of turn-to-turn faults in power transformer winding: an experimental study, in *Proceedings 2019 International Aegean Conference on Electrical Machines and Power Electronics, ACEMP 2019 and 2019 International Conference on Optimization of Electrical and Electronic Equipment, OPTIM 2019*, (2019), pp. 199–204. https://doi.org/10.1109/ACEMP-OPTIM44294.2019.9007169
37. S. Zhang, Y. Wang, M. Liu, Z. Bao, Data-based line trip fault prediction in power systems using LSTM networks and SVM. IEEE Access **6**, 7675–7686 (2018). https://doi.org/10.1109/ACCESS.2017.2785763
38. M. Mohammad Taheri, H. Seyedi, M. Nojavan, M. Khoshbouy, B. Mohammadi Ivatloo, High-speed decision tree based series-compensated transmission lines protection using differential phase angle of superimposed current. IEEE Trans. Power Deliv. **33**(6), 3130–3138 (2018). https://doi.org/10.1109/TPWRD.2018.2861841
39. J.J.Q. Yu, Y. Hou, V.O.K. Li, Online false data injection attack detection with wavelet transform and deep neural networks. IEEE Trans. Ind. Inform. **14**(7), 3271–3280 (2018). https://doi.org/10.1109/TII.2018.2825243
40. A. Al-Abassi, H. Karimipour, A. Dehghantanha, R.M. Parizi, An ensemble deep learning-based cyber-attack detection in industrial control system. IEEE Access **8**, 83965–83973 (2020). https://doi.org/10.1109/ACCESS.2020.2992249
41. D. Djenouri, R. Laidi, Y. Djenouri, I. Balasingham, Machine learning for smart building applications. ACM Comput. Surv. **52**(2), 1–36 (2019). https://doi.org/10.1145/3311950
42. I.H. Witten, E. Frank, M.A. Hall, *Data Mining: Practical Machine learning* (Elsevier, New York, 2011)
43. C. Cortes, V. Vapnik, Support-vector networks. Mach. Learn. **20**(3), 273–297 (1995). https://doi.org/10.1023/A:1022627411411

44. A. Zendehboudi, M.A. Baseer, R. Saidur, Application of support vector machine models for forecasting solar and wind energy resources: a review. J. Clean. Prod. **199**, 272–285 (2018). https://doi.org/10.1016/j.jclepro.2018.07.164
45. A. Moradzadeh, S. Zeinal-Kheiri, B. Mohammadi-Ivatloo, M. Abapour, A. Anvari-Moghaddam, Support vector machine-assisted improvement residential load disaggregation, in *2020 28th Iranian Conference on Electrical Engineering (ICEE)*, (2020), pp. 1–6. https://doi.org/10.1109/icee50131.2020.9260869
46. A.G. Ivakhnenko, Polynomial theory of complex systems. IEEE Trans. Syst. Man Cybern. **1**(4), 364–378 (1971). https://doi.org/10.1109/TSMC.1971.4308320
47. I. Ebtehaj, H. Bonakdari, A.H. Zaji, H. Azimi, F. Khoshbin, GMDH-type neural network approach for modeling the discharge coefficient of rectangular sharp-crested side weirs. Eng. Sci. Technol. **18**(4), 746–757 (2015). https://doi.org/10.1016/j.jestch.2015.04.012
48. H. Jafarian, H. Sayyaadi, F. Torabi, Modeling and optimization of dew-point evaporative coolers based on a developed GMDH-type neural network. Energy Convers. Manag. **143**, 49–65 (2017). https://doi.org/10.1016/j.enconman.2017.03.015
49. N. Nariman-Zadeh, A. Darvizeh, A. Jamali, A. Moeini, Evolutionary design of generalized polynomial neural networks for modelling and prediction of explosive forming process. J. Mater. Process. Technol. **164–165**, 1561–1571 (2005). https://doi.org/10.1016/j.jmatprotec.2005.02.020
50. V.N. Vapnik, *The Nature of Statistical Learning Theory* (Springer, New York, 1995)
51. J. Antonanzas, R. Urraca, F.J. Martinez-De-Pison, F. Antonanzas-Torres, Solar irradiation mapping with exogenous data from support vector regression machines estimations. Energy Convers. Manag. **100**, 380–390 (2015). https://doi.org/10.1016/j.enconman.2015.05.028
52. F. Antonanzas-Torres, R. Urraca, J. Antonanzas, J. Fernandez-Ceniceros, F.J. Martinez-de-Pison, Generation of daily global solar irradiation with support vector machines for regression. Energy Convers. Manag. **96**, 277–286 (2015). https://doi.org/10.1016/j.enconman.2015.02.086
53. Specht, Probabilistic neural networks for classification, mapping, or associative memory, in *IEEE International Conference on Neural Networks*, (1988), pp. 525–532. https://doi.org/10.1109/ICNN.1988.23887
54. C.M. Hong, F.S. Cheng, C.H. Chen, Optimal control for variable-speed wind generation systems using general regression neural network. Int. J. Electr. Power Energy Syst. **60**, 14–23 (2014). https://doi.org/10.1016/j.ijepes.2014.02.015
55. Y.W. Huang, M.Q. Chen, Y. Li, J. Guo, Modeling of chemical exergy of agricultural biomass using improved general regression neural network. Energy **114**, 1164–1175 (2016). https://doi.org/10.1016/j.energy.2016.08.090
56. J. Nirmal, M. Zaveri, S. Patnaik, P. Kachare, Voice conversion using general regression neural network. Appl. Soft Comput. **24**, 1–12 (2014). https://doi.org/10.1016/j.asoc.2014.06.040
57. Z. Yu, F. Haghighat, B.C.M. Fung, H. Yoshino, A decision tree method for building energy demand modeling. Energ. Buildings **42**(10), 1637–1646 (2010). https://doi.org/10.1016/j.enbuild.2010.04.006
58. P. Moutis, S. Skarvelis-Kazakos, M. Brucoli, Decision tree aided planning and energy balancing of planned community microgrids. Appl. Energy **161**, 197–205 (2016). https://doi.org/10.1016/j.apenergy.2015.10.002
59. S. Salzberg, *Book Review-C4. 5: Programs for Machine Learning* (Morgan Kaufmann, Burlington, 1993)
60. R. Yan, Z. Ma, Y. Zhao, G. Kokogiannakis, A decision tree based data-driven diagnostic strategy for air handling units. Energ. Buildings **133**, 37–45 (2016). https://doi.org/10.1016/j.enbuild.2016.09.039
61. J.R. Quinlan, Induction of decision trees. Mach. Learn. **1**(1), 81–106 (1986). https://doi.org/10.1023/A:1022643204877
62. K. Benmouiza, A. Cheknane, Forecasting hourly global solar radiation using hybrid k-means and nonlinear autoregressive neural network models. Energy Convers. Manag. **75**, 561–569 (2013). https://doi.org/10.1016/j.enconman.2013.07.003

63. S. Li, H. Ma, W. Li, Typical solar radiation year construction using k-means clustering and discrete-time Markov chain. Appl. Energy **205**, 720–731 (2017). https://doi.org/10.1016/j.apenergy.2017.08.067
64. K. Wang, X. Qi, H. Liu, J. Song, Deep belief network based k-means cluster approach for short-term wind power forecasting. Energy **165**, 840–852 (2018). https://doi.org/10.1016/j.energy.2018.09.118
65. A. Shrestha, A. Mahmood, Review of deep learning algorithms and architectures. IEEE Access **7**, 53040–53065 (2019). https://doi.org/10.1109/ACCESS.2019.2912200
66. Y. Lecun, Y. Bengio, G. Hinton, Deep learning. Nature **521**(7553), 436–444 (2015). https://doi.org/10.1038/nature14539
67. D. Zhang, X. Han, C. Deng, Review on the research and practice of deep learning and reinforcement learning in smart grids. CSEE J. Power Energy Syst. **4**(3), 362–370 (2018). https://doi.org/10.17775/CSEEJPES.2018.00520
68. L. Zhang, J. Lin, B. Liu, Z. Zhang, X. Yan, M. Wei, A review on deep learning applications in prognostics and health management. IEEE Access **7**, 162415–162438 (2019). https://doi.org/10.1109/ACCESS.2019.2950985
69. A. Moradzadeh, K. Pourhossein, Location of disk space variations in transformer winding using convolutional neural networks, in *2019 54th International Universities Power Engineering Conference, UPEC 2019 - Proceedings*, (2019), pp. 1–5. https://doi.org/10.1109/UPEC.2019.8893596
70. A. Moradzadeh, K. Pourhossein, Short circuit location in transformer winding using deep learning of its frequency responses, in *Proceedings 2019 International Aegean Conference on Electrical Machines and Power Electronics, ACEMP 2019 and 2019 International Conference on Optimization of Electrical and Electronic Equipment, OPTIM 2019*, (2019), pp. 268–273. https://doi.org/10.1109/ACEMP-OPTIM44294.2019.9007176
71. P. Li, Z. Chen, L.T. Yang, Q. Zhang, M.J. Deen, Deep convolutional computation model for feature learning on big data in internet of things. IEEE Trans. Ind. Inform. **14**(2), 790–798 (2018). https://doi.org/10.1109/TII.2017.2739340
72. N. Koroniotis, N. Moustafa, E. Sitnikova, Forensics and deep learning mechanisms for botnets in internet of things: A survey of challenges and solutions. IEEE Access **7**, 61764–61785 (2019). https://doi.org/10.1109/ACCESS.2019.2916717
73. J. Han, S. Miao, Y. Li, W. Yang, H. Yin, A wind farm equivalent method based on multi-view transfer clustering and stack sparse auto encoder. IEEE Access **8**, 92827–92841 (2020). https://doi.org/10.1109/ACCESS.2020.2993808
74. Z.A. Khan, S. Zubair, K. Imran, R. Ahmad, S.A. Butt, N.I. Chaudhary, A new users rating-trend based collaborative denoising auto-encoder for top-N recommender systems. IEEE Access **7**, 141287–141310 (2019). https://doi.org/10.1109/ACCESS.2019.2940603
75. W. Wang, X. Du, D. Shan, R. Qin, N. Wang, Cloud intrusion detection method based on stacked contractive auto-encoder and support vector machine. IEEE Trans. Cloud Comput. (2020). https://doi.org/10.1109/TCC.2020.3001017
76. D.A. Clevert, T. Unterthiner, S. Hochreiter, Fast and accurate deep network learning by exponential linear units (ELUs), in *4th International Conference on Learning Representations, ICLR 2016 - Conference Track Proceedings*, (2016)
77. Y. Bengio, P. Simard, P. Frasconi, Learning long-term dependencies with gradient descent is difficult. IEEE Trans. Neural Netw. **5**(2), 157–166 (1994). https://doi.org/10.1109/72.279181
78. S. Hochreiter, J. Schmidhuber, Long short-term memory. Neural Comput. **9**(8), 1735–1780 (1997). https://doi.org/10.1162/neco.1997.9.8.1735
79. K. Wang, X. Qi, H. Liu, Photovoltaic power forecasting based LSTM-convolutional network. Energy **189**, 116225 (2019). https://doi.org/10.1016/j.energy.2019.116225
80. A. Mohamed, G.E. Dahl, G. Hinton, Acoustic modeling using deep belief networks. IEEE Trans. Audio Speech Lang. Process. **20**(1), 14–22 (2012). https://doi.org/10.1109/TASL.2011.2109382

81. C.-Y. Zhang, C.L.P. Chen, M. Gan, L. Chen, Predictive deep Boltzmann machine for multiperiod wind speed forecasting. IEEE Trans. Sustain. Energy **6**(4), 1416–1425 (2015). https://doi.org/10.1109/TSTE.2015.2434387
82. B. Choubin, S. Khalighi-Sigaroodi, A. Malekian, Ö. Kişi, Multiple linear regression, multi-layer perceptron network and adaptive neuro-fuzzy inference system for forecasting precipitation based on large-scale climate signals. Hydrol. Sci. J. **61**(6), 1001–1009 (2016). https://doi.org/10.1080/02626667.2014.966721
83. R. Wang, S. Lu, W. Feng, A novel improved model for building energy consumption prediction based on model integration. Appl. Energy **262**, 114561 (2020). https://doi.org/10.1016/j.apenergy.2020.114561
84. K. Amasyali, N.M. El-Gohary, A review of data-driven building energy consumption prediction studies. Renew. Sustain. Energy Rev. **81**, 1192–1205 (2018). https://doi.org/10.1016/j.rser.2017.04.095
85. Z. Xuan, Z. Xuehui, L. Liequan, F. Zubing, Y. Junwei, P. Dongmei, Forecasting performance comparison of two hybrid machine learning models for cooling load of a large-scale commercial building. J. Build. Eng. **21**, 64–73 (2019). https://doi.org/10.1016/j.jobe.2018.10.006
86. S. Sekhar Roy, R. Roy, V.E. Balas, Estimating heating load in buildings using multivariate adaptive regression splines, extreme learning machine, a hybrid model of MARS and ELM. Renew. Sustain. Energy Rev. **82**, 4256–4268 (2018). https://doi.org/10.1016/j.rser.2017.05.249
87. M. Hossain, M.N. Sulaiman, A review on evaluation metrics for data classification evaluations. Int. J. Data Min. Knowl. Manage. Process **5**(2), 01–11 (Mar. 2015). https://doi.org/10.5121/ijdkp.2015.5201
88. W. Kong, Z.Y. Dong, B. Wang, J. Zhao, J. Huang, A practical solution for non-intrusive type II load monitoring based on deep learning and post-processing. IEEE Trans. Smart Grid **11**(1), 148–160 (2020). https://doi.org/10.1109/TSG.2019.2918330

Chapter 5
Introduction and Literature Review of the Application of Machine Learning/Deep Learning to Control Problems of Power Systems

Samira Sadeghi, Ali Hesami Naghshbandy, Parham Moradi, and Navid Rezaei

5.1 Introduction

Due to the significant change in the structure of the power system and its transformation into a modern power system, which is expected to be more pronounced in the future, the need to find advanced techniques to solve complex problems of the power system has become very prominent [1]. The evolution of the power system can be summarized as follows:

Changes in power generation sources have shifted power generation from large thermal power plants to smaller and distributed generations, such as wind [2] and solar [3], at all levels of transmission and distribution. The presence of small distributed generations and renewable energy sources along with electronic power converters at all levels of the power system causes uncertainty in power generation and reduces system inertia resulting in faster dynamics because the presence of renewable energy sources challenges frequency regulation and control [4].

The existence of new energy storage technologies, including types of large storage devices connected to the transmission network and small storage devices connected to the distribution network and microgrids, are the main factors in identifying a modern power system [5]. These new technologies make it possible

S. Sadeghi · A. Hesami Naghshbandy · N. Rezaei (✉)
Faculty of Engineering, Department of Electrical Engineering, University of Kurdistan, Sanandaj, Iran
e-mail: samira.sadeghi@eng.uok.ac.ir; hesami@uok.ac.ir; n.rezaei@uok.ac.ir

P. Moradi
Faculty of Engineering, Department of Computer Engineering, University of Kurdistan, Sanandaj, Iran
e-mail: p.moradi@uok.ac.ir

M. Nazari-Heris et al. (eds.), *Application of Machine Learning and Deep Learning Methods to Power System Problems*, Power Systems,
https://doi.org/10.1007/978-3-030-77696-1_5

to balance generation and consumption in sensitive and uncertain conditions of power system performance and can act as important control devices in different operating modes [1].

The loads of the power system, which are also considered as a kind of producer, have caused a change in the load profile. These loads include electronic devices that are available in all homes and industrial buildings and are used to meet load demand response and participate in electricity markets and control the power system. New types of electric loads include electric vehicles with charging stations installed anywhere in the power system, which have changed the load profile with the possibility of charging on the load side [6].

Increasing the use of Flexible Alternating Current Transmission Systems (FACTS) and various types of static and dynamic compensators for more efficient control of the power system [7] and also increasing the establishment of High Voltage Direct Current (HVDC) lines for easy voltage transfer are important and essential changes in the power system [8], which have made it necessary to evaluate the power system and check its control.

The creation of micro-energy networks as part of the power system is one of the undeniable factors in evaluating the control of the power system. These controllable microgrids are considered as an independent and single geographical area with the ability to produce and consume together, which includes a variety of small, renewable and distributed sources [9]. The existence of microgrids with the ability to operate as connected to the grid and separate from that has a significant role in creating flexibility in controlling the system and after the occurrence of faults and during network restoration [4].

Finally, all the changes in the power network have led to the modernization of the power system and the creation of the smart grid. Outstanding technologies for smart grids include advanced communication and measurement infrastructure, Phasor Measurement Units (PMU) in transmission and distribution systems [10], and smart meters in distribution networks. Thus, the emergence of loads with the ability to produce and consume, a variety of Internet of Things (IOT)-based equipment, advanced telecommunications equipment, etc. have caused the smart grid to face a variety of data and many complexities [11]. Therefore, the need for complex calculations and the introduction of new methods in controlling the modern power grid at any time and in any mode of operation have become important [12].

In power system control problems, due to the large amount of data (big data), the need of algorithms based on data analysis is essential. These types of algorithms that are based on data measurement are called machine learning algorithms. In other words, machine learning methods are applied to analyzing, processing, predicting, and categorizing big data in complex problems [13]. The purpose of using these types of algorithms is to extract patterns and order in the data. In fact, learning means improving behavior based on past experience. Therefore, the reason for turning to machine learning in the field of power systems includes the following [14]:

- Strengthening computing environments and producing sufficient information data.

- Complexity and nonlinearity of power system security issues.
- Generation of sufficiently rich security information databases with acceptable responses.
- Contestations of power companies in order to apply the approach of their proposed power system and meeting and responding the practical needs.
- Rapid and dynamic changes in the organization of the power system that will require the use of regular approaches in assessing dynamic security to maintain reliability.
- Recent events and blackouts in various parts of the world have highlighted the use of machine learning techniques to predict various disturbances.
- Generating large volumes of diverse data in various fields of the power system by moving toward smart power networks and its rapid development.

Recently, machine learning methods have been shown in numerous applications, such as power system control and stability studies and Dynamic Security Assessment.

Over the past few decades, with the development of advanced infrastructure and extensive interconnection of modern applications in the power system including Supervisory Control and Data Acquisition (SCADA), Phasor Measurement Unit (PMU), and modern telecommunication and communication technologies, the power grid is increasingly exposed to external damage and threats [15]. Monitoring on the condition of the power system, especially after disruptions and when the loads are changing, has always been one of the most important issues and concerns. Also, connecting renewable energies and developing the network with the ability to produce and consume in one place, increasing the number of consumers and, consequently, transmission lines, have faced the security assessment and stability with many challenges [13].

5.1.1 Reliability and Security

The probability of favorable power system performance and with minimum interruptions over a long period is called reliability. The ability of the power system to deal with impendent disturbances without interrupting services to electricity consumers is called security, which depends on the operating conditions of the system as well as the type of disturbance [16]. Mathematical analysis of the system response to changes made after a fault in the system and the new equilibrium conditions is called system security analysis. This analysis is called Static Security Assessment (SSA) if the analysis only assesses equilibrium conditions after turbulence (steady-state operating point). If the system evaluates the transient performance of the system after it malfunctions, it is called a Dynamic Security Assessment (DSA) [17].

In DSA offline analysis, accurate time domain stability analysis is performed for all valid disturbances and types of operating conditions, there is no severe limit in calculations time, and detailed analysis is performed for a wide range of conditions and probabilities [14]. Online DSA analysis is used to complete and update offline DSA to consider current operating conditions. In DSA online, a rapid screening process is performed to limit the number of occurrences, rapidly assess stability by predicting stability or calculating instability margins, and assess the severity of turbulence. Traditional time domain simulations also involve extensive numerical integrals to detect oscillation paths and voltage changes. In DSA online, calculation time is of particular importance [18].

5.1.2 Stability

The continued smooth operation of the power system performed by the occurrence of a perturbation is called stability, which depends on the operating conditions and the nature of the physical disturbance. Power system stability is divided into three categories: steady-state stability, dynamic stability, and transient stability [19]. Steady-state stability, the ability of different machines in the power system to restore and synchronize after a small and quiet disturbance, is like a gradual change of load. Transient stability refers to the stability after a sudden large disturbance such as a fault, generator failure, switching operation, and large load change. Dynamic stability is a state between steady-state and transient stability and is overcome by voltage regulators, controllers, and governors. In this type of stability, small perturbations of 10–30 s are desired [16]. Types of stability include transient and small signal stability for rotor angle and voltage and frequency stability [19]. Stability assessment methods include power flow analysis, eigenvalue analysis, time domain simulation, numerical integration method, direct or Lyapunov method, probabilistic methods, expert system and metaheuristic methods, and database analysis and machine learning approaches [18].

Therefore, models and various methods in security and stability studies of power system have been presented. Despite the almost satisfactory performance of conventional methods, these techniques are computationally complex, fiscally expensive, inaccurate, and time-consuming. In this regard, recently, machine learning and deep learning techniques have been highly regarded in monitoring complex applications and their modeling [20]. Unlike traditional methods, machine learning methods are computationally robust, principled, and explicitly reliable when used in classification studies. Machine learning methods have the ability of learning and understanding the features of changing different loads, grid data, etc. in normal, emergency, and post-fault conditions. Therefore, these methods are specific to the dynamic smart grid system [21].

5.2 Overview of Machine Learning and Deep Learning

Machine learning (ML) and deep learning (DL) are two topics that are widely used these days in computer science and other engineering disciplines.

Machine learning is used in computational problems in which designing and programming explicit algorithms with appropriate performance is difficult or impossible. The fundamental goal of ML algorithms is to generalize learning beyond the trained examples that is the successful interpretation of data. In ML, systems learn by receiving various inputs and performing statistical analysis to generate outputs in a specific statistical range. Therefore, ML allows computers to automatically perform the decision-making process on new input data after receiving the sample data and modeling it [22]. Deep learning is a subset of machine learning which is able to predict outputs and make decision-making patterns by artificial intelligence and mimicking the function of the human brain. DL refers to the Artificial Neural Network that consists of several layers of learning. This method allows machines to solve complex problems even when using very diverse, unstructured, and interconnected data. The DL algorithm will perform better when it learns more deeply [23]. Each machine learning algorithm is a subset of the field of artificial intelligence. Deep learning is also a subset of machine learning and tries to extract information based on multilevel learning. In fact, the place of artificial intelligence, machine learning, and deep learning are shown in Fig. 5.1. It should be noted that artificial intelligence is any code, technique, or algorithm that enables machines to behave like humans. Now this code may consist of some condition written by the programmer, or it may contain a mathematical relation that produces a fixed value for a particular input [24].

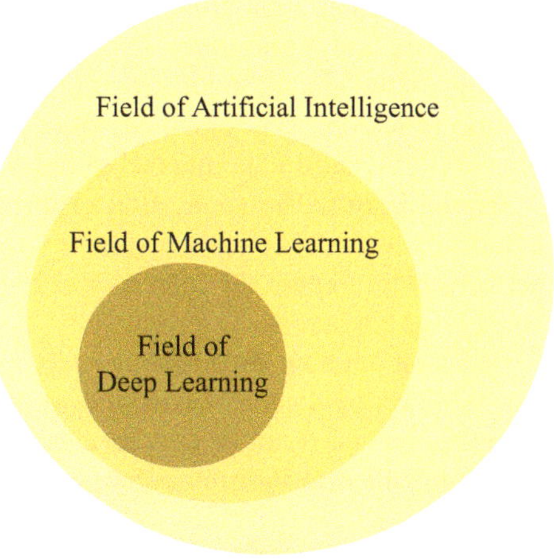

Fig. 5.1 Position of artificial intelligence, machine learning, and deep learning [24]

5.2.1 Machine Learning Technique

Over the past four decades, machine learning has made great strides in terms of theoretical understanding and actual applications in many different fields, the main reason for achieving these important successes being its dramatic increase in the computing potency.

ML is a category of techniques applied in big data processing with the development of algorithms and a category of principles to provide the outcomes needed by users. This method is mostly used for the development of automatic machines by implementing algorithms and a category of predefined principles [22]. In ML, data is used, and a set of rules will be executed by the mentioned algorithm. Therefore, machine learning techniques are classified as automatic instructions to produce the desired results. However, this operation can be done without human role and automatically converts the data into a template and will automatically detect the production problem. Machine learning is a more dynamic and robust method. Also, machine learning algorithms try to minimize the error by defining the error function. Before learning begins, the parameters start randomly and then gradually adapt [23]. Machine learning consists of several parts; first, the data is entered into the system, then the data is preprocessed, and if it has noise or a perturbation, it is first deleted, and then a series of properties are extracted from the raw data. These properties will represent the raw data. This data is then passed to a classification or clustering algorithm that plays a decision-maker role. The characteristics of a machine learning method depend on the following [14]:

Feature Selection: The purpose of this item is to decrease the data and dimensions of the input space, which, by reducing the attributes that have not any beneficial information to forecast the desired output information, the dimensions are reduced.

Model Selection: Usually, models are selected that have the best fit of learning situations in the predefined class. This requires the organized selection of the parameters and building of model using the temporary search technique or optimization approaches appropriate for the type of the desired model.

Exegesis and Validation: This item is synthesized to understand the systemic meaning of the model, and it is very important to determine its validity. Comparison of its information can be obtained from the model with previous allocation and tested on a set of invisible experimental samples.

Model Using: The application of the model is to predict the output using the assumed values of the input parameters. Sometimes, the inversion of the model is performed to prepare the information to change the input parameters in order to attain a certain output.

Machine learning methods are used in practical problems for the purposes of diagnosis or prediction. In diagnosis, the probabilistic values assumed by the attributes are the result of the causality of membership in the classification, but in predicting, the probabilistic values are consistent with the future state of the system.

Data selection is another important issue in machine learning methods for which the following should be considered:

- The data must be a good representation of the problem because it is to be modeled by an algorithm.
- Poor data quality (for example, data that is noisy, out of data, or that not much data is available) cannot lead to a suitable model. The solution here is to remove data noise, outdated data, and unspecified data before learning.
- There are a lot of unrelated and redundant (features that are similar and some of which are not necessary) features for the data.

Practical limitations in implementing machine learning methods include the high cost of collecting databases, complexity of learning according to computational needs in terms of time and CPU memory, and execution cost due to the complexity of the used algorithms. However, challenges of machine learning algorithms in different issues include wrong choice of learning algorithm, inadequate and incorrect selection of data for the training process, and overlearning of algorithm.

5.2.2 Deep Learning Technique

Deep learning is a type of neural network which attracts, processes, and calculates metadata as input through some nonlinear conversion layers and returns it as output data. This algorithm has a unique feature which is the feature of automatic extraction. This means that the algorithm understands the required and related attributes to solve the problem. Also, deep learning algorithm reduces the tasks of programmers to explicitly select features. This algorithm is even used to solve supervised or unsupervised challenges [23]. In DL, each hidden layer teaches a set of special and unique attributes that the performance of each layer is based on the output of the former layer. This type of hierarchical learning also converts low-level features into high-level features. In doing so, the DL is applied to solve the intricate subjects involving multiple nonlinear layers [22]. Large neural networks are called deep learning. The core of deep learning is based on having fast computers and enough data to train large neural networks. In deep learning, the model may also achieve greater accuracy than human accuracy [23]. Neural networks could have two or three layers, but today, Deep Neural Networks can have as many as 150 layers. Each neuron tries to store the lowest layer of information and transfer it to the next layers of neurons. This is done in such a way that the information inside the neurons is extracted along with a hidden layer of neuronal information. With this account, it can be concluded that the data goes from the lowest layer to the highest layer and collects information [25].

Deep learning models require large volumes of labeled data and neural network architecture. These models extract features automatically and do not require manual feature extraction. One of the most common models of Deep Neural Networks is the Convolutional Neural Networks, abbreviated CNN or Conv Net. In deep learning, the feature extraction and classification sections do not exist separately, and the data is applied directly to the algorithm and is obtained at the output of the data label. In fact, the classification section and the feature extraction section are merged [22].

5.2.3 Categorization of ML and DL

Deep learning is a special mode of machine learning. In machine learning, work begins with the manual extraction of features, and features are used to build a model that does the job of categorization, but in deep learning, the feature extraction process is not manual and is done automatically. Another difference is that in deep learning, the model expands as the amount of data increases, but in machine learning, the model expands to a certain extent and then does not change as the data increases. One of the key features of deep learning is that it improves with increasing data [22]. The most general categorizations for machine learning and deep learning algorithms include supervised learning, unsupervised learning, and reinforcement learning [23]. Another categorization of machine learning algorithms include semilogarithmic learning, offline learning, online learning, sample-based learning, and model-based learning [25].

Supervised Learning: The model is trained using specific inputs and outputs and consists of two general categories [23]:

Countable (discrete) outputs: that use classification algorithms.

Uncountable (continuous) outputs: that use regression algorithms.

Some of the most important supervised classification methods are as follows: Linear Regression, Logistic Regression, Support Vector Machine (SVM), k-Nearest Neighbor (k-NN), Decision Tree (DT), Random Forest (RF), Naive Bayes, Concept Learning, Neural Network, and Gradient Boosting.

Unsupervised Learning: The model is trained using a series of data whose output is not known and includes two general categories as follows [25]:

Clustering: Only similarities are used for clustering.

Association rules (dimensional reduction algorithms): This category includes the analysis of input data and the extraction of relationships between features and attributes.

Some of the most important unsupervised clustering methods are as follows: Hierarchical Agglomerative Clustering (HAC), *k*-Means, *C*-Means, Density-Based Gaussian (DBSCAN), Community Detection, Mean-Shelf Clustering, and Gaussian Mixture Models (GMM).

Reinforcement Learning: This is one of the machine learning trends that is inspired by behavioral psychology. This method focuses on the behaviors that the machine must perform to maximize its reward. Here, the algorithm learns the optimal behavior based on the feedback (reward or penalty) it receives from the environment. Reinforcement learning has two major differences from supervised learning: The first is that it does not have the right input and output pairs, and the dysfunctional behaviors are not corrected from the outside, and the second is that there is a strong focus on live performance that requires finding the proper balance between discovering new things and utilizing stored knowledge [26].

Also, it should be noted that the choice of machine learning method to solve the problem depends on factors such as the nature and complexity of the problem, the type and nature of the required data, the limitations and scope of learning, and the expected outcome and results of the problem. In this regard, the tasks of machine learning techniques include forecasting, modeling, categorization, and classification. In addition, most research contains machine learning algorithms based on methods for selecting, extracting, reducing, and optimizing the features of problem data. Some of important selection, extraction, and reduction methods include Principal Component Analysis (PCA), Forward Feature Selection (FFS), Backward Feature Selection (BFS), Wavelet Transform (WT), Partial Least Squares Regression (PLSR), and Factor Analysis (FA). Furthermore, data preprocessing and optimization methods in most cases include metaheuristic algorithms such as Genetic Algorithm (GA), Artificial Bee Colony (ABC), Particle Swarm Optimization (PSO), and Ant Colony Optimization (ACO).

According to the above categories, many machine learning methods are classified in Fig. 5.2. Also, most of the scholars' research is devoted to hybrid machine learning methods or multiple strategies.

5.3 Overview of Power System Control Problems

The power system is the most important vital infrastructure in the modern world. Thus, this system with proper design and operation must meet the following basic requirements:

- It must be able to meet the growing demand for active and reactive powers and should maintain and control the spinning reserve requirements.
- The quality of the supplied power must satisfy the minimum required standards according to the frequency stabilization, voltage stabilization, and reliability level.
- The system must provide energy at the lowest cost [16].

Also, preventing partial and total blackouts caused by various disturbances in the power system are the important goals in controlling the system against all kinds of faults. In assessing the security of the power system, different operational modes are investigated [13]. To meet the above needs, different levels of control are used that includes a complex set of equipment. Figure 5.3 shows different parts of a power system with their corresponding controls. The controllers create the proper operation of the system by maintaining the voltage, frequency, and other variables of the system within the allowable range. These controllers have critical impacts on the dynamical performance of the power system. The goals of controlling a power system depend on its operating conditions. Typically, the aim is to operate the

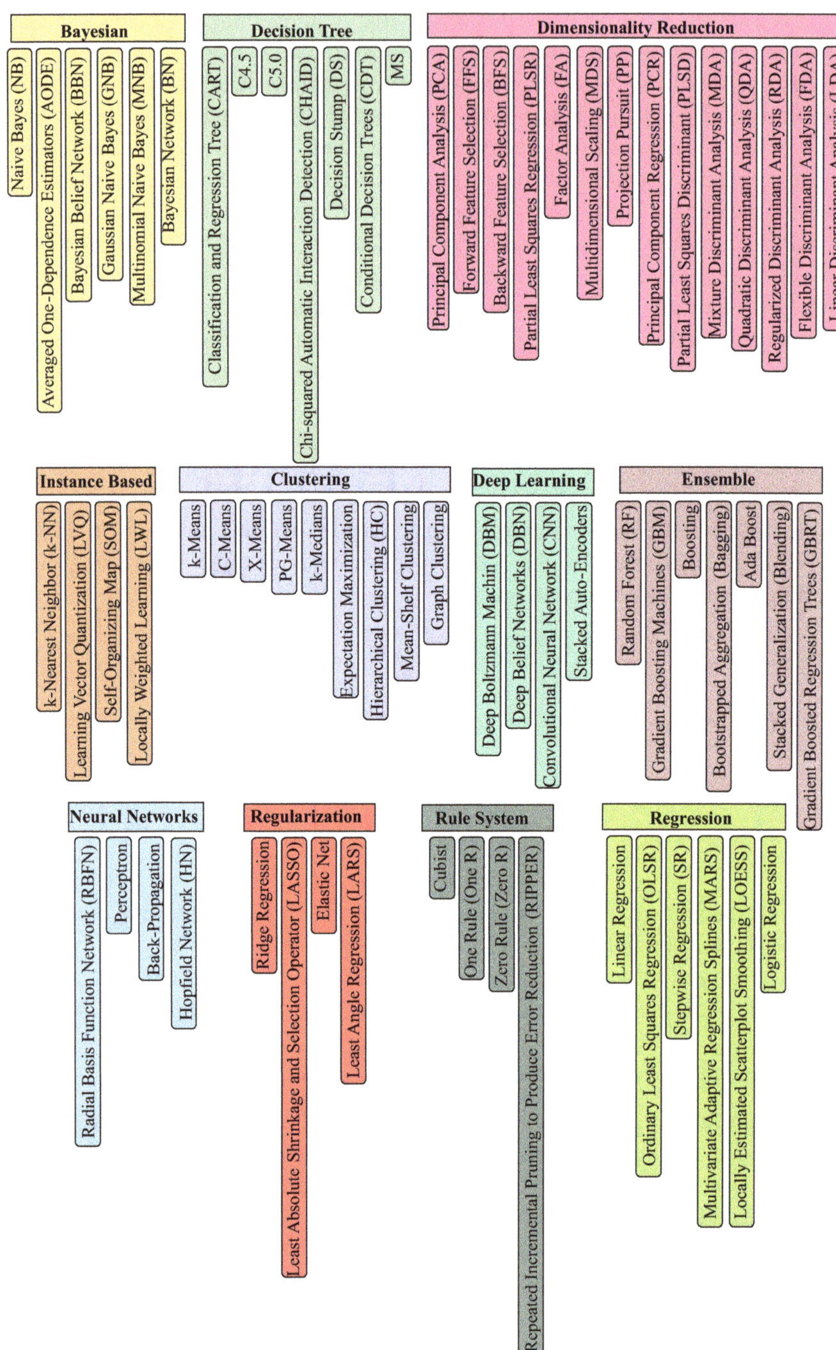

Fig. 5.2 Machine learning methods categorization [14, 22, 23, 25]

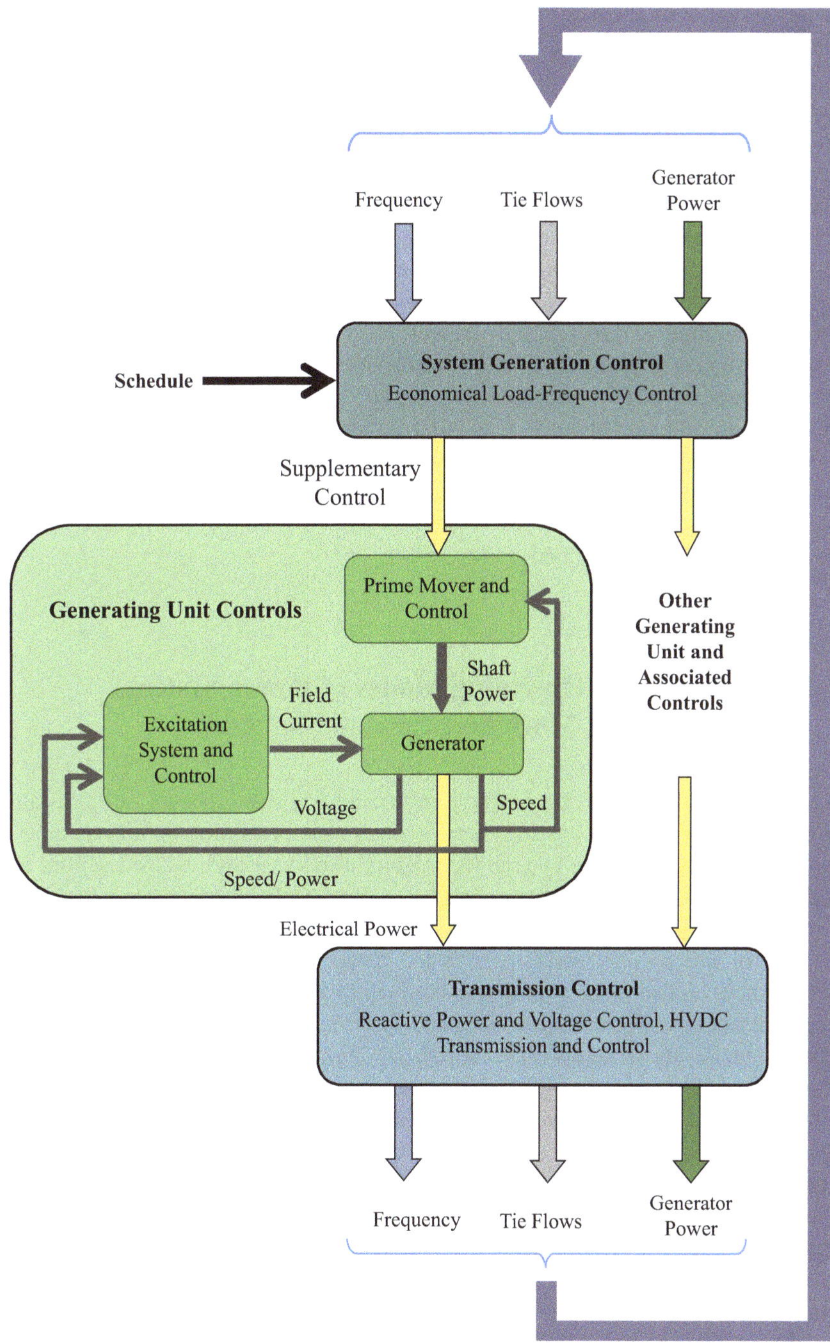

Fig. 5.3 Subsystems of a power system and their control [16]

power system with the best possible efficiency, while the voltage and frequency are close to nominal values [18].

When an unusual situation occurs, new goals must be defined to get the system back to the normal condition. It is rare for a single disturbance in a power system to lead to a major breakdown and the collapse of a seemingly secure system. Such interruptions usually occur by combining a set of events that put pressure on the system beyond its capabilities. Under normal circumstances, all power system variables are within the allowable range. The goal of controllers of a power system is to restore a disrupted system to the normal situation. Hence, it should be noted that dynamical performance of a large interconnected power system is affected by a wide range of equipment failures and outages [16]. In relation to controlling various issues, it should be noted that most control functions such as automatic generation control are performed in control centers located in the power plant. Substations also allow the introduction of some local control functions such as control of distributed generation sources and load management. The control process is often hierarchically and through the exchange of information and decision-making and control signals. The control process must be set instantaneously, accurately, and in coordination between the control commands [14].

5.4 Evaluation of Operating Modes of Power System Control Using Machine Learning Methods

It is an undeniable fact that a power system cannot always operates in its sustained state because in a power system, there is always the possibility of various types of faults such as three-phase to ground, single-phase to ground, etc. Therefore, after identifying the behavior of the system when these faults occur, appropriate control strategies should be established [18].

Control and stability assessment of the power system is in the face of various disturbances and suggests appropriate treatment measures to resolve its main weaknesses at any time. Perturbations and faults may be due to outer or inner occurrences. For example, these faults are caused by lightning (external) and switching (internal) [14].

Before a disturbance occurs, the necessary preventives must be taken to deal with any type of disturbance so that the system can continue to operate. Also, generation and demand of energy should be balanced, and all variables of the power system should be in their operational range. When a disturbance occurs and in an emergency state, the necessary arrangements must be adopted to maintain synchronicity and protect the power system. After a disturbance occurs, when a part of the network or the entire network is removed, the necessary operations must be performed to quickly restore the power system [20]. Therefore, the control system is examined in preventive, normal, emergency, and restorative operating modes, which most studies in controlling the power system are done in two control modes including

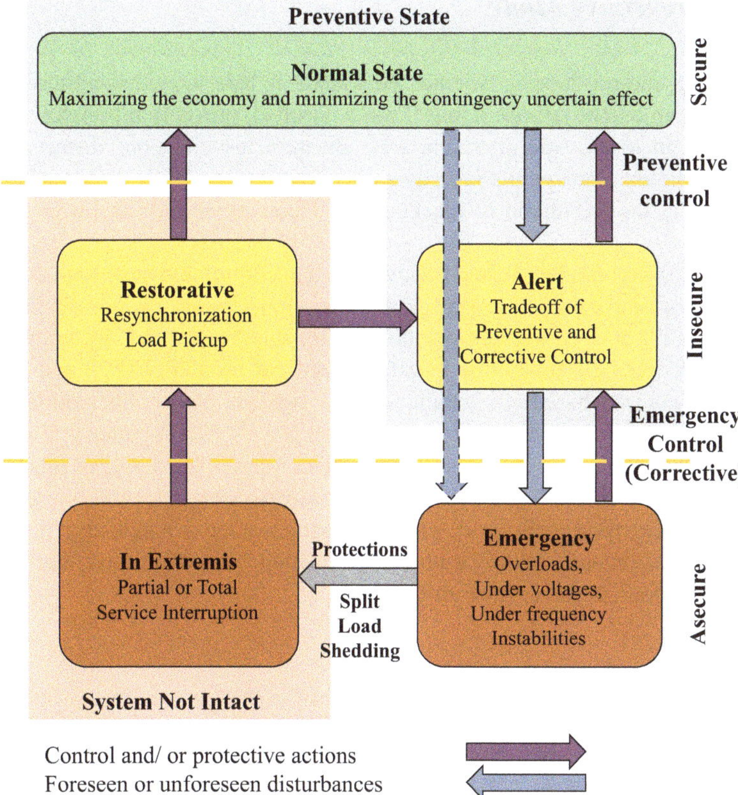

Fig. 5.4 Operating states and transitions [14]

prevention and emergency. Figure 5.4 displays a detailed explanation of the scheme provided by Dy Liacco, which defines the operating modes of a power system [14].

The use of PMUs in modern power systems has led to significant advances in researches based on measurable data. Nevertheless, main subjects such as big data related to PMUs, uncertainties related to measurement faults, and nonlinearity of loads in future and current power system, and computationally complexity of them, expose the constraints of traditional control methods. Therefore, the need for fast, reliable, advanced, scanning, calibrated, and dynamic methods in controlling the modern power system and in the face of continuous blackouts and other threats to the power system has become essential [10, 27].

Some fast system controls include protective relays, turbine governors, and automatic voltage regulators on generators that operate in fractions of seconds, while subsystem level controls such as secondary voltage control and automatic generation control during act for a few seconds. Other types of controls (preventive and emergency) are activated by detecting system limitations and disturbances [26].

5.4.1 Preventive Mode

In this case, the question is whether the system is able to withstand any possible disturbance in its normal operation. If not controlled, prevention transfers this state of the system to the safe operating area. Because instantaneous disturbances are difficult to predict, preventive security assessments are primarily aimed at balancing and reducing the likelihood of blackout and inconsistencies with lower operating costs [20].

Some of the researches related to power system control that have been done using machine learning methods in preventive operational mode include line trip fault prediction [28] at device level, maintaining frequency stability [29], voltage instability prediction [30], predicting post-fault transient instability [31], and proactive frequency control based on anticipation of sudden power fluctuation [32] at subsystem level and predicting transient stability [33, 34] in smart grid and wide area level, which these preventions avoid potential future hazards. For this purpose, in Table 5.1, for preventive control state in power systems, several levels (device, subsystem, microgrid, smart grid, wide area) are considered, and at each level, some papers have examined various control issues, which researchers have used machine learning methods to solve these problems.

Table 5.1 Summary of applications of machine learning methods in preventive control mode

Control level	Control problem	Learning algorithm	Refs
Device	Define fault current direction	Decision Trees (DT) and Support Vector Machine (SVM)	[35]
	Line trip fault prediction	Long Short-Term Memory (LSTM) Networks and Support Vector Machine (SVM)	[28]
Subsystem	Maintaining frequency stability	Extreme Learning Machine (ELM)	[29]
	Voltage instability prediction	Deep Recurrent Neural Network (DRNN)	[30]
	Transient stability	Classification and Regression Tree (CART) and Multilayer Perceptron (MLP)	[36]
	Predicting post-fault transient instability and develop emergency generator-shedding control	Ensemble Online Sequential Learning Machine (E-OS-ELM)	[31]
	Proactive frequency control based on anticipation of sudden power fluctuation	Extreme Learning Machine (ELM)	[32]
Smart grid	Predicting transient stability	Extreme Learning Machine (ELM)	[33]
Wide area	Determine the transient stability aspects	Radial Basis Function Neural Network (RBFNN)	[34]
	Real-time transient stability assessment	Least Square Support Vector Machine (LS-SVM)	[37]

5.4.2 *Normal Mode*

Due to small and continuous changes in the power generation of generators and loads in the power system, it is necessary to control the system in normal operating state. In normal operating mode, the behavior of all system loads and the amount of generated energy by the generators at all control levels produce a certain frequency and voltage without interruption, and the system does not see any physical or security faults. Also, all nominal limitations in each power device are met [20].

Some of the important control problems of different levels that are investigated under normal conditions, in this part, include issues such as improvement in the performance indices of relays [38] and controlling maximum power point tracking (MPPT) for photovoltaic systems [3] at the device and local level; voltage stability margin monitoring [39], automatic generation control (AGC) [40], damping low-frequency oscillations by power system stabilizer (PSS) [41], power quality disturbances control [42], and online rotor angle stability prediction [43] at the subsystem level; load response [44], dynamic energy management [9], and droop control, automatic generation control, and economic dispatch [45] at the microgrid and smart grid level; and transient stability assessment [46] at the wide area level.

Table 5.2 summarizes some of the researches that examine various problems of normal power system control. In this table, different machine learning methods have

Table 5.2 Summary of applications of machine learning methods in normal control mode

Control level	Control problem	Learning algorithm	Refs
Device	Improving performance indices of protection relays	Extreme Learning Machine (ELM)	[38]
	Maximum power point tracking (MPPT)	Markov Decision Process (MDP)	[3]
Subsystem	Voltage stability margin (VSM) monitoring	Local Regression	[39]
	Automatic generation control (AGC)	Long Short-Term Memory Recurrent Neural Network	[40]
	Damping low-frequency oscillations	Artificial Neural Networks (ANN)	[41]
	Power quality disturbances control	Deep Neural Network (DNN)	[42]
	Online rotor angle stability prediction	Ensemble Decision Tree	[43]
Microgrid/ smart grid	Demand response	Deep Neural Network (DNN)	[44]
	Dynamic energy management	Decision Tree (DT)	[9]
	Droop control, automatic generation control, and economic dispatch	Extreme Learning Machine (ELM)	[45]
Wide area	Transient stability assessment	Core Vector Machine (CVM)	[46]

been used to investigate various issues at the control levels of the device, subsystem, microgrid, smart grid, and wide area.

5.4.3 Emergency Mode

The purpose of identifying an emergency state is to whether the system seeks to lose integrity after the actual malfunction begins. There is an important interaction here that response time is much more important, while economic considerations are temporarily given secondary priority [20]. Emergency control is performed with the aim of taking urgent measures to prevent partial or complete interruption of services. In emergency control mode, some of the normal operating limits of the system are violated. For example, lines are overloaded, and the frequency is lower than the allowable value. Also, the frequency may exceed the allowable value due to the departure of a large part of the load, or due to various faults or lack of precise control of the reactive power, the voltage may be out of the allowable range [14].

Some important problems of emergency operating control mode examined in some studies are summarized in Table 5.3. These issues have also been investigated using machine learning methods at various levels. These problems include switch PWM fault distinction in rectifier [47] and detecting fault by relays [48] in device level; transient stability [49], online voltage stability monitoring [50], fault detection [51], enhancing electromechanical oscillations damping [52], emergency load-shedding control [53], and frequency control to prevent system collapse [54] in subsystem level; short-term voltage instability [55], fault detection [56], islanding detection approach [57], and mitigating cascading failures, preventing blackout, and adaptive output power regulation of generators via frequency control [58] in smart grid and microgrid level; and enhancing transient stability and damping the inter-area oscillations [2], oscillatory angle stability [59], damping low-frequency oscillation [60], and short-term voltage stability [61] in wide area level.

5.4.4 Restoration Mode

When both preventive and emergency control have failed to return system parameters to their inequality limits, local protection devices will operate to protect components of the electrical system from irreparable damage. This may cause further disturbances, system failures, and even partial or complete blackouts.

In restorative mode, some system loads may not be supplied so that partial or total blackouts occurs in this mode, but the operational parts of the system are returned to normal. Restoration at each control level and for each type of load will be different in terms of speed and accuracy. Also, the restorative method in smart grids is different from the current power grid. In this mode, the task of operator is minimizing the

Table 5.3 Summary of applications of machine learning methods in emergency control mode

Control level	Control problem	Learning algorithm	Refs
Device	Open switch fault diagnosis PWM in voltage source rectifier	Least Square Support Vector Machine (LSSVM)	[47]
	Detecting fault by relays	Support Vector Machines (SVM)	[48]
Subsystem	Transient stability	High-Performance Back Propagation Neural Network (HBPNN)	[49]
	Online voltage stability monitoring	Ensemble Ada Boost Classifier	[50]
	Fault detection	Nearest Neighbor (KNN), Decision Trees, and Support Vector Machines (SVM)	[51]
	Enhancing electromechanical oscillations damping	Tree-Based Batch Mode Reinforcement Learning (RL)	[52]
	Emergency load-shedding control	Artificial Neural Network and Analytic Hierarchy Process Algorithm	[53]
	Frequency control to prevent system collapse	Multi Q-Learning—Reinforcement Learning	[54]
Microgrid/ smart grid	Short-Term Voltage instability	Random-Weights Neural Networks	[55]
	Mode detection and fault detection	Artificial Neural Network (ANN), Support Vector Machine (SVM), and Decision Tree	[56]
	Islanding detection approach	Support Vector Machine (SVM)	[57]
	Mitigating cascading failures, preventing blackout, adaptive adjustment of generators' output power through frequency control	Artificial Neural Networks (ANN)	[58]
Wide area	Enhancing transient stability and damping the inter-area oscillations	Reinforcement Learning (RL), Neural Network (NN)	[2]
	Oscillatory angle stability	Actor-Critic Neural Network	[59]
	Damping low-frequency oscillation	Artificial Neural Network (ANN)	[60]
	Short-term voltage stability	Random Forest	[61]

value of unreleased power via recoordinating the lost generation in the shortest feasible time and adding the interrupted load in order of primacy.

In Table 5.4, for restorative control mode in power systems, few numbers of researches are done with machine learning methods. In this state, several levels of subsystem, microgrid, and wide area are considered, and at each level, some papers have examined various control issues. The control problems investigated in this mode include generation rescheduling and load shedding [62], fault-induced delayed voltage recovery [63], and restoration of power grid systems [64] in subsystem level, post-fault restoration [65] in microgrid level, and real-time short-term voltage stability assessment [66] in wide area level.

In some studies, several control modes are examined simultaneously. For example, in [67], two modes of emergency and restoration control are performed using the reinforcement learning method (multi-agent Q-learning) in power distribution

Table 5.4 Summary of applications of machine learning methods in restoration control mode

Control level	Control problem	Learning algorithm	Refs
Subsystem	Generation rescheduling and load shedding	Decision Tree (DT)	[62]
	Fault-induced delayed voltage recovery (FIDVR)	Weighted Kernel Extreme Learning Machine (WKELM)	[63]
	Restoration of power grid systems	Q-Learning	[64]
Microgrid	Post-fault restoration	Feature Selection	[65]
Wide area	Real-time short-term voltage stability (STVS) assessment	Extreme Learning Machine (ELM)	[66]

systems. So the fault location detection is done in an emergency state, and then the isolation and restoration are done.

According to the tables, it can be seen that in most research, machine learning methods may be helpful in restorative state but have focused on preventative, normal, and emergency situations. In fact, machine learning and deep learning methods are focused on predictive modes, and these methods are more effective for preventive and emergency modes than other modes.

5.5 Application of Machine Learning Methods in Evaluating the Security and Stability of the Power System

Since various social, economic, and political activities are linked to the national power system, the safe and sustainable operation of the power system are also determined by governments and public industry stakeholders and have the highest priority. Sometimes, enemies or rivals can access network information and change control commands, thus destabilizing network performance and causing blackouts and financial losses, and national security can be compromised [15]. In addition, the increasing growth of energy demand, the presence of annoying loads, rearrangements and changes in topology and architecture of network, overloading of transmission lines, etc. have led to the power system deviating from its safe range so they cause instability and disruption in power quality [68].

Since the purpose of controlling the power system is to create a stable and secure system in all control modes, this section examines in detail the stability and security of the power system. The main areas of power system security and stability include transient stability, voltage stability, frequency stability, and power quality disturbances, in which machine learning are widely developed [16]. Evaluation of machine learning methods in various areas of power system sustainability concentrates on highlighting methods, attainments, and restrictions in the scheme of classifiers, the production of data sets, and the test systems that are used. Also, machine learning methods are widely used to monitor, detect intrusion, predict, and classify various threats of the power system.

Ensuring the security and stability of the power grid is a major challenge today, especially after being exposed to various pressures and disturbances. Therefore, important issues that play an important role in assessing the stable and safe operation of the power system are as follows:

- Transient stability assessment.
- Voltage stability assessment.
- Power quality disturbances.
- Frequency stability assessment.

The three main categories of transient stability, frequency stability, and voltage stability are recognized as important issues in dynamical security assessment [69].

5.5.1 Transient Stability Assessment

The capability of power system generators to keep coordination after a large perturbation such as a severe external fault, sudden and immediate loss of loads, or generators of the power system is called transient stability. Transient instability is one of the main reasons of power instability, which will lead to island construction and widespread blackouts [70]. Conventional methods for evaluating transient stability in a power system, including time domain simulation, equal areas criterion, transient energy function, and Lyapunov, do not meet its new and modern requirements because evaluating the swing curves of generators, different loads, faults, and fault clearance times requires extensive computational effort [71].

Monitoring and evaluating the stability of the power system is very important for its efficient operation in the margins of its stability. If there is a huge amount of high-speed data collected by PMU, using conventional methods of transient stability assessment alone may not meet the needs of transient stability in real time and will be computationally time-consuming. Thus, machine learning techniques have recently been widely used in power system control issues, including transient stability studies. These methods process the great values of PMU data, then analyze data, and finally classify and detect the stability state of the power system. In machine learning methods, due to the generalizability, explicitly trained data-driven models can accurately predict the stability [46]. Transient stability is assessed using machine learning methods in three stages: feature generation, feature preprocessing and optimization, and classification/prediction.

Feature Generation: The first and most important step in organizing the analysis of reliable transient stability assessment models is the generation of input data set vectors. Synchronous variables of the power system sampled by PMUs based on the wide area measurement system (WAMS) that were collected before or immediately after clearing a fault provide the possibility of performing advanced protect operations a wide area for control and decision [10].

Choosing the proper features is an important criterion for the transient stability assessment. Therefore, generating/extracting feature data through time domain simulation processes is an important issue [72]. The major concerns in transient stability assessment is the selection of appropriate trajectories characteristics [73]. To predict the stability or instability of the system, various procedures such as rotor angles, rotor speed, and voltage amplitude are applied as forecasters [74].

Feature preprocessing and Optimization: Various techniques for reducing, selecting, and optimizing attributes appropriate to the type of each data set and each classification algorithms are proposed to eliminate redundancies, improve classification, and investigate transient instability [75].

Classification/Prediction: Typically, "offline training, online application" methods are used in many transient stability assessment studies based on machine learning methods. This method is that first the model is taught offline and then the transient stability test is done online [72]. In the study of stability with offline training, for example, generators and different load changes are usually modeled with different possible disturbances, including three-phase faults to ground, etc. with different fault clearance times. But despite the innumerable successes of using the "offline training, online application" model, this model is not applicable in the real world because the set of training data generated by the offline method cannot exhibit all the features and characteristics of a modern power system [10].

In the study of transient stability, the results of neural network classifiers can be continuous so that the transient stability margins and boundaries appear softer [36]; decision trees are well interpretable [76]; support vector machine can calculate the distance between a moment and a stable boundary, in which case it is mostly used to define a certainty index [68]; Random Forest algorithms can regularly evaluate feature weights and arrange features according to sorting rate [77]; and Extreme Learning Machine has a fast learning speed [78].

Table 5.5 summarizes the machine learning algorithms and preprocessing/optimization methods adopted in each learning method for some recent research works in the field of transient stability. Different types of power systems have been used in various studies to investigate transient stability. According to Table 5.5, the 39-bus system is used in most articles.

Comparison of machine learning methods in the table is done based on the obtained accuracy criterion, which according to the references, this value is usually between 85% and 100%, and the accuracy criterion with higher percentage shows the superiority of the method.

5.5.2 Voltage Stability Assessment

Transient stability is related to generator synchronization, but voltage stability is related to load dynamics and reactive power management. Voltage stability indicates the capability of the power system to retain the buses voltage at acceptable values after perturbation in a particular operating condition [79]. Reactive power

Table 5.5 A summary of some of the offered machine learning methods to classify transient stability assessment

Refs	Machine learning algorithm	Preprocessing/optimization technique
[10]	Ensemble of Online Sequential-Extreme Learning Machine (EOS-ELM)	Binary Jaya (Bin Jaya)
[36]	Classification and Regression Tree (CART) and Artificial Neural Network (ANN)	–
[46]	Core Vector Machine (CVM)	–
[68]	Ensemble of SVM	Min-max normalization
[72]	Bayesian Multiple Kernels Learning (BMKL)	–
[73]	Twin Convolutional SVM (TWCSVM)	–
[74]	Aggressive SVM (ASVM) and Conservative SVM (CSVM)	–
[75]	Extreme Learning Machine (ELM)	PSO
[76]	Decision Trees (DT)	–
[77]	Random Forest	Recursive Feature Elimination Strategy
[78]	ELM	Kernelized Fuzzy Rough Sets (KFRS) and Memetic Algorithm

management and consequently monitoring and evaluating the voltage stability situation are important issues to prevent various blackouts and maintain the overall stability of the modern power system. When a significant fault occurs in the power system, the voltage profile loses stability, the speed of the induction motors decreases rapidly, and they receive a high reactive current, which ultimately leads to an uncontrollable voltage drop and blackouts. Voltage instability is divided into short term and long term [80].

Most conventional methods do not prepare accurate useful information about stability issues although they require relatively complex calculations, but they ignore the dynamics of the modern power system. Also, calculating a large number of loads bus currents has many limitations in conventional models of assessing P-V and Q-V curves of voltage stability analysis. Therefore, machine learning methods have recently been used to overcome shortcomings and violations of conventional voltage stability assessment methods [81]. The study of machine learning methods in assessing voltage stability is done in two stages, including the stages of generation and selection of features and the classification of voltage stability margin index.

Generation and Selection of Features: If the power system is able to keep admissible stable voltages at all system bases under operating conditions of normal and after a fault has occurred, it will be classified in the voltage stability zone. In this regard, to evaluate the voltage stability using machine learning methods, the data of the Phasor Measurement Units is used that ordinary input vectors are considered as the voltage phasor. Here, as in the case of transient stability, the input vectors for classifier algorithms are usually educated via "offline training," and the generated output vector is considered as the index vector for voltage stability margin [82].

Table 5.6 A summary of some of the offered machine learning methods to classify voltage stability assessment

Refs	Machine learning algorithm	Preprocessing/optimization technique
[27]	Decision Tree (DT)	Principal Component Analysis (PCA) and Correlation Techniques
[55]	Neural Network with Random Weight (NNRW)	Relief Algorithm
[80]	Feed Forward Back Propagation (FFBP)	–
[81]	SVM	*k*-Means
[82]	Artificial Neural Network (ANN)	Imperialist Competitive Algorithm (ICA)
[83]	SVM	Genetic Algorithm (GA)
[84]	ANN	Gram-Schmidt Orthogonalization (GSO)
[85]	Probabilistic Fuzzy Decision Tree (PFDT)	Case-Based Reasoning (CBR)
[86]	SVM	Multi-Objective Biogeography-Based Optimization (MOBBO)
[87]	Feed Forward Back Propagation Network (FFBPN)	Linear Optimization

Voltage stability evaluation models are very complex and nonlinear and involve a large amount of data sets, so the selection and reduction of features in these models is a very important issue. In this case, many features of the power system are not suitable, and it is better not to use them directly as classifier inputs. Therefore, in most models, feature extraction and reduction methods are used [27]. In addition, different optimization methods have been used to increase exactitude, improve the parameters adjustment of machine learning algorithms, and decrease the data training time [83].

Classification of Voltage Stability Margin Index: Using ML algorithms, the nonlinear relationship between input vector and output vector can be extracted and adjusted, in which the input vector is the operating parameters related to the power system, and the output vector is the voltage stability margin. In the problem of voltage stability assessment, some of the features of machine learning algorithms are as follows:

The calculation time of the Artificial Neural Network algorithm is very short, and it predicts the stability margin index very accurately, but this algorithm suffers from overtraining and in which setting parameters can be a concern [84]. DT has simple division rules due to fewer data samples and has excellent performance in online voltage rating classification [27].

Table 5.6 summarizes the machine learning methods for predicting, monitoring, and analyzing voltage stability assessments, in which some machine learning algorithms and data preprocessing/optimization techniques are adopted. Also, in this case, comparison of machine learning methods is done based on the obtained accuracy criterion, and the accuracy criterion with higher percentage shows the superiority of the method.

5.5.3 *Power Quality Disturbances Assessment*

Sudden deviation of the voltage amplitude, frequency, and phase angle from the defined standard rate is called power quality disturbances. These disturbances are often caused by the increased use of switching devices, nonlinear loads, inverters, and rectifiers in the power system. Types of power quality disturbances include harmonic distortion, voltage sag, flicker, interruption, swell, transient, etc. [88]. In each power system, there are different types of disturbances in power quality, so accurate diagnosis and classification of events in the study of disturbances related to power quality are done in three stages of feature extraction/selection, feature optimization, and classification of events [89].

Feature Selection: In the first step of feature selection, the volume of data must be considered because generated feature data with large dimension is not ideal for use as a classifier input. Duplicate data with unrelated features significantly increases the time of calculations and decreases the accuracy of classification. Therefore, extraction of dominant features in the subject of machine learning is usually necessary for classification [90]. In the study of power quality disturbances using machine learning methods, frequency domain techniques such as Fourier Transform (FT) [91] and various time domain signal processing methods such as Empirical Mode Decomposition (EMD) [92] have been used successfully to select features of waveforms. In many studies, the Wavelet Transform (WT) method [88] is used because they provide good time-frequency properties and have a very good ability to analyze local discontinuities of signals.

Feature Optimization: Various feature optimization techniques are used to improve the performance of power quality disturbances events classification. The main purpose of using these methods is to remove additional attributes such as noise in the adopted features and optimize the accuracy of the classifiers. Most attribute selection methods for events can be based on wrapper or based on filter [89]. Filter-based feature selection is fast because it ranks attributes according to their inherent properties, while feature selection based on the wrapper needs more time, but it is a more efficient option. If no feature optimization method is used, classification may need complex and time-consuming simulations, voluminous calculations sources, and more time [89].

Classification of Events: Countless machine learning and deep learning algorithms have been used in the power quality disturbances classification. Among the various types of classifiers, SVM [89] and PNN [88] are the most widely used due to some of the salient features. The PNN technique has a good performance for classification and has high accuracy in signal distance studies. Also, PNN technique does not require initial weight adjustment. The SVM technique is also ranked as a powerful classifier.

Table 5.7 shows some of the machine learning techniques along with the method of extracting and optimizing features in some articles in the field of power quality disturbances. As shown in the table, the most used machine learning tool in power quality disturbances assessment and classification is the SVM technique because it

Table 5.7 A summary of some of the offered machine learning methods to classify power quality disturbances events

Refs	Machine learning algorithm	Feature extraction technique	Feature optimization/reduction technique
[42]	Deep Neural Network (DNN)	Singular Spectrum Analysis (SSA) and Wavelet Transform (WT)	Compressive Sensing (CS)
[88]	Probabilistic Neural Network (PNN)	Wavelet Transform (WT)	Artificial Bee Colony (ABC)
[89]	Support Vector Machine (SVM)	Variational Mode Decomposition (VMD) and S-Transform (ST)	Sequential Forward Selection (SFS), Sequential Backward Selection (SBS), and Gram-Schmidt Orthogonalization (GSO)
[90]	SVM	Wavelet Packet Transform (WPT)	Genetic Algorithm (GA)
[93]	SVM	WT	Particle Swarm Optimization (PSO)
[94]	*k*-Means, Decision Tree (DT), and SVM	ST	Ant Colony Optimization (ACO)
[95]	SVM	WT	–
[96]	Radial Basis Function Neural Network (RBFNN)	Empirical Mode Decomposition (EMD) and Hilbert Transform (HT)	PSO
[97]	SVM	Wavelet Multiresolution Analysis (WMRA)	–
[98]	Convolutional Neural Network (CNN)	Curvelet Transform (CT)	Static Single Assignment

ensures high accuracy and efficiency. Here, too, the percentage of accuracy obtained is considered as a criterion for evaluating the superiority of one method over other methods. So the closer the accuracy to 100%, the superiority of the method.

5.5.4 *Frequency Stability Assessment*

The most important task of a power grid is to generate and transmit electricity to consumers while maintaining the quality standards of electrical power. One of the most important of these criteria is to maintain the network frequency close to the nominal value, in which, in order to maintain the frequency, a balance of production and consumption must be established at all times. Otherwise, the network frequency will change and may become unstable and cause severe damage to network equipment such as turbines, motors, transformers, etc. Therefore, special control and protection operations are used to maintain the balance between production and consumption and to maintain the network frequency close to the nominal value. The primary frequency control is the first control operation used for this purpose, and the secondary and tertiary controls are the backup controls. Primary and secondary frequency control characteristics, including dynamic behavior and their requirements, depend on various parameters (inertia constant, the degree of load dependence on frequency, and the speed and number of units participating in frequency control and their available capacity), which are basically the inherent parameters of the system and are different for various networks. These parameters are different for various networks and play a very important role in the operation and control of network frequency and the allocation of primary and secondary storage. Also, by using spinning reserve with the help of primary frequency control, frequency drop to frequency loading thresholds can be prevented to protect against frequency instability [99]. If, due to major disturbances, such as power plant outages, power imbalances persist and the system frequency deviates significantly from the nominal value, and underfrequency load shedding steps begin, in this regard, in order to increase the accuracy of frequency security assessment of systems that are under disturbances, new methods are needed [53].

Frequency stability assessment is based on the degree of frequency deviation and the rate of change of frequency from the nominal value. Then according to the amount of changes, the necessary measures are taken for stability, and even corrective measures may be taken after instability. In frequency security assessment using machine learning methods, the model is trained offline and used for online applications [100].

Frequency assessment in three issues of automatic generation control (AGC) [32, 40, 101], frequency load control (LFC) [99, 102, 103], and underfrequency load shedding (UFLS) [29, 53, 54, 100] is done by using machine learning methods in two stages of generation and optimization of features and classification based on frequency deviation.

Generation and Optimization of Features: The initial data set for determining the frequency stability margin includes the frequency value and the frequency change rate. A variety of optimization methods are used to reduce redundant features and initial data. The initial training of the data is in the form of measuring the distance of the samples in order to identify the features and group the samples in different categories [100].

Classification Based on Frequency Stability Margin: Compared to traditional and common techniques, machine learning methods have the ability to make quick decisions in choosing the type of strategy, reducing the time of decision and restoration, and improving frequency stability. The relationship between the frequency deviation from the nominal value and other attributes is used to determine stability [100]. Some methods, such as Artificial Neural Network and extreme learning machine, have high speed in teaching and learning process, but extreme learning machine algorithm is faster [53]. ELM also has fewer learning limitations and does not fall into local optimizations [29]. Frequency assessment in multi-area and wide area systems is an important criterion in determining system stability [102].

Table 5.8 illustrates some of the machine learning methods along with the techniques of preprocessing and optimizing features in some articles in the field of frequency stability. As shown, the most used machine learning tool in frequency

Table 5.8 A summary of some of the offered machine learning methods to frequency assessment

Refs	Machine learning algorithm	Preprocessing/optimization technique
[29]	Extreme Learning Machine (ELM)	–
[32]	ELM	–
[40]	Long Short-Term emory Recurrent Neural Network	Backward Propagation with the Gradient Optimizer
[53]	Artificial Neural Network (ANN) and Analytic Hierarchy Process (AHP)	*k*-Means
[54]	Deep Q Network (Multi-Q-Learning) and Deep Reinforcement Learning (DRL)	Deep Deterministic Policy Gradient (DDPG)
[99]	Multi-Agent Reinforcement Learning (MARL)	Genetic Algorithm (GA)
[100]	Support Vector Regression (SVR) and Core Vector Regression (CVR)	Mini-Batch Gradient Descent (MBGD) and Fuzzy *k*-Means
[101]	Artificial Emotional Reinforcement Learning (ERL)	–
[102]	Integral Reinforcement Learning	–
[103]	Multi-Agent Deep Reinforcement Learning (MA-DRL)	Deep Deterministic Policy Gradient (DDPG)
[104]	ELM	–
[105]	Principal Component Analysis (PCA), Regression Trees, and Support Vector Machine (SVM)	–

stability assessment and classification is the reinforcement learning technique. Since reinforcement methods provide the learning with feedback from the environment, they have a special application for online training.

5.6 Challenges, Comparative Discussion, and Future Perspectives

Today, practical complexities in the power system have increased, so control and evaluation of stability have become the most important concern in the power system. In this regard, data-based methods should replace model-based methods because data-based methods have online decision-making capabilities, but model-based methods have the contradiction between efficiency and accuracy.

Much progress has been made in using machine learning and deep learning methods related to power system control and stability studies, but some challenges remain unresolved. The quantity and quality of input data sets and experimental systems used in each research will affect the prediction and accuracy of machine learning methods.

Researchers obtain the data, which they need for their research, from open source data and simulated data sets because real power system data may be unavailable for reasons such as security issues, attacks, overcrowding, and breaches, or the number of available data is very small. Therefore, this inconsistency between the available data and the actual data of the power system will cause inconsistencies in forecasts and classifications. In addition, fine-tuning the parameters and measurements of each control problem is an important issue in the application of the machine learning algorithm, and it can be a time-consuming problem. Also, the use of "offline training, online application" method has become a bit challenging due to inconsistency and possible imbalance crisis of data, measurements, and status of power system variables. However, the robust status of the power system is a point of hope for the use of machine learning methods in preventive control measures, protection of the power system in an emergency situation, and its restoration after a fault and blackout. It should be noted that the "offline training, online application" method is more challenging in examining the state of transient stability than other areas because its evaluation is related only to static data after the fault, and also, simulated training samples produced offline mode may not display the current or future state of the power system correctly.

To control the power system and related issues, the following issues are observed:

Most control considerations relate to control issues in normal and emergency situations, while relatively few considerations related to preventive and restorative controls. However, studies on control issues in the operating mode of preventive using machine learning techniques have been surprisingly abundantly presented. The reason for this is to anticipate possible security and stability disruptions and to prevent and plan as much as possible to take the best action in the event of a disruption.

Different control issues have been discussed at different levels of the wide area, smart grids, microgrids, subsystems, and devices. It should be noted that the move toward smart power grids has highlighted the efficiency of machine learning methods.

The most important security and stability problems that can be examined with the largest volume of studies and researches include transient stability assessment after perturbations, voltage stability assessment, and power quality disturbances assessment. It is worth mentioning that power quality assessment includes frequency, voltage, and angle stability.

The predominant machine learning methods used to solve the control problems in this chapter include Support Vector Machine (SVM), k-Nearest Neighbor (k-NN), Decision Tree (DT), Random Forest (RF), Neural Network, and *k*-Means, which sometimes adopt the same important feature selection techniques such as Factor Analysis, Forward Feature Selection (FFS), Backward Feature Selection (BFS), and Principal Component Analysis (PCA).

A very important area that has received a lot of attention in the last decade, and in which issues related to security and stability of control issues are discussed, is the study and prediction of various types of attacks and injuries in control of SCADA power network infrastructure. The SCADA system enables the automatic coordination of control functions such as monitoring the security of the power system, economic operations of load dispatch among power plants, short-term forecasts, automatic production control, emergency control, and etc. across the power system in a coordinated manner. Control systems include a variety of electronic and processing equipment, the most important of which are actuators, sensors, Programmable Logic Controllers (PLCs), and Remote Terminal Units (RTUs). By accessing the software or hardware code of this equipment, attackers can easily infiltrate the control system and take control of the process. Also, the widespread presence of cyber in the form of advanced communication tools and the Internet of Things in the SCADA network has made the power system vulnerable to attacks and security threats. So safety solutions and measures such as antivirus programs, firewalls, encryption algorithms, authentication, traditional intrusion detection systems, etc. are unable to deal with cyberattacks [11].

Recently, cyberattacks on the SCADA network have increased. Therefore, intrusion detection systems are installed for early detection of attacks. In this regard, the ability of machine learning algorithms in autonomous learning, adaptation to change and performance without any prior planning, has led to their use as valid methods for intelligent and efficient intrusion detection systems [15]. Techniques of machine learning algorithms for investigating power grid attacks in SCADA include three main steps including data set generation, data processing, and classification/detection. SCADA network data is recorded and analyzed using machine learning methods to create training data sets to reduce attacks. SCADA data is not available for real-time security reasons. Therefore, predicting the occurrence of instabilities is one of the important issues in controlling the power system. In order to prevent cyberattacks and injuries on SCADA, machine learning methods are used to identify the types of attacks.

5.7 Conclusions

Control of the system in different operating modes in order to provide electrical energy with high reliability for consumers and create a secure and stable situation in the event of any fault in the power system has been considered by all energy stakeholders, especially operators. However, the move toward smart power system and increasing use of distributed generation (DG) and renewable energy sources and converter-based resources, despite their many benefits, have increased instability at the power system level, in such a way that it has the greatest effect on its frequency and changes.

Increasing complexity in large interconnected power systems requires advanced control techniques that effectively control power systems. Also, the hardware cannot evaluate in real time on large networks, and the use of numerical methods and modeling of the power system is complex and time-consuming. Thus, researchers have turned to use machine learning methods. In machine learning, a model is first trained, and then it is used for predicting the class of newly arrived data. Till now, several machine learning techniques have been used to evaluate the power system control. Embedding the machine learning methods in control schemes of power systems is an important and effective method to create controllers with the capability to learn and update decision-making features.

Machine learning and deep learning methods provide suitable solutions to solve each of the control problems at each control level of the operating modes of the electrical power system. Considerations of control levels vary from local to wide area, and to solve each problem using the machine learning methods in this chapter, these results are obtained, that there is no method that is guaranteed for all or even most control problems. However, there are enough ways to evaluate a challenging issue that are reasonably successful. A comparison method is also used to select the best learning method so that several machine learning methods are used to solve a problem, and finally, the method that can offer the most accurate, the best classification with a sufficient number of features, the least complexity, and the shortest simulation time is selected as the method with the best performance in solving the desired problem. In addition, selecting the best method for extracting, reducing, and optimizing features is also a consideration in machine learning and deep learning algorithms.

References

1. DOE-USA, Chapter 5: Increasing efficiency of buildings systems and technologies, in *The Quadrennial Technology Review: An Assessment of Energy Technologies and Research Opportunities* (2015)
2. R. Yousefian, R. Bhattarai, S. Kamalasadan, Transient stability enhancement of power grid with integrated wide area control of wind farms and synchronous generators. IEEE Trans. Power Syst. (2017). https://doi.org/10.1109/TPWRS.2017.2676138
3. P. Kofinas, S. Doltsinis, A.I. Dounis, G.A. Vouros, A reinforcement learning approach for MPPT control method of photovoltaic sources. Renew. Energy **108**, 461–473 (2017). https://doi.org/10.1016/j.renene.2017.03.008
4. A. Rosato, M. Panella, R. Araneo, A. Andreotti, A Neural Network Based Prediction System of Distributed Generation for the Management of Microgrids. IEEE Trans. Ind. Appl. **55**, 922 (2019). https://doi.org/10.1109/TIA.2019.2916758
5. X. Xu, Y. Xu, M.-H. Wang, J. Li, Z. Xu, S. Chai, Y. He, Data-driven game-based pricing for sharing rooftop photovoltaic generation and energy storage in the residential building cluster under uncertainties. IEEE Trans. Ind. Informat. (2020). https://doi.org/10.1109/tii.2020.3016336
6. H. Jahangir, S.S. Gougheri, B. Vatandoust, M.A. Golkar, A. Ahmadian, A. Hajizadeh, Plug-in electric vehicle behavior modeling in energy market: a novel deep learning-based approach with clustering technique. IEEE Trans. Smart Grid (2020). https://doi.org/10.1109/tsg.2020.2998072
7. R. Badar, M.Z. Khan, M.A. Javed, MIMO adaptive bspline-based wavelet neurofuzzy control for multi-type facts. IEEE Access **8** (2020). https://doi.org/10.1109/ACCESS.2020.2969387
8. S. Lan, M.J. Chen, D.Y. Chen, A novel HVDC double-terminal non-synchronous fault location method based on convolutional neural network. IEEE Trans. Power Deliv. **34** (2019). https://doi.org/10.1109/TPWRD.2019.2901594
9. G.K. Venayagamoorthy, R.K. Sharma, P.K. Gautam, A. Ahmadi, Dynamic energy management system for a smart microgrid. IEEE Trans. Neural Netw. Learn. Syst. (2016). https://doi.org/10.1109/TNNLS.2016.2514358
10. Y. Li, Z. Yang, Application of EOS-ELM with binary Jaya-based feature selection to real-time transient stability assessment using PMU data. IEEE Access **5** (2017). https://doi.org/10.1109/ACCESS.2017.2765626

11. J. Gao, L. Gan, F. Buschendorf, L. Zhang, H. Liu, P. Li, X. Dong, T. Lu, Omni SCADA intrusion detection using deep learning algorithms. IEEE Internet Things J. (2020). https://doi.org/10.1109/jiot.2020.3009180
12. E. Hossain, I. Khan, F. Un-Noor, S.S. Sikander, M.S.H. Sunny, Application of big data and machine learning in smart grid, and associated security concerns: a review. IEEE Access **7** (2019)
13. O.A. Alimi, K. Ouahada, A.M. Abu-Mahfouz, A review of machine learning approaches to power system security and stability. IEEE Access **8** (2020)
14. L.A. Wehenkel, Automatic learning techniques in power systems (1998)
15. M. Kalech, Cyber-attack detection in SCADA systems using temporal pattern recognition techniques. Comput. Secur. **84** (2019). https://doi.org/10.1016/j.cose.2019.03.007
16. P. Kundur, *Power System Stability and Control* (McGraw-Hill, New York, 1993)
17. A. Dissanayaka, U.D. Annakkage, B. Jayasekara, B. Bagen, Risk-based dynamic security assessment. IEEE Trans. Power Syst. **26** (2011). https://doi.org/10.1109/TPWRS.2010.2089809
18. L. Wang, P. Pourbeik, Assessment of power system stability and dynamic security performance, in *Power System Stability and Control*, 3rd edn. (CRC Press, 2017)
19. P. Kundur, J. Paserba, V. Ajjarapu, G. Andersson, A. Bose, C. Canizares, N. Hatziargyriou, D. Hill, A. Stankovic, C. Taylor, T. Van Cursem, V. Vittal, Definition and classification of power system stability. IEEE Trans. Power Syst. **19** (2004). https://doi.org/10.1109/TPWRS.2004.825981
20. M. Glavic, (Deep) reinforcement learning for electric power system control and related problems: a short review and perspectives. Annu. Rev. Control **48**, 22–35 (2019)
21. X. Wang, X. Luo, M. Zhang, X. Guan, Distributed detection and isolation of false data injection attacks in smart grids via nonlinear unknown input observers. Int. J. Electr. Power Energy Syst. **110** (2019). https://doi.org/10.1016/j.ijepes.2019.03.008
22. C.C. Aggarwal, Neural networks and deep learning (2018)
23. E. Alpaydin, *Introduction to Machine Learning*, 3rd edn. (2014)
24. Miraftabzadeh SM, Foiadelli F, Longo M, Pasetti M (2019) A survey of machine learning applications for power system analytics. In: Proceedings - 2019 IEEE International Conference on Environment and Electrical Engineering and 2019 IEEE Industrial and Commercial Power Systems Europe, EEEIC/I and CPS Europe 2019
25. P. Mehta, M. Bukov, C.-H. Wang, A.G.R. Day, C. Richardson, C.K. Fisher, D.J. Schwab Review machine learning. arXiv:180308823 [cond-mat, physics:Physics, stat]. (2018). https://doi.org/arXiv:1803.08823v1
26. M. Glavic, R. Fonteneau, D. Ernst Reinforcement learning for electric power system decision and control: past considerations and perspectives. IFAC-Papers OnLine. (2017) https://doi.org/10.1016/j.ifacol.2017.08.1217
27. H. Mohammadi, M. Dehghani, PMU based voltage security assessment of power systems exploiting principal component analysis and decision trees. Int. J. Electr. Power Energy Syst. **64** (2015). https://doi.org/10.1016/j.ijepes.2014.07.077
28. S. Zhang, Y. Wang, M. Liu, Z. Bao, Data-based line trip fault prediction in power systems using LSTM networks and SVM. IEEE Access **6** (2017). https://doi.org/10.1109/ACCESS.2017.2785763
29. Y. Dai, Y. Xu, Z.Y. Dong, K.P. Wong, L. Zhuang, Real-time prediction of event-driven load shedding for frequency stability enhancement of power systems. IET Gener. Transm. Distrib. **6** (2012). https://doi.org/10.1049/iet-gtd.2011.0810
30. H. Hagmar, L. Tong, R. Eriksson, L.A. Tuan, Voltage instability prediction using a deep recurrent neural network. (2019). https://doi.org/10.1109/TPWRS.2020.3008801
31. H. Yang, W. Zhang, F. Shi, J. Xie, W. Ju, PMU-based model-free method for transient instability prediction and emergency generator-shedding control. Int. J. Electr. Power Energy Syst. **105** (2019). https://doi.org/10.1016/j.ijepes.2018.08.031

32. S. Wen, Y. Wang, Y. Tang, Y. Xu, P. Li, Proactive frequency control based on ultra-short-term power fluctuation forecasting for high renewables penetrated power systems. IET Renew. Power Gener. **13** (2019). https://doi.org/10.1049/iet-rpg.2019.0234
33. Y. Xu, Z.Y. Dong, K. Meng, R. Zhang, K.P. Wong, Real-time transient stability assessment model using extreme learning machine. IET Gener. Transm. Distrib. **5** (2011). https://doi.org/10.1049/iet-gtd.2010.0355
34. B.P. Soni, A. Saxena, V. Gupta, S.L. Surana, Identification of generator criticality and transient instability by supervising real-time rotor angle trajectories employing RBFNN. ISA Trans. (2018). https://doi.org/10.1016/j.isatra.2018.08.008
35. J. Morales, E. Orduña, H. Villarroel, J.C. Quispe, High-speed directional protection without voltage sensors for distribution feeders with distributed generation integration based on the correlation of signals and machine learning. Electr. Power Syst. Res. **184** (2020). https://doi.org/10.1016/j.epsr.2020.106295
36. Y.J. Lin, Comparison of CART- and MLP-based power system transient stability preventive control. Int. J. Electr. Power Energy Syst. **45** (2013). https://doi.org/10.1016/j.ijepes.2012.08.066
37. B.P. Soni, A. Saxena, V. Gupta, S.L. Surana, Transient stability-oriented assessment and application of preventive control action for power system. J. Eng. **2019** (2019). https://doi.org/10.1049/joe.2018.9353
38. R. Dubey, S.R. Samantaray, B.K. Panigrahi, An extreme learning machine based fast and accurate adaptive distance relaying scheme. Int. J. Electr. Power Energy Syst. (2015). https://doi.org/10.1016/j.ijepes.2015.06.024
39. S. Li, V. Ajjarapu, M. Djukanovic, Adaptive online monitoring of voltage stability margin via local regression. IEEE Trans. Power Syst. **33** (2017). https://doi.org/10.1109/tpwrs.2017.2698065
40. S. Wen, T. Zhao, Y. Wang, Y. Tang, Y. Xu, P. Li, A deep learning method for power fluctuation identification from frequency fluctuations, in *IEEE Power and Energy Society General Meeting*, (2019)
41. D.K. Chaturvedi, O.P. Malik, Generalized neuron-based adaptive PSS for multimachine environment. IEEE Trans. Power Syst. **20** (2005). https://doi.org/10.1109/TPWRS.2004.840410
42. H. Liu, F. Hussain, Y. Shen, R. Morales-Menendez, M. Abubakar, S. Junaid Yawar, H.J. Arain, Signal processing and deep learning techniques for power quality events monitoring and classification. Electr. Power Components Syst. (2019). https://doi.org/10.1080/15325008.2019.1666178
43. Y. Chen, M. Mazhari, C.Y. Chung, S.O. Faried, B.C. Pal, Rotor angle stability prediction of power systems with high wind power penetration using a stability index vector. IEEE Trans. Power Syst. (2020). https://doi.org/10.1109/tpwrs.2020.2989725
44. R. Lu, S.H. Hong, Incentive-based demand response for smart grid with reinforcement learning and deep neural network. Appl. Energy (2019). https://doi.org/10.1016/j.apenergy.2018.12.061
45. A unified time scale intelligent control algorithm for micro grid based on extreme dynamic programming. CSEE J. Power Energy Syst. (2019). https://doi.org/10.17775/cseejpes.2019.00100
46. B. Wang, B. Fang, Y. Wang, H. Liu, Y. Liu, Power system transient stability assessment based on big data and the core vector machine. IEEE Trans. Smart Grid **7** (2016). https://doi.org/10.1109/TSG.2016.2549063
47. T. Shi, Y. He, T. Wang, B. Li, Open switch fault diagnosis method for PWM voltage source rectifier based on deep learning approach. IEEE Access (2019). https://doi.org/10.1109/ACCESS.2019.2917311
48. M. Tasdighi, M. Kezunovic, Preventing transmission distance relays maloperation under unintended bulk DG tripping using SVM-based approach. Electr. Power Syst. Res. **142** (2017). https://doi.org/10.1016/j.epsr.2016.09.024

49. Y. Liu, Y. Liu, J. Liu, M. Li, T. Liu, G. Taylor, K. Zuo, A MapReduce based high performance neural network in enabling fast stability assessment of power systems. Math. Probl. Eng. (2017). https://doi.org/10.1155/2017/4030146
50. S.S. Maaji, G. Cosma, A. Taherkhani, A.A. Alani, T.M. McGinnity, On-line voltage stability monitoring using an ensemble AdaBoost classifier, in *2018 4th International Conference on Information Management*, (ICIM, 2018, 2018)
51. R.A. Sowah, N.A. Dzabeng, A.R. Ofoli, A. Acakpovi, K.M. Koumadi, J. Ocrah, D. Martin, Design of power distribution network fault data collector for fault detection, location and classification using machine learning, in *IEEE International Conference on Adaptive Science and Technology, ICAST*, (2018)
52. D. Wang, M. Glavic, L. Wehenkel, Trajectory-based supplementary damping control for power system electromechanical oscillations. IEEE Trans. Power Syst. **29** (2014). https://doi.org/10.1109/TPWRS.2014.2314359
53. T.N. Le, N.A. Nguyen, H.A. Quyen, Emergency control of load shedding based on coordination of artificial neural network and analytic hierarchy process algorithm, in *Proceedings - 2017 International Conference on System Science and Engineering, ICSSE 2017*, (2017)
54. C. Chen, M. Cui, F.F. Li, S. Yin, X. Wang, Model-free emergency frequency control based on reinforcement learning. IEEE Trans. Ind. Informat. (2020). https://doi.org/10.1109/tii.2020.3001095
55. Y. Xu, R. Zhang, J. Zhao, Z.Y. Dong, D. Wang, H. Yang, K.P. Wong, Assessing short-term voltage stability of electric power systems by a hierarchical intelligent system. IEEE Trans. Neural Netw. Learn. Syst. (2016). https://doi.org/10.1109/TNNLS.2015.2441706
56. M. Manohar, E. Koley, S. Ghosh, Enhancing the reliability of protection scheme for PV integrated microgrid by discriminating between array faults and symmetrical line faults using sparse auto encoder. IET Renew. Power Gener. **13** (2019). https://doi.org/10.1049/iet-rpg.2018.5627
57. M.R. Alam, K.M. Muttaqi, A. Bouzerdoum, Evaluating the effectiveness of a machine learning approach based on response time and reliability for islanding detection of distributed generation. IET Renew. Power Gener. **11** (2017). https://doi.org/10.1049/iet-rpg.2016.0987
58. S. Zarrabian, R. Belkacemi, A.A. Babalola, Real-time smart grids control for preventing cascading failures and blackout using neural networks: experimental approach for N-1-1 contingency. Int. J. Emerg. Electr. Power Syst. (2016). https://doi.org/10.1515/ijeeps-2016-0039
59. R. Yousefian, S. Kamalasadan, Energy function inspired value priority based global wide-area control of power grid. IEEE Trans. Smart Grid (2018). https://doi.org/10.1109/TSG.2016.2555909
60. S.S. Jhang, H.J. Lee, C.N. Kim, C.H. Song, W.K. Yu, ANN Control for damping low-frequency oscillation using deep learning, in *Australasian Universities Power Engineering Conference, AUPEC 2018*, (2018)
61. J.D. Pinzón, D.G. Colomé, Real-time multi-state classification of short-term voltage stability based on multivariate time series machine learning. Int. J. Electr. Power Energy Syst. (2019). https://doi.org/10.1016/j.ijepes.2019.01.022
62. I. Genc, R. Diao, V. Vittal, S. Kolluri, S. Mandal, Decision tree-based preventive and corrective control applications for dynamic security enhancement in power systems. IEEE Trans. Power Syst. (2010). https://doi.org/10.1109/TPWRS.2009.2037006
63. Q. Li, Y. Xu, C. Ren, A hierarchical data-driven method for event-based load shedding against fault-induced delayed voltage recovery in power systems. IEEE Trans. Ind. Informat. (2020). https://doi.org/10.1109/tii.2020.2993807
64. D. Ye, M. Zhang, D. Sutanto, A hybrid multiagent framework with Q-learning for power grid systems restoration. IEEE Trans. Power Syst. (2011). https://doi.org/10.1109/TPWRS.2011.2157180

65. M. Al Karim, J. Currie, T.T. Lie, A feature based distributed machine learning for post fault restoration of a microgrid under different stochastic scenarios, in *2017 IEEE Innovative Smart Grid Technologies - Asia: Smart Grid for Smart Community, ISGT-Asia 2017*, (2018)
66. Y. Zhang, Y. Xu, Z.Y. Dong, R. Zhang, A hierarchical self-adaptive data-analytics method for real-time power system short-term voltage stability assessment. IEEE Trans. Ind. Informat. **15** (2019). https://doi.org/10.1109/TII.2018.2829818
67. M.J. Ghorbani, M.A. Choudhry, A. Feliachi, A multiagent design for power distribution systems automation. IEEE Trans. Smart Grid **7** (2016). https://doi.org/10.1109/TSG.2015.2453884
68. Y. Zhou, J. Wu, Z. Yu, L. Ji, L. Hao, A hierarchical method for transient stability prediction of power systems using the confidence of a SVM-based ensemble classifier. Energies (2016). https://doi.org/10.3390/en9100778
69. A. Sharifian, S. Sharifian, A new power system transient stability assessment method based on Type-2 fuzzy neural network estimation. Int. J. Electr. Power Energy Syst. **64** (2015). https://doi.org/10.1016/j.ijepes.2014.07.007
70. J.J.Q. Yu, D.J. Hill, A.Y.S. Lam, J. Gu, V.O.K. Li, Intelligent time-adaptive transient stability assessment system. IEEE Trans. Power Syst. **33** (2018). https://doi.org/10.1109/TPWRS.2017.2707501
71. J.J.Q. Yu, A.Y.S. Lam, D.J. Hill, V.O.K. Li, Delay aware intelligent transient stability assessment system. IEEE Access **5** (2017). https://doi.org/10.1109/ACCESS.2017.2746093
72. X. Gu, Y. Li, Bayesian multiple kernels learning-based transient stability assessment of power systems using synchronized measurements, in *IEEE Power and Energy Society General Meeting*, (2013)
73. A.B. Mosavi, A. Amiri, H. Hosseini, A learning framework for size and type independent transient stability prediction of power system using twin convolutional support vector machine. IEEE Access **6** (2018). https://doi.org/10.1109/ACCESS.2018.2880273
74. W. Hu, Z. Lu, S. Wu, W. Zhang, Y. Dong, R. Yu, B. Liu, Real-time transient stability assessment in power system based on improved SVM. J. Mod. Power Syst. Clean. Energy **7** (2019). https://doi.org/10.1007/s40565-018-0453-x
75. Y. Zhang, T. Li, G. Na, G. Li, Y. Li, Optimized extreme learning machine for power system transient stability prediction using synchrophasors. Math. Probl. Eng. (2015). https://doi.org/10.1155/2015/529724
76. M. He, J. Zhang, V. Vittal, Robust online dynamic security assessment using adaptive ensemble decision-tree learning. IEEE Trans. Power Syst. **28** (2013). https://doi.org/10.1109/TPWRS.2013.2266617
77. C. Zhang, Y. Li, Z. Yu, F. Tian, Feature selection of power system transient stability assessment based on random forest and recursive feature elimination, in *Asia-Pacific Power and Energy Engineering Conference, APPEEC*, (2016)
78. Y. Li, G. Li, Z. Wang, Rule extraction based on extreme learning machine and an improved ant-miner algorithm for transient stability assessment. PLoS One **10** (2015). https://doi.org/10.1371/journal.pone.0130814
79. R. Zhang, Y. Xu, Z.Y. Dong, P. Zhang, K.P. Wong, Voltage stability margin prediction by ensemble based extreme learning machine, in *IEEE Power and Energy Society General Meeting*, (2013)
80. H.H. Goh, Q.S. Chua, S.W. Lee, B.C. Kok, K.C. Goh, K.T.K. Teo, Evaluation for voltage stability indices in power system using artificial neural network, in *Procedia Engineering*, (2015)
81. S.M. Pérez-Londoño, G. Olivar-Tost, J.J. Mora-Florez, Online determination of voltage stability weak areas for situational awareness improvement. Electr. Power Syst. Res. **145** (2017). https://doi.org/10.1016/j.epsr.2016.12.026
82. K.S. Sajan, V. Kumar, B. Tyagi, ICA based artificial neural network model for voltage stability monitoring, in *IEEE Region 10 Annual International Conference, Proceedings/TENCON*, (2016)

83. K.S. Sajan, V. Kumar, B. Tyagi, Genetic algorithm based support vector machine for on-line voltage stability monitoring. Int. J. Electr. Power Energy Syst. **73** (2015). https://doi.org/10.1016/j.ijepes.2015.05.002
84. A.R. Bahmanyar, A. Karami, Power system voltage stability monitoring using artificial neural networks with a reduced set of inputs. Int. J. Electr. Power Energy Syst. **58** (2014). https://doi.org/10.1016/j.ijepes.2014.01.019
85. S.R. Nandanwar, M.L. Kolhe, S.B. Warkad, N.P. Patidar, V.K. Singh, Voltage security assessment by using PFDT and CBR methods in emerging power system, in *Energy Procedia*, (2018)
86. H. Mohammadi, G. Khademi, M. Dehghani, D. Simon, Voltage stability assessment using multi-objective biogeography-based subset selection. Int. J. Electr. Power Energy Syst. **103** (2018). https://doi.org/10.1016/j.ijepes.2018.06.017
87. V. Jayasankar, N. Kamaraj, N. Vanaja, Estimation of voltage stability index for power system employing artificial neural network technique and TCSC placement. Neurocomputing **73** (2010). https://doi.org/10.1016/j.neucom.2010.07.006
88. S. Khokhar, A.A. Mohd Zin, A.P. Memon, A.S. Mokhtar, A new optimal feature selection algorithm for classification of power quality disturbances using discrete wavelet transform and probabilistic neural network. J. Int. Meas. Confed. **95** (2017). https://doi.org/10.1016/j.measurement.2016.10.013
89. A.A. Abdoos, P. Khorshidian Mianaei, M. Rayatpanah Ghadikolaei, Combined VMD-SVM based feature selection method for classification of power quality events. Appl. Soft Comput. J. **38** (2016). https://doi.org/10.1016/j.asoc.2015.10.038
90. K. Manimala, K. Selvi, R. Ahila, Optimization techniques for improving power quality data mining using wavelet packet based support vector machine. Neurocomputing **77** (2012). https://doi.org/10.1016/j.neucom.2011.08.010
91. U. Singh, S.N. Singh, Application of fractional Fourier transform for classification of power quality disturbances. IET Sci. Meas. Technol. **11** (2017). https://doi.org/10.1049/iet-smt.2016.0194
92. Z. Liu, Y. Cui, W. Li, A classification method for complex power quality disturbances using EEMD and rank wavelet SVM. IEEE Trans. Smart Grid **6** (2015). https://doi.org/10.1109/TSG.2015.2397431
93. Z. Liquan, G. Meijiao, W. Lin, Classification of multiple power quality disturbances based on the improved SVM, in *Proceedings of the 2017 International Conference on Wireless Communications, Signal Processing and Networking, WiSPNET 2017*, (2018)
94. U. Singh, S.N. Singh, A new optimal feature selection scheme for classification of power quality disturbances based on ant colony framework. Appl. Soft Comput. J. **74** (2019). https://doi.org/10.1016/j.asoc.2018.10.017
95. D. De Yong, S. Bhowmik, F. Magnago, An effective power quality classifier using wavelet transform and support vector machines. Expert Syst. Appl. **42** (2015). https://doi.org/10.1016/j.eswa.2015.04.002
96. S. Shukla, S. Mishra, B. Singh, Empirical-mode decomposition with hilbert transform for power-quality assessment. IEEE Trans. Power Deliv. **24** (2009). https://doi.org/10.1109/TPWRD.2009.2028792
97. H. Erişti, Y. Demir, A new algorithm for automatic classification of power quality events based on wavelet transform and SVM. Expert Syst. Appl. **37** (2010). https://doi.org/10.1016/j.eswa.2009.11.015
98. H. Liu, F. Hussain, Y. Shen, S. Arif, A. Nazir, M. Abubakar, Complex power quality disturbances classification via curvelet transform and deep learning. Electr. Power Syst. Res. **163** (2018). https://doi.org/10.1016/j.epsr.2018.05.018
99. F. Daneshfar, H. Bevrani, Load-frequency control: a GA-based multi-agent reinforcement learning. IET Gener. Transm. Distrib. (2010). https://doi.org/10.1049/iet-gtd.2009.0168

100. H. Li, C. Li, Y. Liu, Maximum frequency deviation assessment with clustering based on metric learning. Int. J. Electr. Power Energy Syst. **120** (2020). https://doi.org/10.1016/j.ijepes.2020.105980
101. L. Yin, T. Yu, L. Zhou, L. Huang, X. Zhang, B. Zheng, Artificial emotional reinforcement learning for automatic generation control of large-scale interconnected power grids. IET Gener. Transm. Distrib. **11**, 2305–2313 (2017). https://doi.org/10.1049/iet-gtd.2016.1734
102. M. Abouhea, W. Gueaieb, A. Sharaf, Load frequency regulation for multi-area power system using integral reinforcement learning. IET Gener. Transm. Distrib. **13** (2019). https://doi.org/10.1049/iet-gtd.2019.0218
103. Z. Yan, Y. Xu, A multi-agent deep reinforcement learning method for cooperative load frequency control of multi-area power systems. IEEE Trans. Power Syst. (2020). https://doi.org/10.1109/tpwrs.2020.2999890
104. Q. Wang, F. Li, Y. Tang, Y. Xu, Integrating model-driven and data-driven methods for power system frequency stability assessment and control. IEEE Trans. Power Syst. **34** (2019). https://doi.org/10.1109/TPWRS.2019.2919522
105. Z. Hou, J. Folium, P. Etingov, F. Tuffner, D. Kosterev, G. Matthews, Machine learning of factors influencing damping and frequency of dominant inter-area modes in the WECC interconnect, in *2018 International Conference on Probabilistic Methods Applied to Power Systems, PMAPS 2018 - Proceedings*, (2018)

Chapter 6
Introduction and Literature Review of the Application of Machine Learning/ Deep Learning to Load Forecasting in Power System

Arash Moradzadeh, Amin Mansour-Saatloo, Morteza Nazari-Heris, Behnam Mohammadi-Ivatloo, and Somayeh Asadi

Abbreviation

ACE	Average coverage error
ANN	Artificial neural network
APE	Absolute percentage error
AR	Auto-regressive
ARIMA	Auto-regressive integrated moving average
ARMA	Auto-regressive moving average
BGA	Binary genetic algorithm
BPNN	Back-propagation neural network
CNN	Convolution neural network
CV	Coefficient of variance
CWC	Coverage width-based criterion
DA	Direction accuracy
DAME	Daily absolute maximum error
DBN	Deep belief network
DBN	Deep neural network
DC	Directional change

A. Moradzadeh · A. Mansour-Saatloo · B. Mohammadi-Ivatloo (✉)
Faculty of Electrical and Computer Engineering, University of Tabriz, Tabriz, Iran
e-mail: arash.moradzadeh@tabrizu.ac.ir; amin_mnsr97@ms.tabrizu.ac.ir; bmohammadi@tabrizu.ac.ir

M. Nazari-Heris · S. Asadi
Department of Architectural Engineering, Pennsylvania State University, State College, PA, USA
e-mail: nazari@psu.edu; sxa51@psu.edu

M. Nazari-Heris et al. (eds.), *Application of Machine Learning and Deep Learning Methods to Power System Problems*, Power Systems,
https://doi.org/10.1007/978-3-030-77696-1_6

DMD	Dynamic mode decomposition
EMAE	Envelope-weighted mean absolute error
EMD	Empirical mode decomposition
ENN	Elman neural network
ESN	Echo state network
FCRBM	Factored conditional restricted Boltzmann machine
FFN	Feed-forward neural network
GB	Gradient boosting
GBA	Gradient boosting machine
GELM	Generalized extreme learning machine
GRU	Gated recurrent unit
GWO	Gray wolf optimizer
HR	Heat rate
IA	Index of agreement
IWNN	Improved wavelet neural network
LSTM	Long short-term memory
MAAPE	Mean arctangent absolute percentage error
MAE	Mean absolute error
MAPE	Mean absolute percentage error
MedAE	Median absolute error
MFFNN	Multilayer feed-forward neural network
MLP	Multilayer perceptron
MLR	Multiple linear regression
MOD	Mean outside distance
MWPI	Mean width of prediction interval
nMAE	Normalized mean absolute error
NRMSE	Normalized root mean squared error
NYISO	New York independent system operator
PCR	Principal component regression
PICP	Prediction interval coverage probability
PJM	Pennsylvania New Jersey Maryland
PMSE	Prognostication mean square error
QRF	Quantile regression forest
RBFNN	Radial basis function neural network
RF	Random forest
RMSE	Root mean square error
RMSLE	Root mean square logarithmic error
RNN	Recurrent neural network
RVM	Relevance vector machine
VMD	Variational mode decomposition
WMAE	Weighted mean absolute error
WNN	Wavelet neural network
WOA	Whale optimization algorithm
WT	Wavelet transform

6.1 Introduction

So far, many scholars have proposed various methods to improve the accuracy of load forecasting. Electrical load forecasting approaches can be classified into three categories: statistical methods, artificial intelligence methods, and hybrid methods. Statistical methods include time series models such as auto-regressive (AR) [1], auto-regressive moving average (ARMA) [2], auto-regressive integrated moving average (ARIMA) [3], seasonal ARIMA (SARIMA) [4], linear regression methods [5], multiple linear regression methods [6], and exponential smoothing methods [7]. The accuracy of ARIMA-based methods depends on some external variables and can be further improved using exogenous variables [8]. However, all these methods have good performance in linear systems and are insufficient in nonlinear systems. Since the real-world electrical load has nonlinear inherent, some researchers have tried to introduce some models by extending Kalman filter methods to handle this issue [9], but the extended models could not show high accuracy. To overcome the mentioned challenges, artificial intelligence methods have been developed for usage in the electrical load forecasting.

In recent years, artificial intelligence, due to its ability in forecasting and dealing with nonlinear data, has received great significance. The most popular techniques are artificial neural network (ANN) methods such as generalized regression neural network (GRNN) [10], multilayer perceptron (MLP) [11], radial bias function (RBF) recurrent neural network [12], back-propagation neural network (BPNN) [13], support vector machine (SVM) [14], support vector regression (SVR) methods [15], fuzzy logic methods [16], and data clustering methods [17]. These methods are used for electrical load forecasting because of their capacity to learn and handle complex systems. However, artificial intelligence methods still have drawbacks, including random selection of weight parameters, high execution time, overfitting, etc. To this end, hybrid models were developed to enhance artificial intelligence performance. For instance, in [18], linear extrapolation combined with fuzzy SVR, and in [19], the season-specific similarity concept was combined with SVM and firefly algorithm to forecast the seasonal electrical load. Furthermore, many of the artificial intelligence methods are combined with optimization methods, e.g., the SVM method was hybridized with gray wolf optimizer (GWO) in [20] or a hybrid model of genetic algorithm, particle swarm optimization, and back-propagation neural network (GA-PSO-BPNN) was proposed in [21].

This chapter has a review of the artificial intelligence methods in electrical load forecasting. To this end, the existing scholars are scrutinized from two different scopes. The first scope classifies the papers from a forecasting time horizon point of view, i.e., short-term forecasting and long-term forecasting. Reviewing of the utilized techniques and their performance analysis are the targets of the second scope.

6.2 Most Important Challenges in Power Systems Short-, Medium-, and Long-Term Load Forecasting

6.2.1 Short-Term Load Forecasting

Short-term load forecasting (STLF) includes 1 h ahead up to 1 week ahead forecasting, and it is important to the scheduling of the power system since it has a considerable impact on power resources' generation, limitations, usage constraints, spinning reserve, security, and reliability. Underestimation of the load leads to power shortage, and overestimation of the load causes power waste and an insufficient number of power plants.

In [22], binary genetic algorithm (BGA) and Gaussian process regression were used for the feature selection process and fitness score measurement of features, respectively. Multi-kernel algorithm and SVR methods were proposed in [23] for load forecasting application, where kernel functions of SVR machine are optimized via the utilized algorithm. For the distribution system reconfiguration, genetic algorithm was introduced in [24], in which wavelet transform-based ARIMA was employed for wind, solar, and load forecasting. In [25], the deep belief network was utilized for hourly load forecasting, in which the preprocessing of data was conducted using Box-Cox transformation. A probabilistic load forecasting study considering data noise uncertainty was performed in [26] using the wavelet neural network improved by the generalized extreme learning machine. RNN based on LSTM and gated recurrent unit (GRU) was proposed in [27] for STLF, where the RNN consists of two different layers that combined with LSTM and GRU. In [28], an STLF considering the uncertainty of PV, wind, and load was introduced, in which deep wavelet transform was used to feature selection, and deep neural network (DNN) consisting of autoencoder neural networks with a cascade layer was used as the forecasting tool. In [29], wavelet decomposition was used to decompose the data into different resolutions. Then gray and cubic exponential smoothing models were applied to forecast the load. Wavelet transform for feature selection was utilized in [30], and relevance vector machine was proposed to forecast the load. Three different aggregation strategies, i.e., information aggregation, hierarchical aggregation, and model aggregation, were analyzed in [15] for four machine learning algorithms. Full wavelet transform instead of traditional wavelet transform along with multilayer neural network were proposed in [31] for STLF. In [11], a feature selection method, namely, RReliefF, was employed to choose the features, and then, MLP was applied to forecast the load. Different DNN-based techniques using the stacking ensemble method was developed in [32]. In this reference, combination of various STLF methods outperforms single DNN methods. In [33], a supervised Fuzzy-ARTMAP neural network was developed to forecast the various nodes' load, where a reverse training technique was introduced to forecast. Dynamic mode decomposition was used in [34] to introduce an STLF method based on error correction. The introduced algorithm includes three stages, i.e., data selection stage, error forecasting stage, and error correction stage, in which the first stage benefits from the gray relation analysis method. To consider the relation of cost and load in the electrical grid, a hybrid of two-sided wavelet transform and SVM, along with the revised mutual information-based feature selection method, was developed in [14] for STLF. An ensemble framework based on WaveNet learners for STLF was proposed in [35], where multiple techniques such as bootstrapping and stacked generalization algorithms were used for the ensemble aggregation of WaveNet learners. Gravitational search algorithm (GSA) and PSO were hybridized with single forecasting methods such as SVR for STLF in [36]. An STLF of an individual household was performed in [37], where the Bayesian networks were used as the forecasting tool. Two of ANN-based algorithms, i.e., FFNN and echo state network, were compared in [38] to investigate their performance in the commercial buildings STLF. In the same context, in [39], nine different combinations of RNN along with clustering were introduced to commercial buildings STLF. GRNN and GSA were applied to the short-term load and price forecasting in [10], where variational mode decomposition was used to improve the forecasting accuracy. In [40], cross-industry standard process for data mining combined with ARIMA was proposed, where the autocorrelation function was employed to compute the correlation of a series and its time shift. In [20], to overcome the shortcoming of existing methods in the forecasting of special days load, a hybrid of SVM and GWO were proposed. To conduct the probabilistic load forecasting in [41], the probability distribution of loads was discretized by dividing the load for multiple ranges, and then, a convolutional neural network (CNN) was applied for STLF. A deep learning framework, namely, hybrid deep meta-ensemble networks composed of four modules, was proposed in [42], in which the first one is the local forecaster of each series, the second one is the global forecaster, the third one is supervised feature learner, and the last one is combiner of local and global forecasts based on obtained features. A hybrid factored conditional restricted Boltzmann machine (FCRBM) and the genetic wind-driven optimization algorithm was introduced in [43], where a modified mutual information approach was used to feature selection. A deep-ensemble STLF method based on the LASSO quantile forecast combination approach was introduced in [44], where the model is a probabilistic load forecasting, and in this model, there is no need for feature selection. A comparative study of three different STLF models, i.e., random forest, multiple linear regression, and gradient boosting, were conducted in [45]. According to the obtained results, the gradient boosting model outperforms the other two models, and it shows good performance in both low and high load periods. In [46], Takagi-Sugeno-Kang neuro-fuzzy forecasting technique was applied for STLF, in which, via training this technique with a locally linear model tree method, all the parameters such as the number of neurons and functions are adjusted by the algorithm. In [47], to perform an STLF, a variational mode decomposition was used for preprocessing of data. After then, quantile regression forest hybridized with Bayesian optimization algorithm were applied to forecast the decomposed data separately, and kernel density estimation was utilized to reconstruct the forecasting results. A multi-objective deep belief network along with empirical mode decomposition was proposed in [48]. A hybrid model composed of variational mode decomposition and autoencoder methods to extract the sufficient subsignals along with LSTM network for STLF were proposed in [49]. In [50], decision tree and weighted average methods were used to decompose the data based on hourly and daily attributes, respectively. Then, a regression model based on time series method and SVM method were applied for STLF. In [51], whale optimization method was used to optimize the parameter of CNN, where the CNN was restructured via MLP method to extract an efficient STLF algorithm. A hybrid Elman neural network and PSO were proposed in [52] for STLF, where the PSO was used to search for optimal rate of the network. To overcome the overfitting problem of ANN, a hybrid algorithm composed of an evolutionary algorithm to search for the optimal weights and ANN was proposed in [53], and the evolutionary algorithm was integrated with a controlled Gaussian mutation technique to enhance the convergence solutions. In [54], a methodology for time series analysis was introduced to extend feature selection of SVR idea which is optimized by the Kernel-penalized iterative method. In [55], the fuzzy clustering method was used to initially cluster raw data. Then radial basis function neural network and CNN were applied to conduct the load forecasting, in which both of them optimized via Adam optimization method. One-dimensional CNN hybridized with RNN in [56] to overcome the incorrect generation of the hidden state vector and also to calibrate the forecasting time. To eliminate the outlier data effect from training data, differential evolutionary PSO was proposed in [57], where ANN was applied for STLF. In [58], to enhance both forecasting error and time, a multi-objective model combined with the ANN was proposed. In [59], a hybrid STLF method for microgrids composed of SVR and LSTM was introduced, in which the hybrid method outperforms single SVR and LSTM. In [60], a Levenberg-Marquardt algorithm based on RNN was utilized to forecast the maritime microgrids. An overview of the short-term literature is provided in Table 6.1.

6.2.2 Midterm Load Forecasting

Midterm load forecasting (MTLF) includes 1 week up to 12 months ahead of forecasting. This type of load forecasting is important for the maintenance and operation of the power system. In [61], a combination of three different models, i.e., random forest regression (RFR), gradient boosting decision tree (GBDT), and SVR, were proposed for probabilistic MTLF, where the kernel density function was applied to achieve the probability density distribution of the load. To plan to produce and purchase a hybrid ST-MTLF model was presented in [62], where the proposed model employs the multilayer feed-forward neural network (MFFNN) and grasshopper optimization algorithm (GOA) to forecast the load. The transfer learning method, along with the Pearson correlation coefficient to enhance the forecasting performance, was introduced in [63], in which the aim of the algorithm is monthly MTLF of a city. Empirical mode decomposition with the LSTM network was used in [64] to forecast the seasonal and daily load demand of a city. A hybrid of PSO and

Table 6.1 Overview on short-term forecasting literature

Ref.	Year	Data	Time horizon
[22]	2019	Occupancy, weather features, and electricity price	30-minute
[23]	2017	Highest temperature, lowest temperature, and wind	Daily
[24]	2019	IEEE 34-bus and IEEE 123-bus	Hourly
[25]	2019	Temperature, electricity price, humidity, barometric pressure, and wind speed	Day-ahead and week-ahead
[26]	2018	Ontario and Australian electricity markets data	Hourly
[27]	2019	Daily weather data	Daily
[28]	2020	Solar and wind generations	Hourly
[29]	2019		Daily
[30]	2020	Highest and lowest temperature	Hourly
[31]	2020	Temperature, humidity, hours, days, and day types	Hourly
[15]	2020	Weather features	Daily
[32]	2020	Weather features	Daily
[11]	2020	New England independent system operator data	Hourly
[33]	2020	New Zealand consumption data	30-minute
[14]	2020	NYISO, NSW zone in Australia's market, and PJM data	Hourly
[34]	2020	Weather features, renewable generation, and demand response data	Hourly
[35]	2019	Hourly load from Italy	Hourly
[36]	2019	Queensland, Australia data	Hourly
[37]	2020	Commission for Energy Regulation (CER) smart metering project data	Hourly
[38]	2020	University campus in Milan, Italy	Hourly
[39]	2020	Weather data	15-minute
[10]	2020	PJM and Spanish market data	Hourly
[40]	2020	Control signals, operating modes and disturbance signals, the status of the auxiliary units, and physical properties	10-s
[20]	2020	Regional special event days and temperature	Daily
[41]	2020	New England independent system operator data	Hourly
[42]	2020	US utility data	Hourly
[43]	2020	PJM data	Hourly
[44]	2019	CER data	Hourly
[45]	2019	Temperature, meteorological variable, holiday, and solar capacity	Hourly
[46]	2020	Climate data and holiday	Hourly
[47]	2020	Weighted temperature and humidity index (WTHI) and day type	Daily
[48]	2020	Power load data and algorithmic parameters	Weekly
[49]	2020	Hierarchical data	Daily
[50]	2020	Dry-bulb temperature, dew point temperature, relative humidity, moisture content, wind speed, and air enthalpy	Hourly
[51]	2020	Electrical load time series	30-minute
[52]	2019	Particular power consumption	30-minute

(continued)

Table 6.1 (continued)

Ref.	Year	Data	Time horizon
[53]	2019	Electrical load data, temperature, rainfall, wind speed, and holiday	Hourly
[54]	2019	Methodological data	Daily
[55]	2020	Temperature data	Weekly
[56]	2019	Day type, temperature, humidity, wind speed, electric load, off-peak, mid-peak, and on-peak	30-minute
[57]	2019	Weather data, day type, and flag	Daily
[58]	2019	Dry-bulb temperature, dew point temperature, hour of the day, day of the week, holiday/weekend indicator (0 or 1), 168-hr (previous week) lagged load, 24-hr lagged load, the previous 24-hr average load	Hourly
[59]	2020	Household and commercial load consumption	Hourly
[60]	2020	Next three latest days for thrusting demand and environmental distortion of 72 h	Hourly

Table 6.2 Overview on medium-term forecasting literature

Ref.	Year	Data	Time horizon
[61]	2019	Electrical load, natural gas load, and price, the average retail price of electricity, natural gas consumed by the electric power sector, and energy electric power sector CO_2 emissions	Monthly
[62]	2020	The times per hour, day, and month and the temperature	Hourly and monthly
[63]	2020	Calendar data, population data, and weather data	Monthly
[64]	2018	Data of Chandigarh, India	Season, day, and time interval of a day
[16]	2013	Time and hourly load	Hourly, daily, and monthly

fuzzy neural network were used in [16] to conduct hourly, daily, and monthly load forecasting. An overview of the medium-term literature is provided in Table 6.2.

6.2.3 *Long-Term Load Forecasting*

Long-term load forecasting (LTLF) includes 1 year up to the next 50 years ahead of forecasting. LTLF has a significant impact on the planning of the power system and expansion decisions. LTLF is complicated than STLF and MTLF due to the high uncertainty of load and economic factors such as energy resources price values. The Prophet and Holt-Winters methods were used in [65] for LTLF of Kuwait for the next decade. Three different configuration methods were employed in [66] to achieve a multi-timestep forecasting problem from a multiyear LTLF. Moreover, an unsupervised learning technique was used to group the feeders. In [12], a fuzzy

Table 6.3 Overview on long-term forecasting literature

Ref.	Year	Data	Time horizon
[65]	2020	Load data	Yearly
[66]	2020	The urban distribution system of Canada	Yearly
[12]	2020	Temperature series and holidays	Yearly
[67]	2020	Iran National Grid data	Yearly
[68]	2020	Brazilian energy utility (Cemig distribution)	Yearly

neural network along with robust-type fuzzy rules was developed for LTLF, where a selection method for weather conditions was proposed to improve the accuracy of the forecasting. A hybrid model consisting of ARIMA, ANN, and SVR were introduced in [67], where the PSO algorithm was used to find the optimal value of parameters. In [68], a spatial load forecasting using the spatial convolution operator was conducted, in which the utilized convolution method acts as a low-pass filter in the frequency domain and in the space defines the relation of nodes. An overview of the long-term literature is provided in Table 6.3.

6.3 Machine Learning and Deep Learning Applications in Load Forecasting of Power System

Today, machine learning and deep learning methods have found a special place in applications related to energy and power systems. Especially in issues of power systems that are related to data processing, the effectiveness of the machine learning and deep learning methods is multiplied. Meanwhile, as mentioned in the previous sections, load forecasting is one of the issues that mainly deals with big, historical, and time series data. So far, there are many methods for load forecasting in power grids, each of which has provided different forecasts according to their capabilities. The load used in power grids follows a variety of parameters such as temperature, wind and solar information, date and time data of the week, etc., which are highly interrelated. However, those methods that can accurately estimate and model the relationship between input variables and load data can be highly effective. Most machine learning and deep learning methods have this capability, and they have been able to perform high-performance load forecasting in a variety of projects. It should be noted that because the data related to the load are of continuous type, regression applications are suitable and ideal for doing this process. In the continuation of this section, different methods of machine learning and deep learning that have been used to forecast the load will be introduced.

6.3.1 Machine Learning and Deep Learning Algorithms Used for Load Forecasting

So far, many methods of machine learning and deep learning have been developed as a tool for processing and predicting load data. Among the machine learning algorithms, the applications of SVR, GRNN, ELM, random forest, and decision tree in load prediction can be mentioned. Each of these methods uses a specific pattern for data processing. The SVR can be considered as a regression version of the support vector machine (SVM) [59]. The ELM is also derived from feed-forward neural networks that have a very high convergence performance with a structure similar to the MLP neural network. In addition to regression applications, this method is used for discrete data and classification. The ELM can be used to evaluate big data, and this is one of the advantages of this method [69]. Meanwhile, the GRNN method is also seen as one of the most widely used algorithms in solving regression problems. The GRNN is an improved technique in neural networks based on nonparametric regression. This method in particular can be used as a powerful tool to solve problems related to online dynamic systems. High training speed and fast convergence against big data are the prominent features of this technique [70]. As mentioned, most machine learning algorithms have a neural network-based structure and often focus on improving the training process so that they can estimate the close relationship between input and output parameters.

With the introduction of deep learning techniques, the use of machine learning algorithms was somewhat reduced so that deep learning methods were able to improve most of the problems associated with machine learning algorithms. By focusing on the literature review, methods such as the LSTM, deep belief network (DBN), Bi-LSTM, autoencoder, and CNN can be mentioned, which have been used to forecast the load. Each of these methods has a unique structure that can have different functions based on its original structure. The LSTM method has been introduced as an algorithm for improving the structure of RNNs. This algorithm was able to eliminate the vanishing gradient problem in the structure of the RNNs [71]. The LSTM is a deep learning technique that is used as a powerful tool for time series data. The proper performance of this method and its high ability to analyze the big, high-dimension, and time series data has introduced this method as one of the most suitable methods of load forecasting [59]. The DBN is one of the deep learning networks that with one deep neural network in its structure, consisting of several variable layers and with connections between layers, learns in-depth training of data structure. This network can also be used without supervision, in which case it may be able to reconstruct its inputs. High-processing speed, data reconstruction for better output, and acceptable performance for different volumes of data are the obvious advantages of this network [72]. Autoencoders are one of the deep learning applications used for unsupervised training of data. These networks have a structure similar to FNNs and consist of two components, including an encoder and a decoder. Minimizing the mean reconstruction loss is one of the most important issues to consider in the training of the autoencoders. These networks are mainly used in

Table 6.4 Overview on load forecasting categories

Method	Algorithm	References
ANN	MLP	[11–13, 15, 16, 38, 40, 42, 51, 53, 58, 62, 67]
	RBF	[10, 12, 55]
	BPNN	[12, 16, 21, 36, 52]
	Fuzzy	[16, 18, 33, 46, 55, 76]
	WNN	[31, 35, 36, 42]
Machine learning	SVR	[12, 14, 15, 18, 50, 51, 59, 61, 67]–[20, 23, 30, 36, 49]
	GRNN	[10, 51, 52]
	ELM	[26]
	Random forest	[15, 45, 47, 61, 77]
	Decision tree	[61]
Deep learning	LSTM	[12, 27, 39, 42, 49, 59, 64, 66, 78]–[80]
	RNN	[49, 56]
	DBN	[25, 48, 49]
	DNN	[28, 32]
	Autoencoder	[39, 49]
	CNN	[39, 41, 55, 56, 68]

applications such as face recognition, fingerprinting, and image processing. The high performance of these networks in processing large-scale data is one of the advantages of this method. Autoencoders have so far shown significant results in load forecasting applications [73, 74]. In addition to most of the deep learning algorithms mentioned in the introduction and performance of each, the CNN is one of the most powerful and widely used methods of deep learning, which is mainly used for classification applications. Layer-to-layer structure and feature extraction from input data in several stages are prominent features of the CNN [41, 75]. This technique is mainly used in many industrial applications and scientific projects due to its high ability to extract data features and detect behavioral patterns.

By reviewing most of the machine learning and deep learning techniques and observing the structure of each, it is inferred that the first basis of most of these methods is ANNs. In presenting each of the machine learning and deep learning techniques, an attempt has been made to improve one of the ANN algorithms and to present a new structure. Accordingly, a review of the literature reveals the dramatic applications of various ANN methods in load forecasting programs. Among the ANN algorithms, networks such as MLP, BPNN, RBF, WNN, and fuzzy network, have been used mainly for load forecasting. Table 6.4 categorizes the load forecasting studies based on the proposed method for each paper, which is based on the ANNs, machine learning, and deep learning algorithms.

6.3.2 *Performance Assessment of Algorithms*

Data processing and estimating the relationship between input variables and the target can be done by different types of learning methods. Each method provides performance based on its structure and formulation. Using machine learning and deep learning applications for forecasting issues, a variety of results are obtained, whereas the superiority and effectiveness of each method are achieved after evaluating and comparing the results. Comparison of results related to learning methods is done by various statistical evaluation metrics. Mean squared error (MSE), root mean squared error (RMSE), mean absolute error (MAE), mean absolute percentage error (MAPE), coefficient of variation of root mean squared error (CVRMSE), and correlation coefficient (R) are the most widely used statistical metrics that have been used to evaluate the results of load forecasting by deep learning and machine learning methods [58, 78, 81]. Each of these metrics refers to a specific concept. Thus, the lower the error values and the higher the correlation value, the more accurate and close the prediction results are to the actual values. In recent years, researchers have dramatically developed a combination of a variety of machine learning and deep learning methods that primarily aim to increase prediction accuracy and reduce results-related errors. Therefore, it is observed that the evaluation of results and the use of statistical performance evaluation metrics are of great importance. Table 6.5 categorizes the load forecasting studies based on the evaluation metrics used in each paper.

6.4 Conclusions

This chapter aimed at providing an updated review on application of machine/deep learning methods and artificial intelligence approaches in electrical load forecasting. Accordingly, this study focused on two viewpoints of the application of machine learning/deep learning to load forecasting in power system in terms of a: time horizon and b: utilized techniques and their performance analysis. In the first

Table 6.5 Load forecasting categories based on various evaluation metrics

Statistical evaluation metric	References
MSE	[59, 61, 79, 82]
RMSE	[10, 14, 40, 47]–[24, 25, 27, 52, 58, 59, 62, 64, 66, 67, 78, 80]–[29, 36, 38, 39]
MAE	[10–13, 15, 22, 27, 28, 30, 36, 46, 47, 50, 53, 58, 59, 62, 67, 78, 80]
MAPE	[10, 11, 25, 27, 28, 30, 31, 33, 36, 38]–[12, 40, 46]–[14–16, 19, 20, 22, 24, 53, 58, 59, 62, 66, 67, 78, 80]
CVRMSE	[78]
R	[10, 15, 21, 35, 36, 39, 40, 45, 50, 58, 59, 66, 78, 80]

viewpoint, short-term, midterm, and long-term forecasting strategies were basically defined and investigated in terms of the pros and cons of machine learning/deep learning methods. Various methods used in the forecasting of electrical load, the data used, and the exact time horizon of the selected papers from the literature showed successful application and high performance of such methods. In the second part, the applied techniques and their performance analysis were performed that demonstrated the focus of most machine learning algorithms in improving the training process. Also, various statistical evaluation metrics that are used for evaluating the performance of the forecasting methodologies were discussed, and the popularity of each evaluation metrics in the literature were analyzed. The current chapter can be useful for researchers of energy systems with a focus on in the area of load forecasting and machine/deep learning studies.

References

1. A. H. Vahabie, M. M. R. Yousefi, B. N. Araabi, C. Lucas, and S. Barghinia, Combination of singular spectrum analysis and autoregressive model for short term load forecasting. *2007 IEEE Lausanne Power Tech*, (2007), pp. 1090–1093
2. S.-J. Huang, K.-R. Shih, Short-term load forecasting via ARMA model identification including non-Gaussian process considerations. IEEE Trans. Power Syst. **18**(2), 673–679 (2003)
3. X. Wang and Y. Liu, ARIMA time series application to employment forecasting. In *2009 4th International Conference on Computer Science & Education*, (2009), pp. 1124–1127
4. V. Debusschere, S. Bacha, One week hourly electricity load forecasting using neuro-fuzzy and seasonal ARIMA models. IFAC Proceedings Volumes **45**(21), 97–102 (2012)
5. K.-B. Song, Y.-S. Baek, D.H. Hong, G. Jang, Short-term load forecasting for the holidays using fuzzy linear regression method. IEEE Trans. Power Syst. **20**(1), 96–101 (2005)
6. N. Amral, C. S. Ozveren, and D. King, Short term load forecasting using multiple linear regression. In *2007 42nd International universities power engineering conference*, (2007), pp. 1192–1198
7. J.W. Taylor, Short-term load forecasting with exponentially weighted methods. IEEE Trans. Power Syst. **27**(1), 458–464 (Feb. 2012). https://doi.org/10.1109/TPWRS.2011.2161780
8. N. Elamin, M. Fukushige, Modeling and forecasting hourly electricity demand by SARIMAX with interactions. Energy **165**, 257–268 (2018)
9. H. Takeda, Y. Tamura, S. Sato, Using the ensemble Kalman filter for electricity load forecasting and analysis. Energy **104**, 184–198 (2016)
10. A. Heydari, M.M. Nezhad, E. Pirshayan, D.A. Garcia, F. Keynia, L. De Santoli, Short-term electricity price and load forecasting in isolated power grids based on composite neural network and gravitational search optimization algorithm. Appl. Energy **277**, 115503 (2020)
11. A. Rafati, M. Joorabian, and E. Mashhour, An efficient hour-ahead electrical load forecasting method based on innovative features. Energy, p. 117511, 2020
12. Z. Wen, L. Xie, Q. Fan, H. Feng, Long term electric load forecasting based on TS-type recurrent fuzzy neural network model. Electr. Pow. Syst. Res. **179**, 106106 (2020)
13. A. Ganguly, K. Goswami, A. Mukherjee, and A. K. Sil, Short-term load forecasting for peak load reduction using artificial neural network technique. In *Advances in Computer, Communication and Control*, Springer, 2019, pp. 551–559
14. E. Zhao, Z. Zhang, and N. Bohlooli, Cost and load forecasting by an integrated algorithm in intelligent electricity supply network. *Sustainable Cities and Society*, p. 102243, 2020

15. C. Feng, J. Zhang, Assessment of aggregation strategies for machine-learning based short-term load forecasting. Electr. Pow. Syst. Res. **184**, 106304 (2020)
16. R.-J. Wai, Y.-C. Huang, Y.-C. Chen, Y.-W. Lin, Performance comparisons of intelligent load forecasting structures and its application to energy-saving load regulation. Soft. Comput. **17** (10), 1797–1815 (2013)
17. Z. Deng, B. Wang, Y. Xu, T. Xu, C. Liu, Z. Zhu, Multi-scale convolutional neural network with time-cognition for multi-step short-term load forecasting. IEEE Access **7**, 88058–88071 (2019)
18. C. Sun, J. Song, L. Li, P. Ju, Implementation of hybrid short-term load forecasting system with analysis of temperature sensitivities. Soft. Comput. **12**(7), 633–638 (2008)
19. M. Barman, N.B. Dev Choudhury, Season specific approach for short-term load forecasting based on hybrid FA-SVM and similarity concept. Energy **174**, 886–896 (2019). https://doi.org/10.1016/j.energy.2019.03.010
20. M. Barman and N. B. D. Choudhury, A similarity based hybrid GWO-SVM method of power system load forecasting for regional special event days in anomalous load situations in Assam, India. *Sustainable Cities and Society*, p. 102311, 2020
21. Y. Hu et al., Short term electric load forecasting model and its verification for process industrial enterprises based on hybrid GA-PSO-BPNN algorithm—A case study of papermaking process. Energy **170**, 1215–1227 (2019)
22. A.T. Eseye, M. Lehtonen, T. Tukia, S. Uimonen, R.J. Millar, Machine learning based integrated feature selection approach for improved electricity demand forecasting in decentralized energy systems. IEEE Access **7**, 91463–91475 (2019)
23. L. Limei and H. Xuan, Study of electricity load forecasting based on multiple kernels learning and weighted support vector regression machine. In *2017 29th Chinese control and decision conference (CCDC)*, (2017), pp. 1421–1424
24. P. Gangwar, A. Mallick, S. Chakrabarti, S.N. Singh, Short-term forecasting-based network reconfiguration for unbalanced distribution systems with distributed generators. IEEE Trans. Indust. Inform **16**(7), 4378–4389 (2019)
25. T. Ouyang, Y. He, H. Li, Z. Sun, S. Baek, Modeling and forecasting short-term power load with copula model and deep belief network. IEEE Trans. Emerging Top. Comput. Intelligence **3**(2), 127–136 (2019)
26. M. Rafiei, T. Niknam, J. Aghaei, M. Shafie-Khah, J.P.S. Catalão, Probabilistic load forecasting using an improved wavelet neural network trained by generalized extreme learning machine. IEEE Trans. Smart Grid **9**(6), 6961–6971 (2018)
27. X. Tang, Y. Dai, T. Wang, Y. Chen, Short-term power load forecasting based on multi-layer bidirectional recurrent neural network. IET Generation, Transmission & Distribution **13**(17), 3847–3854 (2019)
28. M. Alipour, J. Aghaei, M. Norouzi, T. Niknam, S. Hashemi, and M. Lehtonen, A novel electrical net-load forecasting model based on deep neural networks and wavelet transform integration. *Energy*, p. 118106, 2020
29. H.-A. Li et al., Combined forecasting model of cloud computing resource load for energy-efficient IoT system. IEEE Access **7**, 149542–149553 (2019)
30. J. Ding, M. Wang, Z. Ping, D. Fu, and V. S. Vassiliadis, An integrated method based on relevance vector machine for short-term load forecasting. Eur. J. Oper. Res. (2020)
31. M. El-Hendawi, Z. Wang, An ensemble method of full wavelet packet transform and neural network for short term electrical load forecasting. Electr. Pow. Syst. Res. **182**, 106265 (2020)
32. J. Moon, S. Jung, J. Rew, S. Rho, and E. Hwang, Combination of short-term load forecasting models based on a stacking ensemble approach. *Energy and Buildings*, p. 109921, 2020
33. A.J. Amorim, T.A. Abreu, M.S. Tonelli-Neto, C.R. Minussi, A new formulation of multinodal short-term load forecasting based on adaptive resonance theory with reverse training. Electr. Pow. Syst. Res. **179**, 106096 (2020)
34. X. Kong, C. Li, C. Wang, Y. Zhang, J. Zhang, Short-term electrical load forecasting based on error correction using dynamic mode decomposition. Appl. Energy **261**, 114368 (2020)

35. G.T. Ribeiro, V.C. Mariani, L. dos Santos Coelho, Enhanced ensemble structures using wavelet neural networks applied to short-term load forecasting. Eng. Appl. Artif. Intel. **82**, 272–281 (2019)
36. R. Wang, J. Wang, Y. Xu, A novel combined model based on hybrid optimization algorithm for electrical load forecasting. Appl. Soft Comput. **82**, 105548 (2019)
37. M. Bessani, J.A.D. Massignan, T.M.O. Santos, J.B.A. London Jr., C.D. Maciel, Multiple households very short-term load forecasting using bayesian networks. Electr. Pow. Syst. Res. **189**, 106733 (2020)
38. M. Mansoor, F. Grimaccia, S. Leva, and M. Mussetta, Comparison of echo state network and feed-forward neural networks in electrical load forecasting for demand response programs. *Mathematics and Computers in Simulation*, (2020)
39. G. Chitalia, M. Pipattanasomporn, V. Garg, S. Rahman, Robust short-term electrical load forecasting framework for commercial buildings using deep recurrent neural networks. Appl. Energy **278**, 115410 (2020). https://doi.org/10.1016/j.apenergy.2020.115410
40. B. Dietrich, J. Walther, M. Weigold, E. Abele, Machine learning based very short term load forecasting of machine tools. Appl. Energy **276**, 115440 (2020)
41. Q. Huang, J. Li, and M. Zhu, An improved convolutional neural network with load range discretization for probabilistic load forecasting. *Energy*, p. 117902, 2020
42. S. Ma, A hybrid deep meta-ensemble networks with application in electric utility industry load forecasting. Inform. Sci. **544**, 183–196
43. G. Hafeez, K.S. Alimgeer, I. Khan, Electric load forecasting based on deep learning and optimized by heuristic algorithm in smart grid. Appl. Energy **269**, 114915 (2020)
44. Y. Yang, W. Hong, S. Li, Deep ensemble learning based probabilistic load forecasting in smart grids. Energy **189**, 116324 (2019)
45. N. Zhang, Z. Li, X. Zou, S.M. Quiring, Comparison of three short-term load forecast models in Southern California. Energy **189**, 116358 (2019)
46. M. Malekizadeh, H. Karami, M. Karimi, A. Moshari, M.J. Sanjari, Short-term load forecast using ensemble neuro-fuzzy model. Energy **196**, 117127 (2020)
47. F. He, J. Zhou, L. Mo, K. Feng, G. Liu, Z. He, Day-ahead short-term load probability density forecasting method with a decomposition-based quantile regression forest. Appl. Energy **262**, 114396 (2020)
48. C. Fan, C. Ding, J. Zheng, L. Xiao, Z. Ai, Empirical mode decomposition based multi-objective deep belief network for short-term power load forecasting. Neurocomputing **388**, 110–123 (2020)
49. J. Bedi, D. Toshniwal, Energy load time-series forecast using decomposition and autoencoder integrated memory network. Appl. Soft Comput. **93**, 106390 (2020). https://doi.org/10.1016/j.asoc.2020.106390
50. Y. Chu et al., Short-term metropolitan-scale electric load forecasting based on load decomposition and ensemble algorithms. Energ. Buildings **225**, 110343 (2020)
51. X. Ma and Y. Dong, An estimating combination method for interval forecasting of electrical load time series. *Expert Systems with Applications*, p. 113498, 2020
52. K. Xie, H. Yi, G. Hu, L. Li, and Z. Fan, Short-term power load forecasting based on elman neural network with particle swarm optimization. *Neurocomputing*, (2019)
53. P. Singh, P. Dwivedi, V. Kant, A hybrid method based on neural network and improved environmental adaptation method using controlled Gaussian mutation with real parameter for short-term load forecasting. Energy **174**, 460–477 (2019)
54. S. Maldonado, A. González, S. Crone, Automatic time series analysis for electric load forecasting via support vector regression. Appl. Soft Comput. **83**, 105616 (2019)
55. G. Sideratos, A. Ikonomopoulos, N.D. Hatziargyriou, A novel fuzzy-based ensemble model for load forecasting using hybrid deep neural networks. Electr. Pow. Syst. Res. **178**, 106025 (2020)
56. J. Kim, J. Moon, E. Hwang, P. Kang, Recurrent inception convolution neural network for multi short-term load forecasting. Energ. Buildings **194**, 328–341 (2019)

57. D. Sakurai, Y. Fukuyama, T. Iizaka, T. Matsui, Daily peak load forecasting by artificial neural network using differential evolutionary particle swarm optimization considering outliers. IFAC-PapersOnLine **52**(4), 389–394 (2019)
58. P. Singh, P. Dwivedi, A novel hybrid model based on neural network and multi-objective optimization for effective load forecast. Energy **182**, 606–622 (2019)
59. A. Moradzadeh, S. Zakeri, M. Shoaran, B. Mohammadi-Ivatloo, F. Mohamamdi, Short-term load forecasting of microgrid via hybrid support vector regression and long short-term memory algorithms. Sustainability (Switzerland) **12**(17), 7076 (Aug. 2020). https://doi.org/10.3390/su12177076
60. M. Mehrzadi et al., A deep learning method for short-term dynamic positioning load forecasting in maritime microgrids. Applied Sciences **10**(14), 4889 (2020)
61. S. Wang, S. Wang, D. Wang, Combined probability density model for medium term load forecasting based on quantile regression and kernel density estimation. Energy Procedia **158**, 6446–6451 (2019)
62. M. Talaat, M.A. Farahat, N. Mansour, A.Y. Hatata, Load forecasting based on grasshopper optimization and a multilayer feed-forward neural network using regressive approach. Energy **196**, 117087 (2020)
63. S.-M. Jung, S. Park, S.-W. Jung, E. Hwang, Monthly electric load forecasting using transfer learning for Smart Cities. Sustainability **12**(16), 6364 (2020)
64. J. Bedi, D. Toshniwal, Empirical mode decomposition based deep learning for electricity demand forecasting. IEEE Access **6**, 49144–49156 (2018)
65. A.I. Almazrouee, A.M. Almeshal, A.S. Almutairi, M.R. Alenezi, S.N. Alhajeri, Long-Term Forecasting of Electrical Loads in Kuwait Using Prophet and Holt–Winters Models. Applied Sciences **10**(16), 5627 (2020)
66. M. Dong, J. Shi, Q. Shi, Multi-year long-term load forecast for area distribution feeders based on selective sequence learning. Energy **206**, 118209 (2020)
67. M.-R. Kazemzadeh, A. Amjadian, and T. Amraee, A hybrid data mining driven algorithm for long term electric peak load and energy demand forecasting. *Energy*, p. 117948, 2020
68. D.A.G. Vieira, B.E. Silva, T.V. Menezes, A.C. Lisboa, Large scale spatial electric load forecasting framework based on spatial convolution. International Journal of Electrical Power & Energy Systems **117**, 105582 (2020)
69. S. Kumar, S.K. Pal, R.P. Singh, A novel method based on extreme learning machine to predict heating and cooling load through design and structural attributes. Energ. Buildings **176**, 275–286 (2018). https://doi.org/10.1016/j.enbuild.2018.06.056
70. A. Mansour-Saatloo, A. Moradzadeh, B. Mohammadi-Ivatloo, A. Ahmadian, A. Elkamel, Machine learning based PEVs load extraction and analysis. Electronics (Switzerland) **9**(7), 1–15 (Jul. 2020). https://doi.org/10.3390/electronics9071150
71. W. Kong, Z. Y. Dong, Y. Jia, D. J. Hill, Y. Xu, and Y. Zhang, Short-term residential load forecasting based on LSTM recurrent neural network. *IEEE Transactions on Smart Grid*, (2019), doi: https://doi.org/10.1109/TSG.2017.2753802
72. A. Yu et al., Accurate fault location using deep belief network for optical Fronthaul networks in 5G and beyond. IEEE Access **7**, 77932–77943 (2019). https://doi.org/10.1109/ACCESS.2019.2921329
73. Z.A. Khan, S. Zubair, K. Imran, R. Ahmad, S.A. Butt, N.I. Chaudhary, A new users rating-trend based collaborative Denoising auto-encoder for top-N recommender systems. IEEE Access **7**, 141287–141310 (2019). https://doi.org/10.1109/ACCESS.2019.2940603
74. J. Han, S. Miao, Y. Li, W. Yang, H. Yin, A wind farm equivalent method based on multi-view transfer clustering and stack sparse auto encoder. IEEE Access **8**, 92827–92841 (2020). https://doi.org/10.1109/ACCESS.2020.2993808
75. A. Moradzadeh and K. Pourhossein, Location of disk space variations in transformer winding using convolutional neural networks. In *2019 54th International Universities Power Engineering Conference, UPEC 2019 - Proceedings*, (2019), pp. 1–5, doi: https://doi.org/10.1109/UPEC.2019.8893596

76. R.D. Rathor, A. Bharagava, Day ahead regional electrical load forecasting using ANFIS techniques. J Instit. Engineers (India): Series B **101**(5), 475–495 (2020). https://doi.org/10.1007/s40031-020-00477-2
77. G. Dudek, Short-term load forecasting using random forests. Advances in Intelligent Systems and Computing **323**, 821–828 (2015)
78. N. Son, S. Yang, J. Na, Deep neural network and long short-term memory for electric power load forecasting. Appl Sci (Switzerland) **10**(18), 6489 (Sep. 2020). https://doi.org/10.3390/APP10186489
79. M. Tan, S. Yuan, S. Li, Y. Su, H. Li, F.H. He, Ultra-short-term industrial power demand forecasting using LSTM based hybrid ensemble learning. IEEE Trans. Power Syst. **35**(4), 2937–2948 (Jul. 2020). https://doi.org/10.1109/TPWRS.2019.2963109
80. S. Pei, H. Qin, L. Yao, Y. Liu, C. Wang, J. Zhou, Multi-step ahead short-term load forecasting using hybrid feature selection and improved long short-term memory network. Energies **13**(6), 4121 (Aug. 2020). https://doi.org/10.3390/en13164121
81. A. Moradzadeh, A. Mansour-Saatloo, B. Mohammadi-Ivatloo, A. Anvari-Moghaddam, Performance evaluation of two machine learning techniques in heating and cooling loads forecasting of residential buildings. Applied Sciences (Switzerland) **10**(11), 3829 (2020). https://doi.org/10.3390/app10113829
82. S. Tzafestas, E. Tzafestas, Computational intelligence techniques for short-term electric load forecasting. Journal of Intelligent and Robotic Systems: Theory and Applications **31**(1–3), 7–68 (2001). https://doi.org/10.1023/A:1012402930055

Chapter 7
A Survey of Recent Particle Swarm Optimization (PSO)-Based Clustering Approaches to Energy Efficiency in Wireless Sensor Networks

Emrah Hancer

7.1 Introduction

Thanks to the advanced developments in wireless communication and microelectromechanical system (MEMS), wireless sensor networks (WSNs) have gained wide popularity for various fields, such as disaster management, industrial automation, military reconnaissance, smart buildings, etc. [1]. WSNs can therefore establish the connection between the environment, the computational world, and the human society. Typically, WSNs are built of low-price and low-energy sensor nodes. The positions of nodes do not need to be specifically determined, i.e., their positions can be randomly indicated. In a network, all the sensor nodes are able to sense, process, and communicate. Through these characteristics, data acquisition and distribution processes are carried out. The network is operated by a base station or sink node that may be located close to the network framework.

The overall goal of all sensor nodes is to transfer data to the base station or sink node which is the final destination for the acquired data. After the data transmission process, the sink node carries out some required processes (e.g., data transfer to a task manager over the Internet or satellite network). The main distinction between standard and sink nodes is energy usage. While a sink node owns an unlimited power battery, standard sensor nodes own in-rechargeable restricted power batteries. Accordingly, sensor nodes should carefully consider energy consumption as soon as possible.

The most challenging issue in WSNs to be taken into consideration is the consumption of energy since it affects the sensor node lifetime and thereby the

E. Hancer (✉)
Department of Software Engineering, Mehmet Akif Ersoy University, Burdur, Turkey
e-mail: ehancer@mehmetakif.edu.tr

M. Nazari-Heris et al. (eds.), *Application of Machine Learning and Deep Learning Methods to Power System Problems*, Power Systems,
https://doi.org/10.1007/978-3-030-77696-1_7

whole network lifetime. There also exists a trade-off between the energy consumption and the number of sensor nodes. When the number of sensors within the network exceeds the capacity, the usual direct routing process requires higher energy consumption and so inversely affects the network lifetime. The fundamental techniques to enhance the network lifetime are as follows [2]: (1) energy efficiency scheduling, (2) energy efficiency node transmission power tuning, (3) energy efficiency routing, and (4) energy efficiency clustering. Among such techniques, clustering and routing have been widely used for WSNs since bringing several advantages, such as scalability, efficient communication, and fault tolerance [3]. In this chapter, the motivation will be on clustering techniques.

Clustering helps WSNs to manage efficient energy usage by reducing the number of sensor nodes that are far from the sink node and dividing current consumption evenly among sensor nodes. Each separated group of sensor nodes is represented by a cluster head node that gathers data within the group and then transfers it to the sink node. Through this process, it is possible to reduce data transmission traffic, enhance resource allocation, and improve bandwidth reusability. The most well-known clustering approaches to address energy efficiency in WSNs are LEACH [4], LEACH-C [5], SEP [6], EECS [7], and PEGASIS [8]. Although such clustering approaches have obtained promising results, energy efficiency is still an open issue.

Various surveys have been published in the literature to outline the profile of clustering approaches proposed for WSNs. Abbasi and Younis [9] described some important conventional clustering approaches and then classified them in terms of the convergence time. Mamalis et al. [10] considered clustering approaches in the case of probability. While the most popular clustering approaches such as LEACH, HEED, and EEHC were evaluated in the category of probabilistic approaches, other approaches were categorized as weight-based, graph-based, and biologically inspired. Liu [11] reviewed 16 well-known clustering methods and introduced a taxonomy of the corresponding methods in the basis of cluster characteristics. Jiang et al. [12] reviewed some popular clustering approaches in terms of selectivity, count variability, and existence and then introduced a comparative study of the reviewed approaches. Kumarawadu et al. [13] considered clustering approaches in four groups: probabilistic, biologically inspired, neighborhood-based, and identity-based. The criteria to analyze and compare approaches in this work were energy efficiency, clock synchronization, and load balancing. Deosarkar et al. [14] introduced a survey on cluster head selection, where all the related works were categorized into four categories: deterministic, combined, adaptive, and hybrid. Aslam et al. [15] considered LEACH and its variants and introduced a comparative study of the related approaches. Afsar and Tayarani-N [3] considered a large number of clustering approaches from different perspectives, such as load balancing, fault tolerance, scalability, and connectivity. Sambo et al. [16] investigated computational-based clustering methods in terms of 10 criteria, including scalability, radio model, data aggregation, energy efficiency, and nature. Other related surveys can be found in [17–19].

When considering the aforementioned works, it is not possible to find a work except [16] that specifically focuses on particle swarm optimization (PSO)-based clustering approaches. Although [16] considered PSO-based approaches, most of the works were missing. This issue motivated us to perform this comprehensive survey on PSO-based clustering approaches.

The rest of the chapter is as follows: We first provide a general background on the WSNs framework, PSO algorithm, and conventional clustering approaches. We then review the related works with discussions from different perspectives. Finally, we conclude the chapter with current drawbacks and future trends.

7.2 Background

In this section, we will first explain the overall structure of WSNs and particle swarm optimization. We then define the problem.

7.2.1 Structure of WSNs

Sensitive cells in a sensor node can transform measured physical values into an electrical signal to be used in many applications, such as temperature, noise levels, and pressure. As shown in Fig. 7.1, a typical sensor node includes the following components: (1) Power supply, also called battery unit, is built of a tiny battery and AC-DC converter, which provides suitable voltages for the electronic circuits of a sensor node. (2) Physical sensing device is used to measure or observe physical events (e.g., sounds, temperature, earthquakes) in the environment. (3) AD-DC converter transforms analog signals obtained by the sensing device to digital signals. (4) RF communication unit transmits data to another sensor node through radio communication. (5) Processor unit is a microprocessor system that controls and monitors the whole activities of the sensor node. Sensor nodes are located in the target area to gather data concerning related physical events and transmit such data to

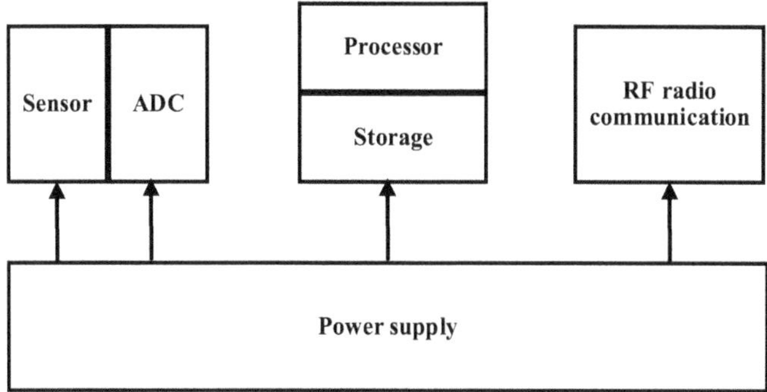

Fig. 7.1 Overall structure of a sensor node [1]

the sink node. Accordingly, they require energy to properly carry out the tasks of WSN. The communication task is treated as the main source of energy consumption. In addition to communication task, there also exists energy loss inactive states due to the following reasons [1]: (1) Idle listen: As it is possible to receive data from its neighbors to a sensor, it must listen the messages over the medium. (2) Overhearing: When a data is sent from a sensor node, all of its neighbors receive this data even though it is sent to only one of them. (3) Interference: A sensor node receives a packet between transmitting and interference ranges but cannot decode the packet. (4) Collision: When a collusion exists during the transmission, the energy arranged for this task is lost. (5) Control packet overhead: A small number of control packets should be used to support transmission tasks. In summary, due to their small batteries, sensor nodes need to address limited energy consumption. This case become more harmful, especially when manual recharge is not possible for the sensor nodes which are generally located in unreachable area. Accordingly, energy efficiency has been considered as an important parameter in the design of WSNs to extend the lifetime of network.

7.2.2 Particle Swarm Optimization

Swarm intelligence is the discipline of evolutionary computation in artificial intelligence that investigates the collective behavior of decentralized and self-organized systems. In particular, the source of collective behaviors is the interactions of individuals within the swarm and with the environment. Ants and termites, schools of fish, bird flocks, and honeybees are representative examples of the swarm intelligence. It is not possible to consider all kinds of individuals in swarm intelligence. To be considered as swarm intelligence, a swarm of individuals should have the following properties:

1. Self-organization: This is an important characteristic of natural systems, which means that it is possible to carry out required processes without the requirement of any centralized authority. With the help of self-organization, swarm agents can build the best work coordination with each other and so pursue their task at high speed and fault tolerance. The main characteristics of self-organization are the following: a) Positive feedback: More agents take part in the same work to promote smart solutions. b) Negative feedback: Agents avoid getting involved in the same state to counterbalance the effect of positive feedback. c) Fluctuation: Randomness allows agents to occur random changes in the system. d) Multiple interactions: Agents tend to interact with each other and so learn from each other to create intelligent behaviors within the swarm.
2. Division of labor: Thanks to the cooperation of specialized agents, it is possible to simultaneously perform a variety of tasks in the swarm rather than carrying out sequential tasks. Accordingly, there exists a more effective and efficient task management system without a doubt.

Since first introduced in 1995, swarm intelligence algorithms have gained overwhelming interest by researchers and so have been applied to a variety of fields from numeric optimization to machine learning to power systems [20]. The earliest swarm intelligence algorithm, particle swarm optimization (PSO) [21] has received a growing-up interest by the researchers, resulting in a variety of applications in various fields. PSO mimics the behaviors of social groups, like fish and birds. The algorithm seeks the optimal solution through agents, called particles, each of which is represented by its position in the possible solution space. The algorithm iteratively updates the particle positions using the information of its local best and the swarm best position in a randomized weighted manner by Eqs. 7.1 and 7.2. After the process of updating the particle positions, the swarm best position is redefined for the next generations. One of the advantages of PSO over evolutionary algorithms, like genetic algorithms (GA) and evolutionary strategies, is a smaller number of parameters to be tuned by a user. Furthermore, the algorithm does not tend to face stagnation problems as fluently as GA. Besides a variety of successful applications in various fields from numerical problems to energy systems [22, 23], PSO has also been frequently used in WSNs.

$$v_i(t+1) = wv_i(t) + c_1r_1(\widehat{x}_i(t) - x_i(t)) + c_2r_2(gbest(t) - x_i(t)) \tag{7.1}$$

$$x_i(t+1) = x_i(t) + v_i(t+1) \tag{7.2}$$

where $\boldsymbol{x_i(t)} = \{\boldsymbol{x_{i,1}}, \boldsymbol{x_{i,2}}, \ldots, \boldsymbol{x_{i,D}}\}$ is the position of the ith particle at time t, $\widehat{\boldsymbol{x}}_i(\boldsymbol{t})$ is its best solution as of time t, $\boldsymbol{gbest(t)}$ represents the best position in the swarm as of time t, w is the inertial weight which keeps the particle moving in the same direction it was originally heading and is generally selected between 0.8 and 1.2, c_1 and c_2 are cognitive and social coefficients which are generally set to 2, and r_1 and r_2 are randomly generated numbers between 0 and 1.

7.2.3 *Problem Definition*

Efficient energy consumption is a crucial issue in WSNs to enhance the lifetime of a network. The most intensive task for energy usage is the data transfer from sensor nodes to the sink node. The required energy for the transmission process varies according to the distance between the sink and sensor nodes. To be specific, energy consumption exponentially increases proportionally to the transmission distance, and so the lifetime of WSNs largely depends on the efficient energy management of the transmission process. Clustering is one of the most widely applied techniques to efficiently manage the transmission process from sensor nodes to the sink node. By clustering, a set of sensor nodes within a WSN are grouped into clusters based on predefined similarity metrics (e.g., distance). Each cluster is represented by its cluster head, which is responsible for acquiring data from sensor nodes within its

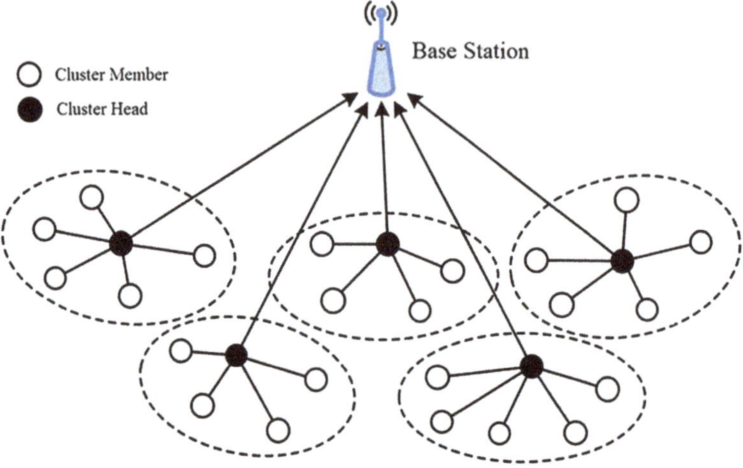

Fig. 7.2 Transmission process in WSNs

cluster and then forward the acquired data to the sink node as shown in Fig. 7.2. A cluster head that is assigned for a long duration prematurely exhausts its power supply. The selection of a cluster head therefore plays a crucial role in the performance and longevity of the network. The well-known conventional clustering approaches are presented as follows:

1. *LEACH* [4]*:* The selection of a cluster head is very crucial in LEACH. If the cluster head is optimally selected, it is possible to increase the energy efficiency and address data transmission problems. In the initial stage, all the sensors within the cluster can be chosen as a cluster head, but LEACH introduces some additional definitions for the cluster head selection. The protocol operation is divided into rounds. The processes such as detection of all sensors, selection of cluster head, and data transmission are carried out in a certain plan in each round. Each sensor node is selected at least once due to the statistical formulation used for the cluster head selection. This is because a sensor acting as a cluster head consumes more energy, and so each sensor node must be selected at least once as a cluster head to evenly divide energy consumption among sensor nodes.
2. *LEACH-C* [5]*:* Since the number of cluster heads is not clear and low-energy sensor nodes are possible to be selected as a cluster head in LEACH, the researchers developed LEACH-C, which carries out a two-layer cluster head selection. In detail, all sensor nodes inform the sink node about their position and energy level. Using this information, a threshold value is determined, and the sensor nodes which are smaller than the determined threshold value are not selected as the cluster head. LEACH-C outperforms LEACH in terms of the cluster quality.
3. *EECS* [7]*:* The algorithm shows similar characteristics with LEACH. All the clusters are combined to only one cluster head, and this cluster head directly

makes a connection with the sink node. To detect the position of sensor nodes, the sink node sends a "hello" packet, and the response time determines the approximate position of sensor nodes. The selection of a cluster head is carried out in a probabilistic manner for each cluster. The energy levels of sensor nodes are evaluated using the *COMPETE-HEAD-MSG* message, and then the node with the highest energy level is selected as the cluster head.

4. *PEGASIS* [8]*:* The algorithm is a hierarchical protocol which is based on a greedy algorithm and a chain-based approach. Sensor nodes come together to build a chain. If any sensor node dies within the chain, the chain is then rebuilt by skipping the dead node. The requests transmitted from other nodes to the leader node are then forwarded to the sink node. Using PEGASIS protocol in a network with many sensor nodes may cause delays in data transmission. To alleviate this drawback, one of the ways is to use multiple leader nodes in the network. When compared to LEACH, PEGASIS was proved to be much more efficient.
5. *H-PEGASIS* [24]*:* The algorithm has been developed to address time delays in PEGASIS due to data collisions. In H-PEGASIS, simultaneous data transfer is introduced to prevent data collisions using CDMA encoding and spatially separated sensors. A chain-based sensor based on CDMA, which acts like a tree structure, transmits data from sublayers to the sink node. Accordingly, delays are reduced while data is simultaneously transmitted.

Other popular conventional clustering approaches to address energy efficiency in WSNs are SEP [6], DEEC [25], HEED [26], and EEHC [27]. More information concerning conventional clustering approaches proposed for WSNs can be found in [28].

7.3 PSO-Based Approaches

Clustering in WSNs is an NP-hard problem due to the selection of m optimal cluster heads among n sensor nodes leading m $\times$ n possibilities. Thanks to their effective search characteristics, swarm intelligence algorithms have strongly been treated as good NP-hard problem-solving tools. Due to its historical and successful background, PSO is maybe the most applied swarm intelligence algorithm in WSNs. The considered PSO-based clustering approaches in this work are as follows:

PSO-C [29]: The approach owns a centralized mechanism that is operated at the sink node or base station. In the initial stage, the base station receives information from sensor nodes concerning their energy level and locations. The base station then calculates the average energy level of all nodes using the received information. The nodes with sufficient energy levels are determined as candidate cluster heads. In the second stage, PSO is applied to determine K best cluster heads among candidate cluster heads by optimizing the objective function which is the weighted combination of the intra-cluster distance and the energy efficiency. After the determination of cluster heads and cluster members, each cluster head builds a TDMA schedule to prevent collisions during data transmission. In this scheme, sensor nodes are only activated during their transmission time to enhance the energy efficiency within the network. The cluster heads gather data from its allocated sensor nodes within its cluster and then forward the gathered data to the base station. According to the results, PSO-C is far superior to LEACH and LEACH-C.

PSO-HC [30]: The approach is another centralized protocol that aims to enhance the network lifetime and maximize the network scalability by minimizing the average energy consumption and building two-hop communication within the cluster. In the first stage, each sensor node sends a hello packet with its ID. When a sensor node receives this packet, it updates its neighbor table with the RRSI value included in the packet. After all sensors determine their neighbors, each sensor node transmits the data including ID, residual energy, and its neighbor table to the base station using the flooding method. The base station then calculates the average energy level of all sensor nodes based on the received data. If any sensor in the network owns a higher energy level than the average, it is determined as a candidate cluster head. In the second stage, PSO is applied to select K optimal cluster heads among the candidate ones. Each particle is represented as a sequence of cluster head IDs. After obtaining the set of cluster heads from the particle, the base station constructs two-tier clusters to improve the network scalability and lifetime. The first tear clusters are constructed by assigning each sensor node to a cluster node based on the RSSI value. The second tear clusters are constructed using all non-clustered sensor nodes from the first tier. According to the results, PSO-HC is more efficient than LEACH, LEACH-C, and PSO-C.

PSO-SD [31]: The approach is not based on a centralized mechanism, which makes it a semi-distributed approach. The approach tries to find the locations of cluster heads by optimizing the following objectives in a weighted manner: the intra-cluster distance, the residual energy, the node degree, and the head count. Each particle represents the locations of cluster heads. The impact of the packet retransmission size along the estimated path to the cluster head is also considered in the approach. According to the results, it performs better than PSO-C and LEACH-C.

PECC [32]: The approach uses a centralized mechanism to form clusters. The sink node sends info collection messages to sensor nodes. After receiving messages, sensor nodes transmit information concerning the location, id, energy level, and energy loss to the sink node. Then the sink node forms clusters by optimizing the average distance and the average energy level in a weighted manner using PSO. For each cluster, a cluster head is selected by optimizing the within-cluster distance and the number of sensor nodes in a weighted manner using PSO. Finally, a multi-hop communication protocol is applied to carry out the data transmission process from sensor nodes to the cluster head and from the cluster head to the sink node. Data is gathered by the cluster head in each cluster to save the residual energy. If the distance from the cluster head to the sink node does not exceed the predefined threshold value, the cluster head transmits data to the sink node through the single-hop transmission. Otherwise, the cluster head seeks for new hop based on the residual energy and distance.

EMPS [33]: Like PSO-C, PSO-HC, and PECC, the approach is based on a centralized mechanism. In this approach, sensor nodes share information with the sink node concerning their position and energy level. Then PSO first splits the network into subregions. Each particle is represented as a region boundary, including the (x,y) coordinates of the point line and the angles of the point line to the X and Y axes. After the split process of subregions, a cluster head is determined for each region based on the distance from the sensor node to the center of gravity and the residual energy. EMPS consists of three characteristic messages to carry out the transmission process: a) The hello packet detects the cluster region that transmits data to the sink node. b) The message-s packet transmits data to the sink node. c) The message-h packet transmits data to the cluster head. From the results, it can be indicated that it efficiently manages energy usage compared to the conventional clustering approaches.

PSO-ECHS [34]: The approach aims to select the optimal cluster heads on the network to improve the network efficiency using PSO. Each particle represents the possible locations of the cluster heads in the network space. The objective function handles the following objectives in a weighted manner: the average intra-cluster distance and the average sink distance. Different from the approaches where sensor nodes are assigned to a cluster based on the distance to the cluster head, the approach assigns sensor nodes to the cluster heads based on a weighted function which comprises of the residual energy, the distance from the cluster head to the sink node, the distance from the sensor node to the cluster head, and the degree of the cluster head. According to the various experiments on a variety of scenarios, the approach outperforms a variety of well-known approaches, such as LEACH, LEACH-C, and PSO-C.

PSO-HSA [35]: The approach follows a two-way hybridized methodology to manage energy usage in the network. In the first stage, clustering is applied using PSO, and 10% of the available sensor nodes are selected as the cluster heads. In the second phase, harmony search [36] is applied to carry out the transmission process. A gateway node is positioned between the sink node and the cluster head. The gateway nodes are responsible for improving energy efficiency during the data transmission process to the sink node. The cluster head first transmits the obtained data to the gateway node based on the distance between the cluster head and the sink node. The gateway nodes are selected based on the distance with the cluster head using harmony search.

SCE-PSO [37]: The approach assumes that the cluster heads (called gateways) and sensor nodes are randomly located in a given area. Any sensor node can connect to any cluster head if the sensor node is within the communication range of the cluster head. If the number of sensor nodes within a cluster exceeds the cluster capacity, its load is divided between other clusters. A position of each particle represents the assigned cluster of the corresponding sensor node. Thus, the dimensionality of each particle is equal to the number of sensor nodes in the network. The objective function consists of three fundamental components: the load of the cluster heads, the intra-cluster distance, and the number of heavily loaded cluster heads. It can be extracted from the objective function that the assignment process of sensor nodes depends on the cluster head load. This enhances the lifetime of the cluster head and so maximizes the whole network lifetime. Once particles are evaluated using the objective function, they are sorted according to the objective value and partitioned into groups, named complexes. Then each complex is individually evaluated using PSO and is updated using the newly generated particles.

PUDCRP [38]: The cluster heads closer to the sink node are more likely to participate in data transmission processes in multi-hop routing. This leads to the end of nodes near the sink node area, referred to as the hotspot problem. Different from the aforementioned approaches, the approach considers the hotspot problem as well as grouping sensor nodes. To alleviate the hotspot problem, the approach divides the network into different-sized circles according to the distribution of sensor nodes. The determination of cluster heads is then considered as a multi-objective problem. It should be notified that the optimal number of cluster heads is dynamically determined without the requirement of any user-specified parameters. From the results, it can be revealed that PUDCRP performs better than a variety of recently introduced approaches, including PSO-SD and PSO-ECHS.

GA-PSO [39]: The approach involves two fundamental stages. In the first stage, the cluster heads are determined using GA based on the weighted objective function which considers the distance from nodes to the cluster head, the distance from the cluster head to the base station, the distance from a non-cluster head to the cluster head, and the total system energy. After the determination of the cluster heads, the second stage is carried out. In this stage, PSO is applied to manage the routing process. The relay nodes are put into the transmission traffic to increase the efficiency of the cluster head. In particular, the cluster head first search for the closest relay nodes or base station to transfer the aggregated data. The method performed better than LEACH variants, but it did not compare with recent PSO-based clustering approaches.

PSO-UFC [40]: The approach first determines the cluster heads by optimizing the following objectives in a weighted manner: (1) average intra-cluster distance, (2) average inter-cluster distance, and (3) residual energy. After the selection of the cluster heads, clusters are formulated in a such way that the clusters near to base station have a smaller size to keep their energy for the inter-cluster relay transmission process. Then a multi-hop routing three is constructed among the selected cluster head. To deal with the fault tolerance task, a surrogate cluster head is determined for each cluster head. According to the experiments, PSO-UFC outperforms various approaches such as LEACH, PSO-C, and EBUC.

PSO-ECSM [41]: The sink node is placed at the middle of the WSN for collecting data from nodes. In other words, the approach is built on a heterogeneous model. The approach selects the cluster heads by considering the following factors: residual energy, node degree, energy distance, average energy, and energy consumption. The approach also resolves the data traffic in a multi-hop network by applying sink mobility. According to a number of experiments, the approach performs better than a variety of clustering approaches in terms of stability, robustness, and network lifetime.

Table 7.1 Comparison of PSO-based approaches

Method	Data agg.	Efficiency	Scalability	Nature	Network	Multi hop	Multi path
PSO-C [29]	Yes	Average	Low	Centralized	Homogeneous	No	No
PSO-HC [30]	–	Average	High	Centralized	Homogeneous	Yes	–
PSO-SD [31]	Yes	High	High	Distributed	Homogeneous	–	Yes
PECC [32]	Yes	Average	Medium	Distributed	Homogeneous	Yes	–
EMPS [33]	Yes	High	High	Centralized	Homogeneous	–	Yes
PSO-ECHS [34]	No	High	High	Centralized	Homogeneous	–	–
PSO-HSA [35]	–	Average	Medium	Centralized	Homogeneous	No	–
SCE-PSO [37]	–	Average	High	Centralized	Homogeneous	No	–
PUDCRP [38]	–	High	High	Centralized	Homogeneous	Yes	Yes
GA-PSO [39]	–	Average	Medium	Centralized	Homogeneous	–	–
PSO-UFC [40]	–	Average	Medium	Centralized	Homogeneous	Yes	Yes
PSO-ECSM [41]	Yes	High	High	Centralized	Heterogeneous	Yes	–

Table 7.1 summarizes the presented PSO-based clustering approaches from different perspectives. According to Table 7.1, a centralized mechanism is much more preferred than a distributed mechanism by PSO-based approaches. It can also be extracted from Table 7.1 that all the PSO-based approaches can improve the network lifetime, thanks to their efficient energy management. Among the compared approaches, PSO-SD, EMPS, PSO-ECHS, PUDCRP, and PSO-ECSM are high-level energy-saving approaches. Furthermore, the scalability of all the approaches can be treated as sufficient except PSO-C.

7.4 Conclusions

In this chapter, we introduced a survey of recently proposed PSO-based clustering approaches to address energy efficiency in WSNs. Among a variety of works in this field, there exists no such work in the literature that tries to reflect the profile of recent PSO-based approaches proposed to deal with energy consumption in WSNs. According to the results, PSO-based approaches can better carry out energy management of the whole network than conventional clustering protocols. In our opinion, the following works can be considered by researchers in the future. First, hybridized frameworks may be an option to improve the efficiency of PSO-based approaches. Second, it is not possible to find a comprehensive comparative analysis of PSO-based approaches in the literature from different perspectives. Third, most of the PSO-based approaches are based on a centralized mechanism, i.e., the impact of distributed mechanism on energy efficiency may be investigated deeply in the future.

References

1. M. A. H. Hussein, *Energy Efficiency in Wireless Sensor Networks*. Master's thesis, (The Graduate School of Natural and Applied Sciences of Cankaya University, 2015)
2. S. Mahfoudh, *Energy Efficiency in Wireless Ad Hoc and Sensor Networks: Routing, Node Activity Scheduling and Cross-Layering*. PhD theses, (Universite Pierre et Marie Curie - Paris VI, 2010)
3. M. Mehdi Afsar, T.-N. Mohammad-H, Clustering in sensor networks: A literature survey. J. Netw. Comput. Appl. **46**, 198–226 (2014)
4. W.B. Heinzelman, A.P. Chandrakasan, H. Balakrishnan, An application-specific protocol architecture for wireless microsensor networks. IEEE Trans. Wirel. Commun. **1**(4), 660–670 (2012)
5. M. Tripathi, M. S. Gaur, V. Laxmi, R. B. Battula, Energy efficient LEACH-C protocol for Wireless Sensor Network. In *Proc. Third International Conference on Computational Intelligence and Information Technology (CIIT)*, 2013
6. G. Smaragdakis, I. Matta, S.E.P. Bestavros, A stable election protocol for clustered heterogeneous wireless sensor networks. In Proc. Second International Workshop on Sensor and Actor Network Protocols and Applications (SANPA), 2004
7. V. Saranya, S. Shankar, G.R. Kanagachidambaresan, Energy efficient clustering scheme (EECS) for wireless sensor network with Mobile sink. Wireless Personal Communications: An Int. J. **100**(4), 1553–1567 (2018)
8. S. Lindsey, C.S. Raghavendra, PEGASIS: power-efficient gathering in sensor information systems. In *Proc. IEEE Aerospace Conference*, (2002)
9. A. Abbasi, M. Younis, A survey on clustering algorithms for wireless sensor networks. Comput. Commun. **30**(14), 2826–2841 (2007)
10. B. Mamalis, D. Gavalas, C. Konstantopoulos, G. Pantziou, *Clustering in Wireless Sensor Networks. RFID and Sensor Networks: Architectures, Protocols, Security, and Integrations* (2009), pp. 323–354
11. X. Liu, A survey on clustering routing protocols in wireless sensor networks. Sensors **12**(8), 11113–11153 (2012)
12. C. Jiang, D. Yuan, Y. Zhao, Towards clustering algorithms in wireless sensor networks-a survey. In *Proc. IEEE Wireless Communications and Networking Conference*, (2009)
13. P. Kumarawadu, D.J. Dechene, M. Luccini, A. Sauer, Algorithms for node clustering in wireless sensor networks: a survey. In *Proc. 4th International Conference on Information and Automation for Sustainability*, (2008)
14. B.P. Deosarkar, N.S. Yadav, R.P. Yadav, Cluster head selection in clustering algorithms for wireless sensor networks: a survey. In *Proc. International Conference on Computing, Communication and Networking*, (2008)
15. M. Aslam, N. Javaid, A. Rahim, U. Nazir, A. Bibi, Z. Khan, A survey of extended LEACH-based clustering routing protocols for wireless sensor networks. In *Proc. 9th IEEE International Conference on Embedded Software and Systems*, (2012)

16. D. Wohwe Sambo, B.O. Yenke, A. Forster, P. Dayang, Optimized clustering algorithms for large wireless sensor networks: A review. Sensors **19**, 1–27 (2019)
17. R.V. Kulkarni, A. Förster, G.K. Venayagamoorthy, Computational intelligence in wireless sensor networks: A survey. IEEE Communications Surveys Tutorials **13**(1), 68–96 (2011)
18. P. Kumari, M.P. Singh, P. Kumar, Survey of clustering algorithms using fuzzy logic in wireless sensor network. In *Proc. International Conference on Energy Efficient Technologies for Sustainability*, (2013)
19. S. Sirsikar, K. Wankhede, Comparison of clustering algorithms to design new clustering approach. In *Proc. 4th International Conference on Advances in Computing, Communication and Control*, (2015)
20. E. Hancer, D. Karaboga, A comprehensive survey of traditional, merge-split and evolutionary approaches proposed for determination of cluster number. Swarm and Evolutionary Computation **32**, 49–67 (2017)
21. J. Kennedy, R. Eberhart, Particle swarm optimization. In *Proc. International Conference on Neural Networks*, (1995)
22. P.H. Mahmoud, N.-H. Morteza, M.-I. Behnam, S. Heresh, A hybrid genetic particle swarm optimization for distributed generation allocation in power distribution networks. Energy **209**, 118218 (2020)
23. N.-H. Morteza, S. Madadi, P.H. Mahmoud, M.-I. Behnam, Optimal distributed generation allocation using quantum inspired particle swarm optimization, in *Quantum Computing: An Environment for Intelligent Large Scale Real Application*, (Springer, Cham, 2018), pp. 419–432
24. S. Lindsey, C. Raghavendra, K.M. Sivalingam, Data gathering algorithms in sensor networks using energy metrics. IEEE Trans. Parallel Distributed Syst. **13**(9), 924–935 (2002)
25. L. Qing, Q. Zhu, M. Wang, Design of a distributed energy-efficient clustering algorithm for heterogeneous wireless sensor networks. Comput. Commun. **29**(12), 2230–2237 (2006)
26. O. Younis, S. Fahmy, HEED: A hybrid, energy-efficient, distributed clustering approach for ad hoc sensor networks. IEEE Trans. Mob. Comput. **3**(4), 366–379 (2004)
27. D. Kumar, T.C. Aseri, R.B. Patel, EEHC: Energy efficient heterogeneous clustered scheme for wireless sensor networks. Comput. Commun. **32**(4), 662–667 (2009)
28. B. Jan, H. Farman, H. Javed, B. Montrucchio, M. Khan, S. Ali, Energy efficient hierarchical clustering approaches in wireless sensor networks: A survey. Wirel. Commun. Mob. Comput. **6457942** (2017)
29. N.M.A. Latiff, T.C. Simonides, B.S. Sharif, Energy-aware clustering for wireless sensor networks using particle swarm optimization. In *Proc. 18th IEEE International Symposium on Personal, Indoor and Mobile Radio Communications*, (2007)
30. R.S. Elhabyan, M.C.E. Yagoub, PSO-HC: Particle swarm optimization protocol for hierarchical clustering in wireless sensor networks. In *Proc. 10th IEEE International Conference on Collaborative Computing: Networking, Applications and Worksharing*, (2014)
31. B. Singh, D.K. Lobiyal, A novel energy-aware cluster head selection based on particle swarm optimization for wireless sensor networks. HCIS **2**(1), 2–13 (2012)
32. C. Vimalarani, R. Subramanian, S.N. Sivanandam, An enhanced PSO-based clustering energy optimization algorithm for wireless sensor network. Scientific World J. (2016)
33. J. Wang, Y. Cao, B. Li, H. Kim, S. Lee, Particle swarm optimization based clustering algorithm with mobile sink for WSNs. Futur. Gener. Comput. Syst. **76**, 452–457 (2017)
34. P.C.S. Rao, P.K. Jana, H. Banka, A particle swarm optimization based energy efficient cluster head selection algorithm for wireless sensor networks. Wirel. Netw **23**(7), 2005–2020 (2017)
35. V. Anand, S. Pandey, Particle swarm optimization and harmony search based clustering and routing in wireless sensor networks. Int. J. Comput. Intelligence Syst. **10**(1), 1252–1262 (2017)
36. Z. Woo, J. Hoon, G.V. Loganathan, A new heuristic optimization algorithm: Harmony search. Simulation **76**(2), 60–68 (2001)
37. D.R. Edla, M.C. Kongara, R. Cheruku, SCE-PSO based clustering approach for load balancing of gateways in wireless sensor networks. Wirel. Netw **25**(3), 1067–1081 (2019)

38. D. Ruan, J. Huang, A PSO-based uneven dynamic clustering multi-hop routing protocol for wireless sensor networks. Sensor Networks **19**, 1835 (2019)
39. D. Anand, S. Pandey, New approach of GA-PSO based clustering and routing in wireless sensor networks. Int. J. Commun. Syst. **33**, e4571 (2020)
40. T. Kaur, D. Kumar, Particle swarm optimization-based unequal and fault tolerant clustering protocol for wireless sensor networks. IEEE Sensors J. **18**, 4614–4622 (2018)
41. B.M. Sahoo, T. Amgoth, H.M. Pandey, Particle swarm optimization based energy efficient clustering and sink mobility in heterogeneous wireless sensor network. Ad Hoc Netw. **106**, 102237 (2020)

Chapter 8
Clustering in Power Systems Using Innovative Machine Learning/Deep Learning Methods

Mohammad Hossein Rezaeian Koochi, Mohammad Hasan Hemmatpour, and Payman Dehghanian

8.1 Introduction

Clustering is the task of dividing a dataset into groups with similar characteristics. These datasets can be made of different types of data such as a set of time-series variations of variables or a set of single data points placed in a two- or more-dimensional space. The choice of the mechanism for assessing such similarities depends on the type of the data. For the case of single data points placed in an p-dimensional space, the Euclidean distance could be a promising approach. For example, Fig. 8.1 shows that a typical set of data points in a 2-D plane can be clustered according to the Euclidean distances between the data points. However, in order to assess the similarity of time-series variations, other criteria and preprocessing tasks rather than Euclidian distance, such as calculating correlation coefficients, applying dynamic time warping (DTW), or using feature extraction techniques such as principle component analysis (PCA) or independent component analysis (ICA), may be needed. Figure 8.2 illustrates how these preprocessing steps are applied to the raw time-series data prior to clustering.

M. H. Rezaeian Koochi (✉)
Department of Electrical Engineering, Shahid Bahonar University of Kerman, Kerman, Iran
e-mail: mh_rezaeian@eng.uk.ac.ir

M. H. Hemmatpour
Department of Electrical Engineering, Jahrom University, Jahrom, Iran
e-mail: m.h.hematpour@jahromu.ac.ir

P. Dehghanian
Department of Electrical and Computer Engineering, School of Engineering and Applied Sciences, George Washington University, Washington, DC, USA
e-mail: payman@gwu.edu

M. Nazari-Heris et al. (eds.), *Application of Machine Learning and Deep Learning Methods to Power System Problems*, Power Systems,
https://doi.org/10.1007/978-3-030-77696-1_8

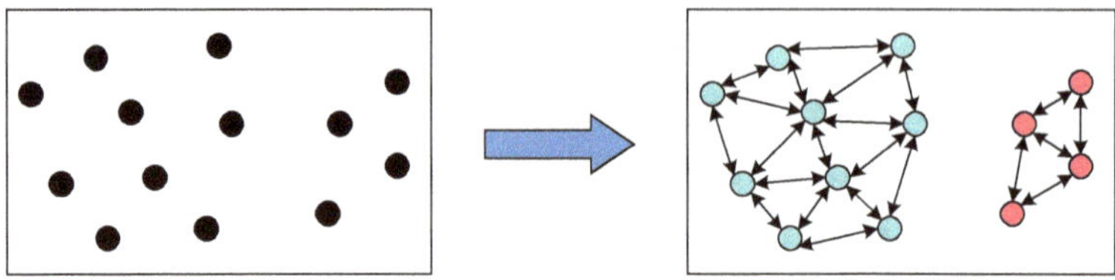

Fig. 8.1 Clustering data points in a 2-D space according to the Euclidean distances

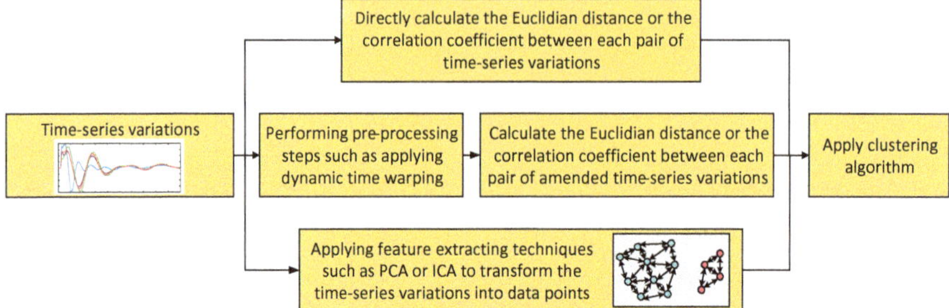

Fig. 8.2 Steps of clustering a dataset of time-series variations

The deployment of advanced metering devices such as smart meters and phasor measurement units (PMUs) in both distribution and transmission systems has broaden the horizons of power system monitoring, operation, and control. In modern power systems, advanced monitoring systems provide a huge amount of raw data including time-series variations of various signals. The scale of these time-series variations can range from less than a minute (related to the post-disturbance variations) to a one-day time frame (related to the daily bids offered by consumers) or even monthly patterns of end users' load variations. In this context, assessing the similarities between the variations and clustering similar signals can be helpful in better operation and planning of the power system. However, since the number of these signals is high and their variation patterns are complicated, the use of advanced feature extraction and clustering techniques is inevitable. Therefore, in the literature, the use of clustering techniques in power system studies including both transmission and distribution networks has been proposed.

Among the works in the related literature, several studies have been dedicated to the use of clustering techniques for grouping the consumers. In [1], the use of clustering techniques for dividing the end users into different categories according to their 24-h bid-offers in a smart grid is proposed. This task could be very helpful in demising flexible incentive rate strategies for demand response programs in power distribution systems. Another class of research in this field relates to the use of clustering techniques for dividing the loads in a smart grid or a micro grid. For example, in [2, 3], the electricity consumption patterns across the households are clustered in order to select the most appropriate household to be incentivized in an incentive-based household demand response program. In doing so, thousands of households are clustered into groups according to the similarity of their energy consumption behaviors. Clustering has also been proposed in [4] for demising local control of distributed energy resources (DERs). In doing so, clustering techniques are used to cluster similar DER characteristic curves based on voltage variations. Furthermore, the use of clustering techniques has been proposed in other applications such as phase identification of smart meters in distribution systems [5] and forecasting of the energy demand [6].

Apart from the above examples on the application of clustering techniques in power system studies, majority of works have been devoted to cluster or, in other words, partition the bulk power system into areas so that the dynamic response of elements in each area is similar. Such power system partitioning is based on a concept called *coherency*, which is defined as the similarity of post-disturbance dynamic response of the system elements [7]. In this regard, those elements revealing similar response to a disturbance are placed in the same cluster and are called coherent. Similar to other applications of clustering in power system studies, here the clustering is carried out on the basis of similarities between time-series signals, which are the post-disturbance observations of the generators' speeds or rotor angles or phase angle of the voltages at all buses across the network.

In general, coherency-based power system partitioning methods can be categorized in two groups, i.e., model-based and measurement-based methods [8]. Model-based methods mostly rely on the slow coherency concept and use the linearized model of the system to distinguish the coherent generators. Such methods are mainly suitable for applications such as dynamic equivalencing or control actions to mitigate the low-frequency oscillations in the system. Figure 8.3 shows how a power system is clustered into areas for dynamic equivalencing. Measurement-based methods, which are also known as data-driven methods, use the data measured by PMUs gathered from all over the system to find the coherent generators. Such methods can also be extended to the buses to determine the boundary of clusters. In such methods, it is assumed that the number of coherent groups and their boundaries may vary for different disturbances with different characteristics. Therefore, measurement-based methods are suitable for online applications such as controlled islanding and special protection systems. Due to the large scale of

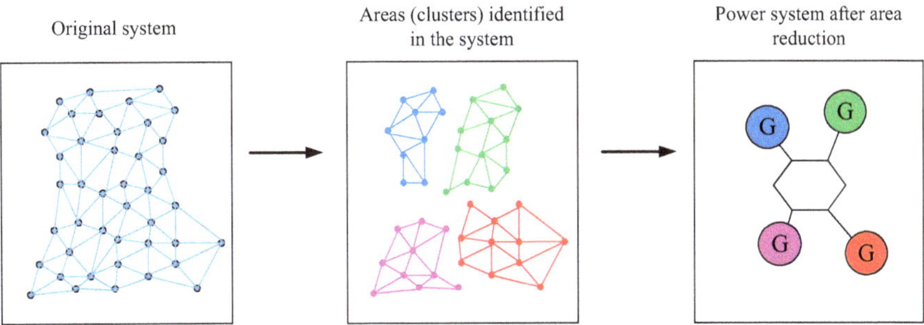

Fig. 8.3 Power system clustering for dynamic equivalencing

interconnected power systems, it is difficult to find the areas in the system quickly following disturbances. Moreover, each bus or generator in the system can be treated as a data point. As a result, the use of machine learning for clustering a power system has been addressed vastly in the literature. In this regard, feature extraction and supervised and unsupervised learning techniques have been used to partition the power system into different areas.

From a measurement-based point of view, the number of areas and their boundaries are not fixed and may change for different disturbances. In fact, factors such as the type of the disturbance and its location, as well as the power system condition at the time of the disturbance occurrence (e.g., the system load level), may affect the dynamic response of the system to the disturbance and cause different power system partitioning schemes. Therefore, the clustering algorithm to be used for power system partitioning is expected to have the following features:

- It should have no dependency on the prior assumption on the number of clusters.
- It should be capable to work well for imbalanced datasets. An imbalanced dataset is a dataset in which both very small and large clusters exist. In power system studies, it is probable that in some cases, a small area along with large areas are formed in the system. Moreover, in unstable cases, usually a single generator starts to lose synchronism and, therefore, forms a small group with one member. The clustering algorithm should be capable to distinguish such small clusters.
- It shouldn't be dependent on the random selection in its procedure (such random selections will be discussed in Sect. 8.4 where unsupervised learning techniques are described). This is necessary to ensure a deterministic solution.
- It is better for the clustering algorithm to feature a few parameters to be tuned. While this is not a necessity, the lower the number of parameters, the simpler the algorithm implementation.

It is noteworthy that similarity is a relative term meaning that two objects that seem to be highly similar from one's view may not be seen as similar objects from the view of someone else. In this regard, the degree of similarity is a better term, especially when dealing with clustering problems. Therefore, it is the user who determines the desired degree of similarity according to the requirements of his clustering problem to set the parameters of the chosen clustering technique. On the other hand, various similarity evaluation criteria are available in the literature for time-series signals. Examples of such criteria are Euclidean distance between the samples of the two signals, Pearson coefficient and semblance coefficient, respectively, defined in (8.1)–(8.3).

$$c_{x,y} = \sqrt{\frac{1}{N}\sum_{k=1}^{N}(x(k) - y(k))^2} \tag{8.1}$$

$$c_{x,y} = \frac{\sum_{k=1}^{N} x(k)y(k)}{\sqrt{\sum_{k=1}^{N} x^2(k) \sum_{k=1}^{N} y^2(k)}} \tag{8.2}$$

$$c_{x,y} = \frac{\sum_{k=1}^{N} (x(k) + y(k))^2}{2\sum_{k=1}^{N} (x^2(k) + y^2(k))} \tag{8.3}$$

where x and y are the two signals whose similarity is to be assessed, and N is the number of samples.

Along with techniques that have been proposed in the literature for power system clustering and will be discussed in the rest of this chapter, a test case will also be used and simulated to numerically evaluate the effectiveness of some of these techniques. The test system used here is the 16-machine, 68-bus test system, which has been introduced for dynamic studies and has been used widely for coherency evaluation and its applications. This system includes two large areas named as New England Test System (NETS) and New York Power System (NYPS) and three reduced areas represented by G14, G15, and G16. The one-line diagram of this system is shown in Fig. 8.4, details of which can be found in [9]. The test case used in this chapter, which is adopted from [10], is characterized by a single line to ground fault applied on the line connecting buses 1 and 2 and close to bus 1 and is cleared after 0.06 s without any line tripping. Note that before applying this fault, the line connecting

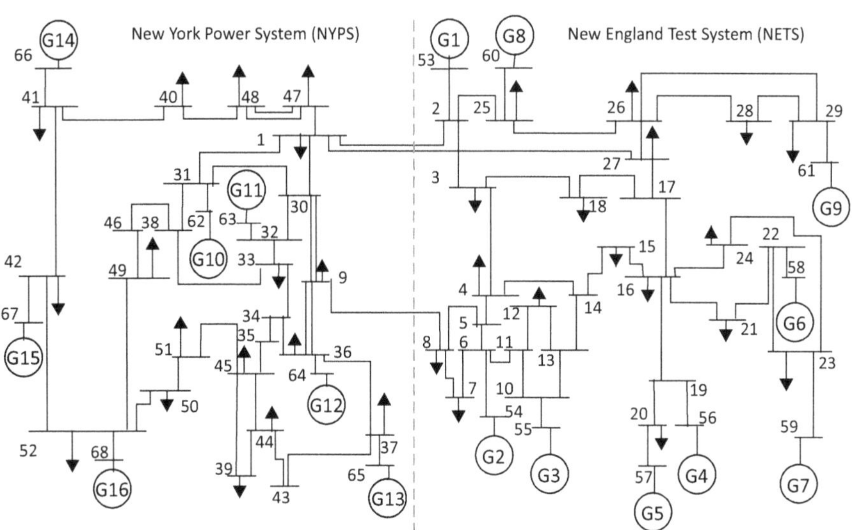

Fig. 8.4 One-line diagram of the 16-machine, 68-bus test system

buses 8 and 9 is removed in order to increase the electrical distance between the NETS and the rest of the system.

In the related literature, different data mining and pattern recognition techniques have been proposed for power system partitioning. Although most effort is on using unsupervised learning approaches, feature extraction techniques and supervised learning approaches have been also proposed in the literature. Therefore, in the rest of this chapter, the use of feature extraction techniques is first discussed. After describing the application of supervised learning approaches, the use of unsupervised learning methods will be then addressed. In addition, the advantages and disadvantages of these techniques will be explored in power system clustering, helping system planners and operators' decision-making in real-world settings.

8.2 Power System Clustering Using Feature Extraction Methods

Apart from various supervised and unsupervised learning methods examined by researchers for finding groups of coherent generators or buses, there are also feature extraction methods which have been proposed for coherency evaluation in power systems. Feature extraction methods are mainly introduced as a preprocessing task for other data mining techniques. To be specific, these methods are mainly introduced for transforming the huge volumes of raw data into more appropriate datasets with lower dimensions. Among such methods are PCA [11, 12] and ICA [13] which have been applied to generators' speed variations to transform these data into a 3-D space for grouping.

8.2.1 *Principal Component Analysis*

PCA is a multivariate analysis technique which is capable of extracting uncorrelated components from the input signals. PCA has been introduced mainly for reducing the dimension of the problems. Assume that there are N observations of M variables where $M \leq N$. By applying PCA, these N variables, which are in fact correlated variables, are transformed into a set of uncorrelated variables. These uncorrelated variables (principal components), which are the result of applying the p largest eigenvectors of the original matrix to the original matrix, are then mapped to a p-dimensional space. The idea behind PCA is to investigate if the first few components account for the variations in the original variables. Mathematically speaking, consider a matrix X with dimension $M \times N$ in which rows are the variables and columns are the observations. By applying PCA, X will be reconstructed as the sum of the orthonormal functions as follows [14]:

$$X = \begin{pmatrix} d_{1,1} \\ \vdots \\ d_{m,1} \end{pmatrix} \omega'_1 + \begin{pmatrix} d_{1,2} \\ \vdots \\ d_{m,2} \end{pmatrix} \omega'_2 + \cdots + \begin{pmatrix} d_{1,m} \\ \vdots \\ d_{m,m} \end{pmatrix} \omega'_m \tag{8.4}$$

where ω'_i is the ith normalized right eigenvector of matrix $X^T X$. In a p-dimensional space ($p \leq M$), the ith component is mapped to a point with coordinates $d_{i,\ 1}$, $d_{i,\ 2}$, . . ., $d_{i,\ p}$. In this regard, since similar components have similar coordinates in the new space, a group of data points corresponding to the original variables with similar variations will form a cluster in the new space.

In power system studies, the use of PCA for power system partitioning has been proposed in several research works. In [11], the efficiency of PCA on clustering the system generators using the post-disturbance variations in the generators' speed signals has been examined. Moreover, PCA has been also applied in [11] for power system partitioning by investigating the similarities between post-disturbance phase angle variations of the voltages at all buses. However, it is not possible in all cases to visually cluster the data points mapped to the reduced p-dimensional space, meaning that further analysis would be required to find the clusters. For example, in [15], after applying PCA, the use of hierarchical clustering for grouping the components in the three-dimensional space has been proposed. One advantage of PCA is that if it works well, there would be no need to have a prior assumption on the number of clusters. This is because, as stated in the Introduction, from a measurement-based point of view, a power system can be partitioned in different number of areas with different boundaries for different disturbances occurring in the system.

Figure 8.5 shows the locations of coefficients of the first three components for each bus in the 3-D space obtained for the test case. As it can be seen from Fig. 8.5, area 1 (cluster 1) has been formed in a far distance from the other areas and, therefore, can be easily distinguished. However, although it seems that the other

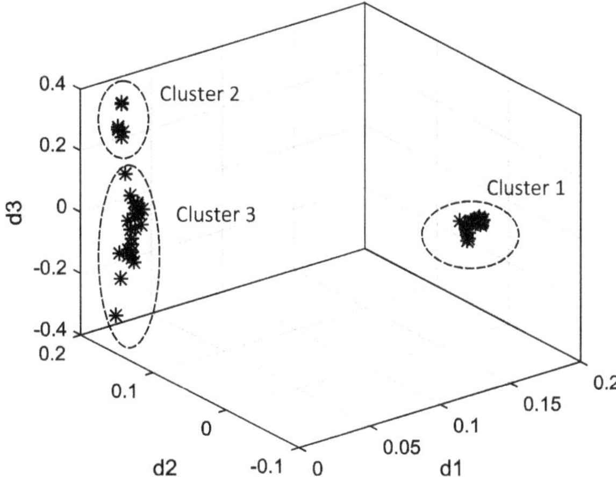

Fig. 8.5 Clusters of buses obtained from applying PCA in a 3-D space

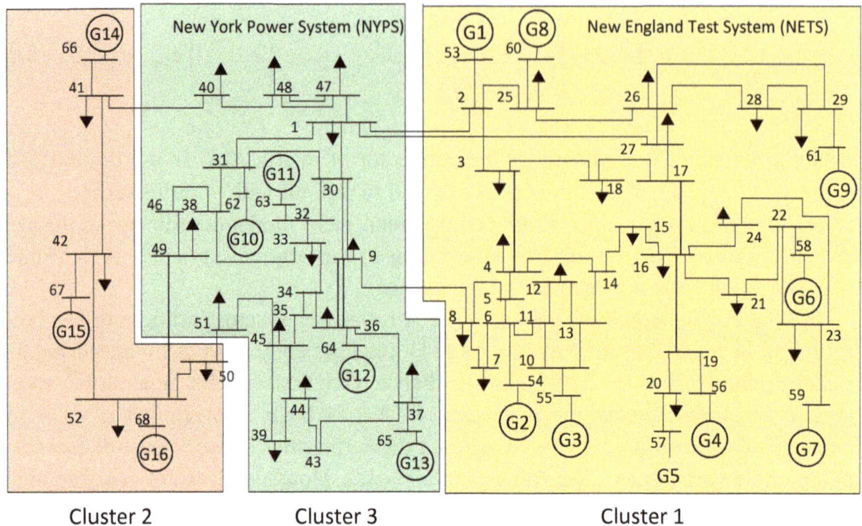

Fig. 8.6 Clusters (areas) obtained for the test system

two clusters can be discriminated visually, in some cases, it will be hard to find the clusters in the 3-D space, and thus, using a clustering technique would be essential. Note that the clustering scheme in this figure is the same as the one obtained by subtractive clustering in [10]. A graphical illustration of these areas is shown in Fig. 8.6.

8.2.2 Independent Component Analysis

ICA is another multivariate technique which is aimed to extract hidden features from raw data [16]. Unlike PCA in which a set of uncorrelated components are extracted from the raw data, ICA tries to transform the data into a set of independent components. More specifically, ICA tries to find components that are independent and non-Gaussian. According to [13], being independent can be an implication of being uncorrelated, but being uncorrelated does not guarantee an independence.

Considering a set of N observations of M variables stored in a matrix X with dimension $M \times N$, ICA will decompose X into independent components as shown in (5) [16].

$$X = \begin{pmatrix} d_{1,1} \\ \vdots \\ d_{m,1} \end{pmatrix} c'_1 + \begin{pmatrix} d_{1,2} \\ \vdots \\ d_{m,2} \end{pmatrix} c'_2 + \cdots + \begin{pmatrix} d_{1,m} \\ \vdots \\ d_{m,m} \end{pmatrix} c'_m \quad (8.5)$$

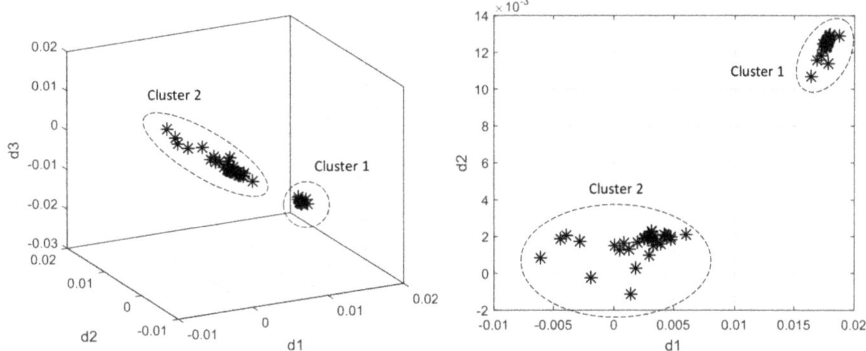

Fig. 8.7 Clusters obtained by ICA for the test case: (**a**) 3-D space and (**b**) 2-D space

Similar to PCA, independent components have to be sorted according to their dominance. Then dominant components, which can be the first two for the 2-D visualization or the first three for the 3-D visualization, will be used for clustering. For example, considering the first three dominant independent components and plotting the first three elements of rows of *d*-vectors, i.e. $d_{i,\ 1}$, $d_{i,\ 2}$, and $d_{i,\ 3}$, it will be seen that the data points corresponding to coherent signals will take place close together in a 3-D space. However, as described for PCA, it is not always possible for ICA to cluster the data points visually or graphically. Therefore, although ICA can extract useful information from raw data, at last, it may be needed to use a clustering technique to find the final clusters. Figure 8.7 shows the results of applying ICA (here fast ICA) to the test case used in this chapter. As it can be seen, ICA partitions the system into two clusters, where the second cluster is formed by combining the clusters 2 and 3 in Fig. 8.5.

8.3 Power System Clustering Using Supervised Learning Methods

Supervised learning approaches, mostly known as classifiers, are another type of learning methods, which are used in various scientific fields such as engineering, medicine, and economics. In these methods, a set of data with already known group labels are used for training the classifiers. In fact, supervised methods try to find the rules that determine the relationships between the already observed data and their associated known labels and then use them to predict the label of the next observation. Therefore, a dataset consisting several attributes and a target is needed for training and testing the classifier where each of these attributes and targets can be of any types of numerical, nominal, or Boolean variables.

In the literature, supervised learning approaches such as decision trees (DTs) and artificial neural networks (ANNs) have been developed to build classifiers for

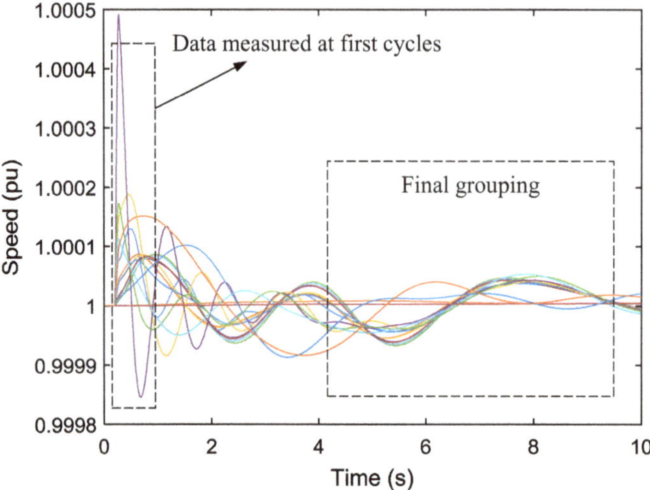

Fig. 8.8 Data window for generators grouping prediction using generators' speed variations

predicting the stable or unstable groups of generators using the first cycles after fault occurrence. In coherency analysis, these approaches are aimed to replace the timely coherency identification methods with a multi-class classifier to predict the coherent groups of generators. Among classification methods, the use of neural networks and decision trees for coherency prediction has been suggested. A typical representation of generators' speed oscillations has been shown in Fig. 8.8. According to this figure, a classifier is intended to predict the final grouping that is going to be formed using the information and patterns hidden in the data measured at the first cycles following the disturbance.

8.3.1 Artificial Neural Networks

In ANNs, several layers, each consists of neurons, are put together to build a classifier. ANNs are vastly used in different fields for predicting and regression applications. In the literature related to power system coherency analysis, the use of ANN has been addressed in few studies. The primary work has been published in [17], where authors have suggested the use of ANN for finding clusters using three samples of speed signals of generators. ANNs have also been proposed to serve as a preprocessing task in generators clustering. In a recent study, Siddiqui et al. [18] used a neural network structure to predict the time-series signals which is the post-disturbance rotor angle trajectories. In their work, they have used the first six cycles' variations following a disturbance to predict the future rotor swing trajectories and then have clustered the generators by evaluating the similarities of these predicted trajectories.

8.3.2 Decision Trees

DTs are a sort of classifiers with a treelike shape forming from a set of decision rules extracted from a training dataset. To extract the rules, every observation in the data must be accompanied with a known target content. A DT consists of two types of nodes, namely, decision and terminal nodes. The input of a DT includes several attributes, which should be tested in their related decision nodes. For example, in the root node, which is the first decision node of a DT, a test on the associated attribute is carried out, and based on the output value of the first test, the next decision node is then determined. This procedure will continue until it reaches to one of the terminal nodes or leaves of the DT. Similar to ANNs, a training dataset along with a testing dataset are used to train and evaluate the accuracy of a DT.

In the literature, several studies have used different types of DTs for generator coherency prediction. For example, authors in [19, 20] have shown that using appropriate DT training techniques, it is possible to build a power system model to predict the clusters of coherent generators using the data measured by PMUs in a very short time period following a disturbance. In another work presented in [21], it has been proposed to use a set of simple DTs to find the most coherent generators following a disturbance. Quantile regression forest is another type of DTs, which has been proposed in [22] to be used for predicting unstable group of coherent generators, which is very helpful for early identification of the necessary remedial actions in the emergency scenarios following disturbances.

In these studies, it is assumed that having a good knowledge of the uncertainties in a power system, it can be possible to find almost all possible clustering schemes that would happen in the system. Then a training dataset is generated in which the target is any of these probable clustering schemes while the attributes are demised characteristics appropriately extracted from the synchrophasor measurements. Finally, a training and testing procedure is carried out to discover the hidden patterns and rules between the values of the attributes and the target. However, one notable issue arising in these studies is that even in these methods, a clustering technique may be needed to establish the target input, if the number of data points (generators or buses) is high.

8.4 Power System Clustering Using Unsupervised Learning Methods

The vast majority of works in the field of power system clustering is, as expected, done using unsupervised learning approaches, which are called as clustering type of data mining techniques. Various clustering methods have been proposed for power systems clustering. However, these methods have their advantages and disadvantages, which should be examined and addressed before being used for power system clustering.

With respect to the nature of their procedures, unsupervised clustering techniques can be divided into two categories. In the first one, which includes algorithms such as k-means (KM) [23] and fuzzy c-means (FCM) [24], the desired number of clusters (N_C) is known, and therefore, the task is to divide the dataset into N_C groups according to the degree of similarities between each pair of data points. However, in the second category are placed algorithms, such as subtractive clustering (SC) [25–27] and density-based spatial clustering of applications with noise (DBSCAN) [28], in which instead of having a prior assumption on the desired number of clusters, a radius indicating the desired highest degree of dissimilarity between each pair of data points in a cluster would be determined by the user, and thus, the only task would be to let the algorithm find the clusters. In other words, in the second category, it would be the algorithm itself that determines the number of clusters. Figure 8.9 illustrates how the clustering works in these two categories.

Subsects. 8.4.1–8.4.7 will describe these methods as well as their cons and pros. In addition, the literature related to the use of each clustering method in the field of power system clustering will be introduced in their respective subsections.

8.4.1 K-Means Clustering Algorithm

K-means clustering is a popular clustering algorithm which uses an iterative procedure to cluster the data points into a predefined number of clusters. In fact, KM is a hard clustering algorithm meaning that in its final solution the membership value of each data point to one cluster is found to be 1 while for other clusters, it is 0. This algorithm searches for the clusters in the dataset so that a cost function is minimized. The cost function, which is defined as the sum of dissimilarities between the member of N_C clusters, is as follows:

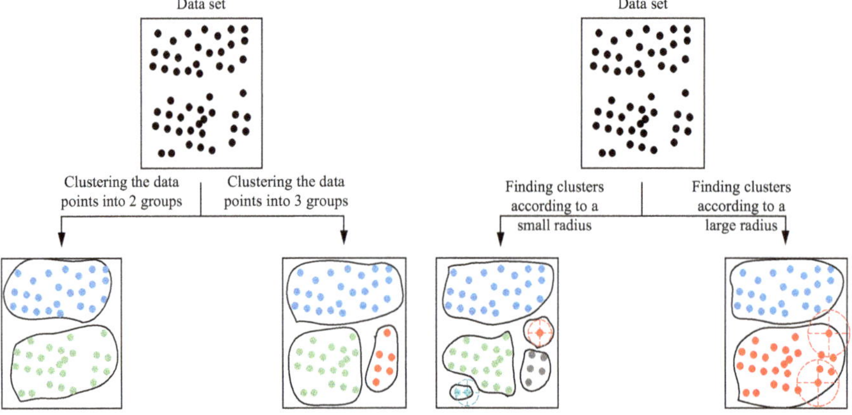

Fig. 8.9 Two types of clustering procedure by clustering techniques

$$J = \sum_{j=1}^{N_C} \left(\sum_{i \in Gj} d_{i,j}{}^2 \right) \tag{8.6}$$

For any data point i in the dataset, a membership function will be defined as in (8.7).

$$u_{i,j} = \begin{cases} 1 & \text{if jth datapoint belongs to ith cluster} \\ 0 & \text{otherwise} \end{cases} \tag{8.7}$$

Note that a data point i is considered to belong to the jth cluster if among all centers, the jth center has the lowest dissimilarity with the data point. At each iteration, cluster centers are updated according to (8.8).

$$z_j = \frac{1}{|G_j|} \sum_{x_i \in G_j} x_i \tag{8.8}$$

In (8.8), c_j is the center of the jth cluster, G_j is the set of data points in the jth cluster, and $|G_j|$ represents its size, while x_i is the ith data point in G_j. It should be noted that as (8.8) shows, in KM algorithm, a cluster center in the final solution may not essentially be a data point. Considering (8.6)–(8.8), the iterative procedure for clustering a dataset will be as follows:

- Step 1: Initialize N_C number of cluster centers selected from the dataset.
- Step 2: Find membership values for all data points using (8.7).
- Step 3: Compute the cost function defined in (8.6).
- Step 4: Update cluster centers using (8.8) and return to step 2.

Two stopping criteria can be used here. In the first one, algorithm will stop if the difference between the value of the cost function obtained in two consecutive iterations becomes less than a threshold, while in the second, algorithm will stop if it has been run for a sufficient number of time. However, the drawbacks of KM are twofold. Firstly, KM needs the number of clusters to be predefined. Secondly, its performance highly depends on the random selection of initial cluster centers. Therefore, it may happen that KM finds a wrong solution especially for a dataset in which both very small and very large clusters exist. To overcome this challenge, one approach is to run KM several times to find better solutions or use a subtractive method to select the initial cluster centers appropriately.

In the related literature, several studies have used KM to find coherent generators and coherent areas. For example, authors in [29, 30] have used KM to cluster generators in predefined number of groups. In so doing, they define a row vector for each generator representing the modal response of the generator and then use the angle between the vectors of two generators as a criterion for obtaining dissimilarities. In another work presented in [31], singular value decomposition (SVD) is first applied to the rotor angle variation signals of generators to construct a matrix called the characteristic coefficient matrix with dimension $k \times N_G$ where N_G is the number of generators, and k is selected to retain the largest k singular values during SVD process. Next, KM is applied on this matrix to cluster N_G generators. It is worthy to note that the characteristic coefficient matrix is considerably smaller than the matrix of the original samples, resulting in KM to be computationally more attractive. However, there are other types of studies in which a combination of KM and evolutionary algorithms is proposed for better partitioning of the power systems [32, 33]. For example, in [33], KM is applied on the generators' rotor samples in conjunction with PSO to contrive a PSO-KM method for better clustering.

In this subsection, the KM algorithm is applied to cluster the buses of the test case. To do that, the Pearson coefficient is first used to assess the similarity of phase angle variations of voltages at buses. Then the distance between the value of Pearson coefficient between two buses and unity value is considered as the dissimilarity between the two buses. Finally, KM algorithm uses these dissimilarity values to cluster the buses. Note that the number of clusters has been set to 3. However, as stated above, the solution obtained by KM depends on the centers selected at the initial iteration. Therefore, KM algorithm is applied to the dataset 1000 times to see how many times the correct solution shown in Fig. 8.6 will be obtained. Results showed that KM algorithm succeeded to find the clusters in 95 runs of the algorithm.

8.4.2 Partitioning around Medoid Algorithm

PAM clustering algorithm, which is known also as k-medoids clustering algorithm, is a clustering algorithm basically similar to KM. However, unlike KM in which cluster centers are not the real data points, in PAM algorithm, cluster centers (medoids) are essentially selected in every iteration from data points in the dataset. In fact, a medoid is a data point whose average dissimilarity to other data points in the cluster is minimal. PAM algorithm needs the initial selection of medoids. In doing so, a building phase, which consists of an iteration-based procedure, has been added for selecting the initial medoids wisely [34]. However, it is essential in PAM to predefine the desired number of clusters. Briefly speaking, the implementation steps of the PAM algorithm are as follows (more details can be found in [35, 36]):

- Step 1: Initialize N_C number of medoids (cluster centers) selected from the dataset.
- Step 2: Find membership values for all data points.
- Step 3: Update medoids according to the dissimilarities of the data points to the medoids selected previously.
- Step 4: Stop updating medoids if the stopping criterion holds. Otherwise, return to step 2.
- Step 5: Assign each data point to the cluster centered by the nearest medoid.

In the field of coherency evaluation, authors in [34, 37] have proposed the use of PAM algorithm. In their works, they have applied PAM algorithm on dissimilarities between rotor angle variations of generators to cluster them for dynamic equivalencing. In order to examine the effectiveness of PAM algorithm, it has been applied on the test case defined in Sect. 8.1. In doing this examination, the building phase has not been considered, and the aim has been to examine the capability of the algorithm in finding clusters. Moreover, the number of clusters has been set to three, and the algorithm has applied 1000 times on the dataset. It has been found that in 274 runs, the PAM algorithm has succeeded to find the correct clusters.

8.4.3 Fuzzy C-Means Clustering Algorithm

Unlike KM and PAM algorithms, in FCM algorithm, an object is allowed to belong to several clusters with different degrees of membership. FCM is an iteration-based clustering algorithm, and its solution is dependent on the initial selection of the membership values. This is because cluster centers in each iteration are determined on the basis of membership values. Moreover, the final solution of the algorithm is dependent on the random selection of the initial memberships and the number of clusters. FCM algorithm utilizes a cost function more general than (8.6), in which membership values are included. In this cost function, which is defined in (8.9) and should be minimized, a weighting exponent ($m \geq 1$) is used to control the fuzziness of the clustering process [24].

$$J = \sum_{j=1}^{N_C} \left(\sum_{i \in Gj} u_{i,j}^m {d_{i,j}}^2 \right) \tag{8.9}$$

Moreover, though a data point belongs to all clusters with different degrees of memberships, the sum of degrees must equal to 1, or:

$$\sum_{j=1}^{N_C} u_{i,j} = 1 \qquad \forall i = 1, \ldots, n \tag{8.10}$$

Similar to (8.8), a set of cluster centers will be updated in each iteration except that in FCM, the fuzziness of the membership values controls the updating process. In addition to updating the centers, membership values should be updated at each iteration as well. In doing so, (8.11) and (8.12) are used, respectively.

$$z_j = \frac{\sum_{i=1}^{n} u_{i,j}^m x_i}{\sum_{i=1}^{n} u_{i,j}^m} \tag{8.11}$$

$$u_{i,j} = \frac{1}{\sum_{k=1}^{N_C} \left(\frac{d_{i,j}}{d_{i,k}} \right)^{2/(m-1)}} \tag{8.12}$$

As stated, FCM algorithm is an iteration-based algorithm in which centers and membership values are updated at each iteration so that the cost function is minimized. The implementation steps of this algorithm are as follows:

- Step 1: Initialize the algorithm by random selection of membership values.
- Step 2: Find cluster centers using (8.11).
- Step 3: Compute the cost function defined in (8.9).
- Step 4: Update the membership values using (8.12) and return to step 2.

Among significant works which have used FCM, we can refer to [38] where FCM has been proposed for grouping the power system generators into coherent areas. To do that, a coherency measure is first defined to assess the similarity of rotor angle variations of generators. Then, FCM algorithm is applied to cluster the generators on the basis of such similarities. In this subsection, FCM algorithm is applied on the test case defined in this chapter. Similar to Subsections 8.4.1 and 8.4.2, the number of clusters has been set to 3, and algorithm has been run for 1000 times. Moreover, in the FCM algorithm, the number of iterations is set to 1000. It has been found that FCM could obtain the correct results in 995 out of 1000 runs.

8.4.4 Fuzzy C-Medoids Clustering Algorithm

FCMd is another soft clustering algorithm which, similar to FCM, simultaneously assigns the data points to different clusters with different values of memberships. FCMd is again an iterative-based clustering algorithm, and the difference between FCMd and the FCM is that FCM focuses on updating the membership values, while in FCMd, the focus is on appropriately upgrading the medoids (cluster centers) [39]. Therefore, instead of initializing the algorithm with random selection of membership values, FCMd starts with random selection of medoids. In its algorithm, FCMd uses a cost function similar to (9) except that $d_{i,\,j}$ will be the dissimilarity between the ith data point in the jth cluster and its medoid. However, here the membership values will be obtained as follows:

$$u_{i,j} = \frac{\left(\frac{1}{d_{i,j}} \right)^{1/(m-1)}}{\sum_{k=1}^{N_C} \left(\frac{1}{d_{i,k}} \right)^{1/(m-1)}} \tag{8.13}$$

Moreover, medoids are updated in each iteration according to the dissimilarities of the data points to the medoids selected in the previous iteration. Mathematically speaking, medoids are obtained using the membership values as follows [40]:

$$q = \arg \min_{1 \le k \le n} \sum_{i=1}^{n} u_{i,j}^{m} d_{k,i} \tag{8.14}$$

$$z_j = x_q \tag{8.15}$$

Accordingly, the implementation steps of the FCMd algorithm are as below:

- Step 1: Initialize N_C number of medoids (cluster centers) selected from the dataset.
- Step 2: Compute membership values for all data points using (8.13).
- Step 3: Update medoids according to (8.14) and (8.15).
- Step 4: Stop updating medoids if the stopping criterion holds. Otherwise, return to step 2.
- Step 5: Assign each data point to the cluster centered by the nearest medoid.

The stopping criteria is that whether for all clusters the new medoid obtained in the current iteration is equal to the medoids obtained in the previous iteration: If the answer is yes, the algorithm will stop. It is noteworthy that FCMd suffers from those shortcomings associated to KM and FCM, but its performance can be better than that of FC, if the initial centers are selected appropriately. In the literature, FCMd has been proposed by the authors in [40, 41] to provide a better solution for large power system partitioning. The reason is that in case where the number of data points is high, using FCM clustering algorithm can be time-consuming. To evaluate the performance of the FDMd algorithm, it has been applied to the test case used in this chapter for 1000 times, and it was seen that FCMd algorithm could find the clustering scheme shown in Fig. 8.6 in 94 runs.

8.4.5 Subtractive Clustering Algorithm

SC is another clustering algorithm that has been primarily introduced as a pre-processing tool for other clustering algorithms such as KM and FCMd. SC is primarily aimed to find cluster centers using an iterative procedure. However, it can itself be used for clustering as well. In this algorithm, a density value D_i is first assigned to a data point x_i using the density measure defined in (8.16).

$$D_i = \sum_{j=1}^{N_B} e^{\left[-\frac{(d_{i,j})^2}{(r_a/2)^2}\right]} \tag{8.16}$$

where $d_{i,j}$ is the dissimilarity between data points i and j, and r_a is a positive constant used to represent the neighboring radius. According to (8.16), the lower the value of $d_{i,j}$ is (the more data point j is similar to data point i), the higher the value of D_i will be. In other words, since the density measure in (8.16) is defined on the basis of sum of similarities between all data points, the data point with maximum density value will be chosen as the first cluster center which is because it has the highest similarity with the highest number of data points. In the next step of the algorithm, all density values will be revised with respect to the density value of the cluster center determined in the previous step. This subtractive revision is done through using the following equation:

$$D_i^p = D_i^{p-1} - D_c^{p-1} e^{\left[-\frac{\left(d_{i,c_{p-1}}\right)^2}{(r_b/2)^2}\right]} \qquad i = 1, 2, \ldots, N_B \tag{8.17}$$

where

$$D_c^{p-1} = \max_i \left\{ D_i^{p-1} \right\} \tag{8.18}$$

In (8.17), r_b is a positive constant aimed to define a neighborhood around the previous center whose members experience a measurable reduction in their density values. Moreover, D_c^{p-1} is the density value of the cluster center selected at iteration p-1, $d_{i,c_{p-1}}$ is the dissimilarity between data point i and the center determined at iteration p-1. Now, the cluster center at iteration p will be selected as the one having the highest density value among the revised values obtained from (8.17). The process of finding cluster centers will be repeated until the rejection criterion rejects the candidate center. Note that various rejection criteria can be defined ranging from a simple one as in [10] to a more complicated criterion. Once the cluster centers are determined, the similarity of each data point with each cluster center should be assessed. Accordingly, a matrix with dimensions $N_C \times N_D$ will be formed as follows [10]:

$$R = \begin{bmatrix} r_{1,1} & \cdots & r_{1,N_B} \\ \vdots & \ddots & \vdots \\ r_{N_C,1} & \cdots & r_{N_C,N_B} \end{bmatrix} \tag{8.19}$$

where

$$r_{i,j} = D_j^1 - D_{c_i}^1 e^{\left[-\frac{\left(d_{j,c_i}\right)^2}{(r_b/2)^2}\right]} \tag{8.20}$$

Finally, assuming that the ith element of the jth column of R has the lowest value in the column, it can be said that the jth data point should be assigned to a cluster whose center is the ith cluster center determined by the algorithm. The performance of SC algorithm for power system partitioning and its effectiveness over FCMd and KM algorithms has been demonstrated in [10]. Particularly, the effectiveness of the SC algorithm in finding the areas for the test case used in this chapter has been shown in [10]. Note that SC is mainly featured with no random selection operation, and therefore, it has to be run once, and its results are deterministic. According to the SC algorithm, two parameters have to be predefined, i.e., a neighborhood radius (r_a) and the squash factor which is defined as the ratio of r_b to r_a. In addition, one or more parameters have to be predefined as well according to the rejection criteria selected by the user.

8.4.6 Density-Based Spatial Clustering of Applications with Noise

DBSCAN is another density-based clustering technique which has been used in different areas of science. In a way similar to subtractive clustering, the basics of DBSCAN rely on the concept of density in an *n*-dimensional space. Considering a 2-D space shown in Fig. 8.10, the following definitions hold [28]:

- A data point p_r is a core data point if at least *Pmin* data points including p_r are within the distance ε of data point p_r.
- If data point q_r is within distance ε of data point p_r, it is said that data point q_r is directly reachable from data point p_r.
- A data point p_r is a noise if it is not directly reachable from any core data point.
- A data point q_r is reachable from data point p_r if there is a path $p_{r1}, \ldots, p_{rn}$ with $p_{r1} = p_r$ and $p_{rn} = q_r$, where each $p_{r(i+1)}$ is directly reachable from p_{ri}.

According to the above definitions, a cluster is a set of at least *Pmin* data points so that each pair of them are directly reachable. For example, considering the example data points shown in Fig. 8.9, if *Pmin* is set to 3, a cluster with eight members (shown in green) will be formed, while the other three data points are found to be noises (shown in yellow).

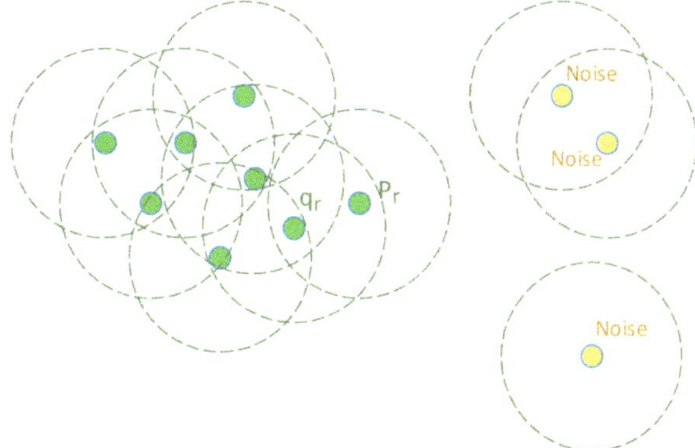

Fig. 8.10 Example of clustering the data points in a 2-D space (*Pmin* = 3)

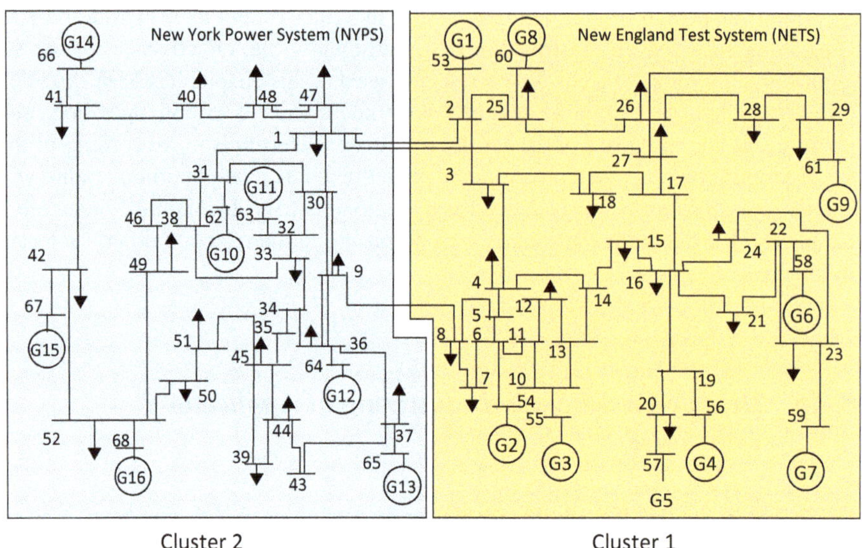

Fig. 8.11 Clusters (areas) obtained for the test system using DBSCAN

In SubSect. 8.4.5, the use of subtractive clustering and its advantages for power system coherency-based clustering are discussed. However, DBSCAN is a clustering algorithm simpler than SC algorithm and is more suitable for power system partitioning due to the following features [42].

- It doesn't need a prior assumption on the number of clusters. This is a very important feature since from a measurement point of view, groups of coherent generators are not fixed for different disturbances.
- It needs fewer parameters to be set.
- It doesn't use any random selection operation, and therefore, its solution is deterministic.
- It works well for imbalanced datasets. It can easily find small clusters and noises which is necessary for online measurement-based power system partitioning.

In [42], the use of DBSCAN for clustering the buses in a power system is proposed and examined. Here, this algorithm is applied to the test case defined in this chapter, and its clustering result is shown in Fig. 8.11. Note that to obtain the clustering scheme, ε is set to 0.1. As it can be seen from this figure, DBSCAN has clustered the buses into two areas instead of the three areas shown in Fig. 8.6. The reason backs to the nature of the two categories of clustering techniques defined in the beginning of Sect. 8.4.

8.4.7 *Support Vector Clustering Algorithm*

Support vector clustering (SVC) is derived from support vector machine on the basis of the fact that when data points in the original dataset are mapped into a new space

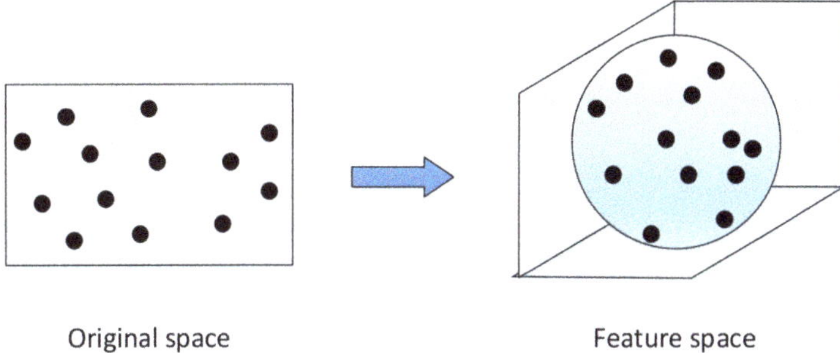

Fig. 8.12 Transforming the data points from the original space into the feature space in the SVC algorithm

with higher dimensions, the minimum sphere in the new space that encloses a group of similar data points is found, and after transforming back into the original space, it can be considered as the cluster boundary (see Fig. 8.12). SVC doesn't need to have a pre-assumption on the number of clusters since it solves an optimization problem to find the optimal clustering. In other words, it can be said that SVC can cluster datasets with arbitrary shapes and is able to estimate the optimum number of clusters. However, SVC needs its input to be in the form of data points in a p-dimension space since it uses the Euclidean distances between data points to transform them into the feature space. Therefore, using PCA or other embedding techniques is necessary to transform the similarity of the time-series data into the Euclidean distances.

A brief mathematical expression of SVC procedure is as follows. In order to prepare the data to be used in this algorithm, SVC employs a nonlinear kernel transformation for mapping the data points to the higher dimensional feature space. In order to find the minimum sphere, an optimization problem in the Lagrangian form is first defined as in (8.21).

$$L = R^2 - \sum_j \left(R^2 + \xi_j - \|\phi(x_j) - a\|^2 \right) \beta_j - \sum_j \xi_j \mu_j + C \sum_j \xi_j \tag{8.21}$$

where $X = x_1, x_2, \ldots, x_N$ is a dataset of N data points, ϕ is a nonlinear transformation function, and R is the smallest enclosing sphere. β_j and μ_j are the Lagrangian multipliers. C is the user defined constant, and $C\sum_j \xi_j$ denotes the penalty term. Moreover, R^2 is the distance of each data point x from the center of the shape which is calculated by (8.22).

$$R^2(x) = K(x, x) - 2\sum_j K(x_j, x)\beta_j + \sum_{i,j} \beta_i \beta_j K(x_i, x_j) \tag{8.22}$$

In (8.22), $K(x_i, x_j) = \phi(x_i).\phi(x_j) = e^{-q\|x_i - x_j\|^2}$ is the Gaussian kernel transformation, and q is the width parameter which controls the boundaries of clusters. The optimization problem defined in (8.21) is then converted to its dual quadratic form to obtain (8.23). Now, the goal would be to obtain the Lagrangian multiplier β_j.

$$W = \sum_{j} K(x_j, x_j)\beta_j - \sum_{i,j} \beta_i \beta_j K(x_i, x_j) \tag{8.23}$$

subject to

$$0 \leq \beta_j \leq C, \sum_{j} \beta_j = 1 \tag{8.24}$$

In the literature, the use of SVC for finding groups of coherent generators is presented in [43, 44] to show its effectiveness over KM and FCM techniques.

8.5 A Discussion on the Use of Machine Learning for Power System Clustering and Possible Future Works

As discussed in Sect. 8.3, several classifiers have been proposed for power system clustering and its predictions. However, a survey of the related literature shows that the number of works in this field is limited. One can note that in addition to the concerns regarding the fact that a clustering technique is still needed to form the schemes in the target variable in the dataset, another important issue that should be taken into account is the sufficiency of the training dataset. This must be addressed as it can affect the efficiency of a classifier for an observation made out of the range of those in the training dataset.

On the other hand, among clustering methods that have been proposed for clustering the generators or buses in a power system, it can be said that each of them suffers from one or more of the following shortcomings, particularly when being considered for online applications; some of these algorithms have more than one parameters to be set, and therefore, their final results depend on the setting of these parameters. Moreover, some of them such as SVC needs more calculation and are time-consuming. In addition, algorithms including KM, FCM, and FCMd use random selection processes in their initial iteration, meaning that their results may change as the initial selection changes. Also, for almost all these algorithms, the number of clusters needs to be predefined.

In order to provide a better vision on the cons and pros of the unsupervised learning clustering techniques described in Sect. 8.4, more evaluations and simulations will be given here. At first, KM, PAM, FCM, and FCMd algorithms have been again applied to the test case for a larger number of runs, which is 10,000 runs in this section. Table 8.1 shows the number of times each algorithm has found the correct clustering scheme among the 1000 and 10,000 runs. It can be seen that the best performance is obtained by FCM clustering algorithm.

As stated in Sect. 8.1, clustering techniques have been proposed to be used in various fields of power system studies. The suitability of a clustering technique for a specific application depends on the requirements of that application. For example, in

Table 8.1 Performance of different clustering algorithms on the test case under different number of simulation runs

Total no. of runs	KM	PAM	FCM	FCMd
10,000	1001	2647	9998	873
1000	95	274	995	94

applications where the number of clusters are given or at least can be easily determined, the use of KM or FCM algorithms could be advisable. In the field of power system clustering, an ideal algorithm for online grouping of coherent generators or buses should be characterized with the following features:

- It has less parameters to be set, i.e., at most one parameter. This is very important particularly if a measurement-based online clustering is desired.
- It uses no random operation to make the results deterministic. It is essential to run the algorithm once and obtain the correct solution. This feature relates to the proper selection of initial values in the first iteration of algorithms such as KM or FCM. In the literature, there have been techniques proposed to be used as a preprocessing for preparing the initial values. However, these preprocessing can add burden in online applications.
- The algorithm itself determines the number of clusters. From a measurement-based point of view, in power system post-disturbance clustering, the number of areas and their boundaries are not fixed and may change for different disturbances. Therefore, it would not be easy to have a prior assumption on the number of areas (clusters).
- It is suitable for both stable and unstable situations since in the latter, one generator starts to lose its synchronization, and therefore, the algorithm must detect it as an individual cluster.

According to the above features needed for the ideal algorithm to be used for power system clustering and considering the features of most well-known clustering techniques proposed in the literature, it can be concluded that SC and DBSCAN algorithms can be the most proper ones. The reasons are that they use no random operation; therefore, their solutions are deterministic, they have low number of parameters to be set, they can find very small clusters in a dataset, and they can themselves find the number of clusters.

8.6 Conclusions

This chapter discussed the use of data mining and machine learning methods in power system clustering applications. Particularly, the use of feature extraction methods as well as supervised and unsupervised learning approaches for power system clustering was discussed. For a better understanding of the performance of the unsupervised approaches, these methods were applied on a test case, and their solutions were examined. These simulations also made it possible to have a more clear vision on the advantages and disadvantages of different clustering methods. Nevertheless, future research is needed to widen the application of advanced clustering techniques, particularly due to the evolving interests in the use of artificial intelligence and deep learning methods in power system applications.

References

1. Z. Luo, S. Hong, Y. Ding, A data mining-driven incentive-based demand response scheme for a virtual power plant. Appl. Energy **239**, 549–559 (2019)
2. T. Teeraratkul, D. O'Neill, S. Lall, Shape-based approach to household electric load curve clustering and prediction. IEEE Trans. Smart Grid. **9**(5), 5196–5206 (2017)
3. S. Dasgupta, A. Srivastava, J. Cordova, R. Arghandeh, Clustering household electrical load profiles using elastic shape analysis. In. *2019 IEEE Milan PowerTech.* (2019), pp. 1-6
4. S. Karagiannopoulos, G. Valverde, P. Aristidou, G. Hug, Clustering data-driven local control schemes in active distribution grids. IEEE Syst. J., 1–10 (2020). https://doi.org/10.1109/JSYST.2020.3004277
5. F. Olivier, A. Sutera, P. Geurts, R. Fonteneau, D. Ernst, Phase identification of smart meters by clustering voltage measurements. In. *2018 Power Systems Computation Conference (PSCC).* (2018), pp. 1-8
6. C. Bock, Forecasting energy demand by clustering smart metering time series. In. *International conference on information processing and Management of Uncertainty in knowledge-based systems.* (Springer, 2018). pp. 431-42
7. M.H.R. Koochi, S. Esmaeili, P. Dehghanian, Coherency detection and network partitioning supported by wide area measurement system. In *2018 IEEE Texas power and energy conference (TPEC).* (2018), pp. 1-6
8. M.H.R. Koochi, S. Esmaeili, G. Ledwich, Taxonomy of coherency detection and coherency-based methods for generators grouping and power system partitioning. IET Generation, Transmission & Distribution. **13**(12), 2597–2610 (2019)
9. G. Rogers, *Power system oscillations.* (Springer Science & Business Media, 2012)
10. M.H. Rezaeian, S. Esmaeili, R. Fadaeinedjad, Generator coherency and network partitioning for dynamic equivalencing using subtractive clustering algorithm. IEEE Syst. J. **12**(4), 3085–3095 (2017)
11. K.K. Anaparthi, B. Chaudhuri, N.F. Thornhill, B.C. Pal, Coherency identification in power systems through principal component analysis. IEEE Trans. Power Syst. **20**(3), 1658–1660 (2005)
12. K. Mandadi, B.K. Kumar, Generator coherency using Zolotarev polynomial based filter bank and principal component analysis. Int. J. Emerging Electric Power Syst. **19**(2), 1–10 (2018)
13. M. Ariff, B.C. Pal, Coherency identification in interconnected power system—An independent component analysis approach. IEEE Trans. Power Syst. **28**(2), 1747–1755 (2012)
14. C. Chatfield, A. Collins, *Introduction to Multivariate Analysis* (CRC Press, Thousand Oaks, 1981)
15. A.M. Almutairi, S.K. Yee, J. Milanovic, Identification of coherent generators using PCA and cluster analysis. In Proc. Power Systems Computation Conference, 1–10 (2008)
16. A. Hyvärinen, J. Karhunen, E. Oja, Independent component analysis, adaptive and learning systems for signal processing, communications, and control. John Wiley & Sons, Inc. **1**, 11–14 (2001)

17. M.-H. Wang, H.-C. Chang, Novel clustering method for coherency identification using an artificial neural network. IEEE Trans. Power Syst. **9**(4), 2056–2062 (1994)
18. S.A. Siddiqui, K. Verma, K. Niazi, M. Fozdar, Real-time monitoring of post-fault scenario for determining generator coherency and transient stability through ANN. IEEE Trans. Ind. Appl. **54**(1), 685–692 (2017)
19. M.H.R. Koochi, S. Esmaeili, R. Fadaeinedjad, New phasor-based approach for online and fast prediction of generators grouping using decision tree. IET Generation, Transmission & Distribution. **11**(6), 1566–1574 (2017)
20. T. Guo, J.V. Milanović, Online identification of power system dynamic signature using PMU measurements and data mining. IEEE Trans. Power Syst. **31**(3), 1760–1768 (2015)
21. M.H.R. Koochi, P. Dehghanian, S. Esmaeili, P. Dehghanian, S. Wang, A synchrophasor-based decision tree approach for identification of most coherent generating units. In: *IECON 2018-44th Annual Conference of the IEEE Industrial Electronics Society,* (2018), pp. 71–6
22. S.M. Mazhari, N. Safari, C. Chung, I. Kamwa, A quantile regression-based approach for online probabilistic prediction of unstable groups of coherent generators in power systems. IEEE Trans. Power Syst. **34**(3), 2240–2250 (2018)
23. J.A. Hartigan, M.A. Wong, Algorithm AS 136: A k-means clustering algorithm. J Royal Statistical Society Series C (Applied Statistics) **28**(1), 100–108 (1979)
24. J.C. Bezdek, *Pattern Recognition with Fuzzy Objective Function Algorithms*. (Springer Science & Business Media, 2013)
25. S. Chiu, Method and software for extracting fuzzy classification rules by subtractive clustering. In: *Proceedings of North American Fuzzy Information Processing*. (IEEE, 1996), pp. 461-5
26. S.L. Chin, An efficient method for extracting fuzzy classification rules from high dimensional data. JACIII. **1**(1), 31–36 (1997)
27. S.L. Chiu, Fuzzy model identification based on cluster estimation. Journal of Intelligent & Fuzzy Systems. **2**(3), 267–278 (1994)
28. M. Ester, H.P. Kriegel, J. Sander, X. Xu, A density-based algorithm for discovering clusters in large spatial databases with noise. In: *KDD Conference*. (1996), pp. 226–31
29. S.K. Joo, C.C. Liu, J.W. Choe, *2001 Power Engineering Society Summer Meeting*. Conference proceedings (cat. No. 01CH37262). (2001), pp. 1811-6
30. S.-K. Joo, C.-C. Liu, L.E. Jones, J.-W. Choe, Coherency and aggregation techniques incorporating rotor and voltage dynamics. IEEE Trans. Power Syst. **19**(2), 1068–1075 (2004)
31. Q. Zhu, J. Chen, X. Duan, X. Sun, Y. Li, D. Shi, A method for coherency identification based on singular value decomposition. In. *2016 IEEE Power and Energy Society General Meeting (PESGM)*. (2016), pp. 1-5
32. E. Cotilla-Sanchez, P.D. Hines, C. Barrows, S. Blumsack, M. Patel, Multi-attribute partitioning of power networks based on electrical distance. IEEE Trans. Power Syst. **28**(4), 4979–4987 (2013)
33. M. Davodi, H. Modares, E. Reihani, M. Davodi, A. Sarikhani, Coherency approach by hybrid PSO, K-means clustering method in power system. In *2008 IEEE 2nd International Power and Energy Conference*. (2008), pp. 1203-7
34. G.-C. Pyo, J.-W. Park, S.-I. Moon, Coherency identification of generators using a PAM algorithm for dynamic reduction of power systems. Energies **5**(11), 4417–4429 (2012)
35. L. Kaufman, P. Rousseeuw, Clustering by means of Medoids in statistical data analysis based on the L1–norm and related methods. (Y. Dodge, Dü.) reports of the Faculty of Mathematics and Informatics. (Delft University of Technology, 1987)
36. H. Spath, *Cluster Analysis Algorithms for Data Reduction and Classification of Objects*. (Ellis Horwood Chichester, 1980)
37. G. Pyo, J. Park, S. Moon, A new method for dynamic reduction of power system using PAM algorithm. In. *IEEE PES General Meeting*. (2012), pp. 1–7
38. S.C. Wang, P.H. Huang, Fuzzy c-means clustering for power system coherency. In. 2005 IEEE International Conference On Systems, Man and Cybernetics. (2005). pp. 2850-5
39. R. Krishnapuram, A. Joshi, O. Nasraoui, L. Yi, Low-complexity fuzzy relational clustering algorithms for web mining. IEEE Trans. Fuzzy Syst. **9**(4), 595–607 (2001)

40. I. Kamwa, A.K. Pradhan, G. Joós, Automatic segmentation of large power systems into fuzzy coherent areas for dynamic vulnerability assessment. IEEE Trans. Power Syst. **22**(4), 1974–1985 (2007)
41. I. Kamwa, A.K. Pradhan, G. Joos, S. Samantaray, Fuzzy partitioning of a real power system for dynamic vulnerability assessment. IEEE Trans. Power Syst. **24**(3), 1356–1365 (2009)
42. F. Znidi, H. Davarikia, M. Arani, M. Barati, Coherency detection and network partitioning based on hierarchical DBSCAN. In. *2020 IEEE Texas Power and Energy Conference (TPEC)*. (2020), pp. 1-5
43. R. Agrawal, D. Thukaram, Support vector clustering-based direct coherency identification of generators in a multi-machine power system. IET Generation, Transmission & Distribution. **7** (12), 1357–1366 (2013)
44. R. Agrawal, D. Thukaram, Identification of coherent synchronous generators in a multi-machine power system using support vector clustering. In. *2011 International Conference on Power and Energy Systems*. (2011), pp. 1-6

Chapter 9
Voltage Stability Assessment in Power Grids Using Novel Machine Learning-Based Methods

Ali Mollaiee, Sasan Azad, Mohammad Taghi Ameli, and Morteza Nazari-Heris

9.1 Introduction

Voltage instability is a crucial issue that has been challenging power system engineers over the past two decades [1, 2]. This issue typically occurs in power systems that suffer increasing load demand and lagged development of the transmission system. As a result, power is delivered near power system limits, and the potential of instability is significantly increased. The voltage collapse can lead to power grid blackouts, which cause significant economic losses and affecting unpredictable impacts on residents' lives and industrial production [3–5]. To mitigate the risk of voltage collapse, stability analysis should be considered during both planning and online operating of power systems. In contrast to planning, online analysis requires online voltage stability assessment to take remedial actions for preventing possible voltage collapse.

A. Mollaiee · S. Azad · M. T. Ameli (✉)
Department of Electrical Engineering, Shahid Beheshti University, Tehran, Iran
e-mail: a.mollaiee@Mail.sbu.ac.ir; sa_azad@sbu.ac.ir; m_ameli@sbu.ac.ir

M. Nazari-Heris
Department of Architectural Engineering, Pennsylvania State University, State College, PA, USA
e-mail: mun369@psu.edu

M. Nazari-Heris et al. (eds.), *Application of Machine Learning and Deep Learning Methods to Power System Problems*, Power Systems,
https://doi.org/10.1007/978-3-030-77696-1_9

9.1.1 Literature Review

The classic voltage stability assessment methods utilize static analysis based on power flow methods such as the Gauss-Seidel or Newton-Raphson method. In reference [6–10], various voltage stability indexes using conventional power flow have been proposed. The main drawback of these techniques is the singularity of the Jacobian matrix at the maximum loading point. To address this issue, the continuation power flow (CPF) method is employed to obtain the voltage stability margin (VSM) [11]. Lee et al. [12] proposed a P-Q-V curve base technique for VSM assessment, which indicates the maximum limits of power demands. Due to the increasing size and complexity of the modern power system, classic power system stability analysis is becoming highly computational and time-consuming. Therefore, the aforementioned techniques are not effective for online applications.

In the last few years, machine learning (ML)-based techniques such as artificial neural networks (ANNs), decision trees (DTs), support vector machines (SVMs), and ensemble methods have been considered in various studies due to their capability to solve nonlinear and complex problems independent of system modeling with desired speed and accuracy [13, 14]. The multilayered perceptron (MLP) neural network trained using the back-propagation algorithm was first employed in [15, 16] to obtain the VSM that utilizes the energy method. Arya et al. [17] proposed applying a radial basis function (RBF) network to approximate the probabilistic risk of voltage instability for several operating conditions. In the proposed work, the database has been generated using the Monte-Carlo simulation. In [18], input feature selection based on neural networks using mutual information is employed to estimate the voltage stability level for several scenarios according to load condition and contingency. Debbie et al. [19] proposed an ANN-based method to estimate the VSM of the power system under normal and N-1 contingency operating conditions (OC). Devaraj et al. [20] proposed a new online monitoring technique using ANN to estimate the VSM of the power system based on synchrophasor measurement under normal and under N-1 contingencies OCs. In [21], various input feature reduction techniques are applied to the RBF network to estimate the voltage stability level with enhanced prediction accuracy. Hashemi et al. [22] use a multi-resolution wavelet transform and principal component analysis for feature extraction of voltage profile along with RBF network to approximate VSM. In [23], a multilayer feedforward artificial neural network (MLFFN) and radial basis function network (RBFN) are employed to assess power system static security assess. This paper uses a composite security index for contingency ranking and security classification. Bahmanyar et al. [24] proposed a new approach to obtain the optimal input variables required to estimate the VSM using ANN. In [25], a Z-score-based bad data processing algorithm is employed to improve the estimation accuracy of the feedforward ANNs. Walter M et al. [26] presented a novel approach for VSM estimation that combines a kernel extreme learning machine (KELM) with a mean-variance mapping optimization (MVMO) algorithm. In [27], the association rules (AR) technique is used to select the most effective loading parameters for the main input of the adaptive neuro-fuzzy inference system (ANFIS).

Generally, ANNs are known as a powerful and flexible method for carrying out nonlinear regression; nevertheless, they have some issues with training time and overfitting. Support vector machine (SVM) is another powerful machine learning technique that uses various kernel functions to perform classification and regression problems. In [28], a multi-class SVM method is applied to classify the power system's security level, either normal, alert, emergency_1, or emergency_2. Moreover, the enhanced multi-class SVM has been employed using the pattern recognition approach for security assessment [29, 30]. Suganyadevi et al. [31] proposed a support vector regression (SVR) model to assess the voltage stability of the power system consolidating flexible alternating current transmission systems (FACTS) devices. In [32], the ν-SVR and ε-SVR models with RBF and polynomial kernel functions have been applied to estimate VSM. Sajan et al. [33] proposed a genetic algorithm-based support vector machine (GA-SVM) approach to estimate the voltage stability margin index (VSMI). In this work, the optimal values of SVM parameters are obtained using the genetic algorithm. In [34], a new least-square SVM using synchrophasor measurements is employed to estimate voltage stability based on online learning.

DT is considered a fast and accurate ML technique for the development of classification and regression models. Furthermore, DT training results' interpretability is a significant advantage over other ML techniques such as SVM and ANN. Firstly, in the subject of power systems, Wehenkel employed DT to assess the transient stability of power systems [35]. Further, DT has been applied for security assessment applications [36, 37]. In [38], DT has been employed for online voltage stability assessment using wide-area measurements. Zheng et al. [39] employed a regression tree to predict the power system stability margin based on the VSM and oscillatory stability margin (OSM). DT can be consolidated with other algorithms such as principal component analysis (PCA) or fuzzy logic (FL) to enhance the training performance. Mohammadi et al. [40] proposed a hybrid model for the online voltage security assessment using a reduced predictor set extracted by PCA. In [41], a contingency grouping method was developed for deriving DTs to assess the power system security considering multiple contingencies. Meng et al. [42, 43] employed participation factor analysis and relief algorithm to select attributes for DT. Recently, ensemble methods have been introduced to improve the accuracy of DTs, and they combine several base models to produce one optimal predictive model. Beiraghi et al. [44] proposed a DT-based method for online voltage security assessment using wide-area measurements. In the proposed work, the ensemble methods such as bagging and adaptive boosting (AdaBoost) are employed to improve the voltage security assessment of the power system performance. Su and Liu [45] proposed a novel online learning framework for monitoring voltage stability using wide-area measurements. In the proposed study, a new enhanced online random forest (EORF) model based on the drift detection and online bagging techniques is implemented, enabling online update of the trained model instead of reconstructing an entire model. In [46], a novel methodology is presented for real-time assessment of short-term voltage stability (STVS) based on multivariate time series. In this study, random forest (RF) and symbolic representation technique is employed to classify the power system stability status in multiple class using the maximal Lyapunov exponent and various dynamic voltage indices. Dharmapala et al. [47] proposed an ML approach to predict the long-term VSM based on loadability margin (LM). In the proposed study, random forest regression (RFS) is employed to estimate LM using different voltage stability indices (VSI).

9.1.2 Contributions and Novelties

In this chapter, the voltage stability assessment using ML technique's voltage stability assessment has been studied. As mentioned above, classic methods utilize CPF to calculate VSM and determine whether the corresponding operating point (OP) is stable or not. The obtained results using analytical methods are deterministic, and CPF is highly computational and can't be applied for online assessment. On the other hand, ML techniques are widely applied in voltage stability assessment according to their potential to train a model for complex systems. ML techniques have a major drawback: Predictive accuracy is related to input data, which can affect actual performance in practical application. Besides, a variety of ML techniques and several approaches in performance enhancement complicated the implementation of an optimal model. To overcome these problems, an online voltage stability assessment framework using ML techniques is proposed to develop the most appropriate ML model for the online voltage stability assessment based on phasor measurement unit (PMU) data.

9.1.3 Structure of Chapter

The rest of the proposed chapter is presented as follows: The problem statement of static voltage stability and some mathematical preliminaries are introduced in Sect. 9.2. Section 9.3 explains the details of the proposed framework for voltage stability assessment. Section 9.4 presents the numerical results of case studies. Finally, the conclusion appears in Sect. 9.5.

9.2 Voltage Stability Problem Statement

The steady-state power system model can be formulated in (9.1). Where x is system state-variable, which is formed by a vector of power grid voltage magnitudes and phase angles. And λ is the loading factor related to load and generator power as follows [11]:

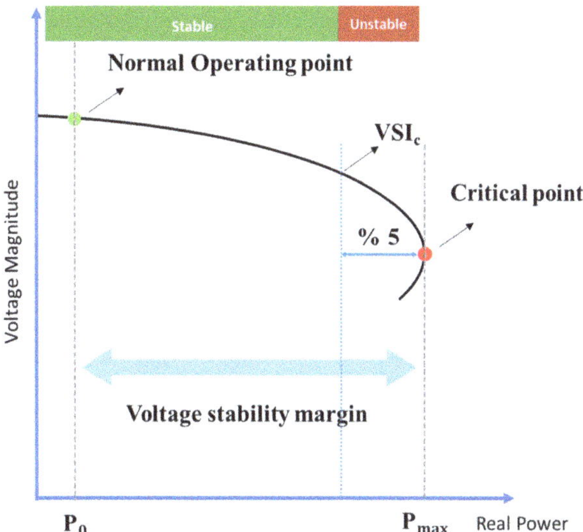

Fig. 9.1 P-V curve for an operation state

$$F(x, \lambda) = 0 \tag{9.1}$$

$$P_L = P_L^0 + \lambda P_L^d \tag{9.2}$$

$$Q_L = Q_L^0 + \lambda Q_L^d \tag{9.3}$$

$$P_G = P_G^0 + \lambda P_G^d \tag{9.4}$$

In the above equation, P_L^0, Q_L^0, and P_G^0 are the base load and generator powers, whereas P_L^d, Q_L^d, and P_G^d are the load and generator power increasing directions. In practice, P_L^d, Q_L^d, and P_G^d can be obtained from load forecasting and generation dispatch, respectively [44].

Figure 9.1 represents a P-V curve for the specific operating condition. P-V curves are employed to visually illustrate VSM, which corresponds to the distance from the current OP to the critical point. For each OC, a P-V curve can be obtained using CPF. The CPF program calculates the maximum loading level starting from a specified initial operating point. The VSM is calculated through (9.5); therefore, the voltage stability index (VSI) can be defined as the percentage of the maximum loading level at the critical point.

$$VSM = P_{max} - P_0 \tag{9.5}$$

$$VSI = \frac{VSM}{P_{max}} \tag{9.6}$$

In the proposed study, the long-term voltage stability margin for the power system is defined using the voltage stability criterion [48]. According to the WECC criteria, the voltage stability index criteria is set at 7% for normal operating conditions, while s is set at 5% for single-contingency conditions. Subsequently, each operating point is labeled as "*Stable*" (S) if $VSI < VSI_c$; otherwise, it is labeled as "*Unstable*" (U).

9.3 The Proposed Framework

This chapter proposes a novel methodology in the online voltage stability assessment using the data mining and machine learning approach. The proposed methodology demonstrates the performance and effectiveness of ML techniques in the online voltage stability assessment. As shown in Fig. 9.2, the framework involves 4 stages: (1) database generation, (2) ML techniques training, (3) performance evaluation, and (4) online application. The OCs are characterized using nominated topology scenarios and load variation patterns (e.g., residential, industrial, or agricultural load) to cover contingency situations and various load types. Various

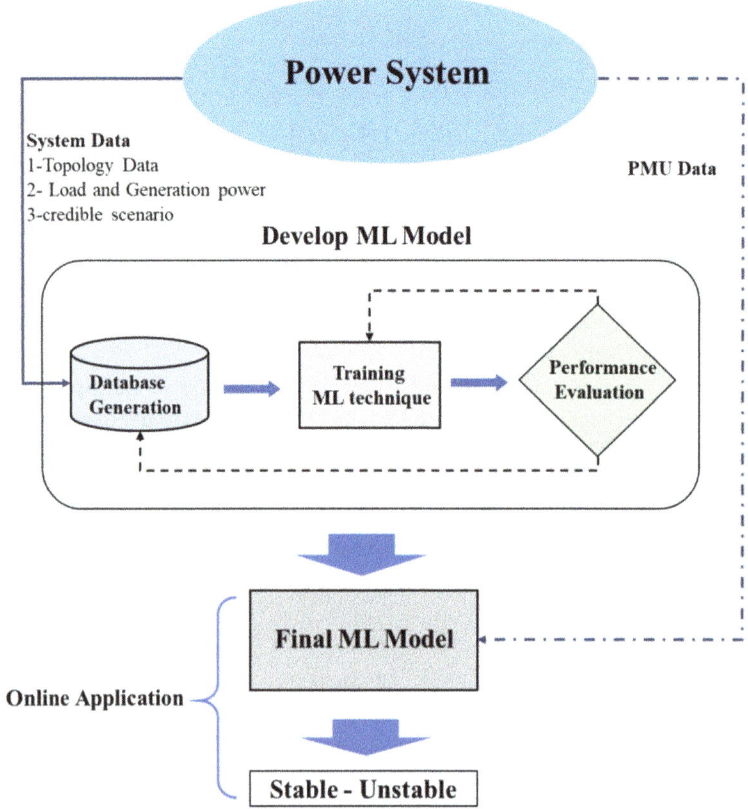

Fig. 9.2 Flowchart of the proposed framework

predictor sets are obtained using feature selection and feature extraction methods based on power system variables. Furthermore, an importance sampling approach is employed to enhance the speed and the accuracy of training. The utilized ML technique is optimized using hyperparameters tuning based on the grid search. Several metrics are conducted to evaluate the performance of the implemented model. In addition to numerical metrics, the confusion matrix is also employed to visually describe the classification model's performance. Also, the PMU measurements infected with noisy and missing data are employed to assess the robustness of the framework in the presence of PMU uncertainty. It is remarkable to mention that all ML-implemented models validated using the K-Fold cross-validation. These stages are explained in detail in the following sections.

9.3.1 Database Generation

ML techniques employ datasets to build a model and predict results. Therefore, achieving a comprehensive database is crucial to obtain an efficient and accurate model. Either historical data or synthetic data can be applied to build a database of the power system OPs for ML applications. Notwithstanding the historical data involving actual information of the power system, it can be lacked details to represent system status, precisely. Therefore, the synthetic data based on prior knowledge can be employed to obtain a sufficient database. In the proposed framework, various topology scenarios, load variation patterns, and the random sampling method are applied to generate an eligible database, which spots all possible operating points. Various combinations of the power system's primary variables (V, δ, P, Q) are considered predictors to indicate the power system's state in each OP. Eventually, importance sampling is employed to remove useless samples and enhance training performance.

9.3.1.1 Sampling

As illustrated in Fig. 9.3, database generation approach consists of three loops to obtain required samples for training:

1. **Load Variation Patterns:** There are various load types in realistic power systems that each type is indicated with specified increasing characteristics. Therefore, loads of the whole system can't increase at the same rate. To address this issue, the system is divided into some areas to define the load variation pattern [44]. In each load variation pattern, the loads in different areas increase at different rates, and all the generators balance the incremental load. The dividing approach can be based on the similarity of load types (e.g., residential and industrial) or the same geographical region.

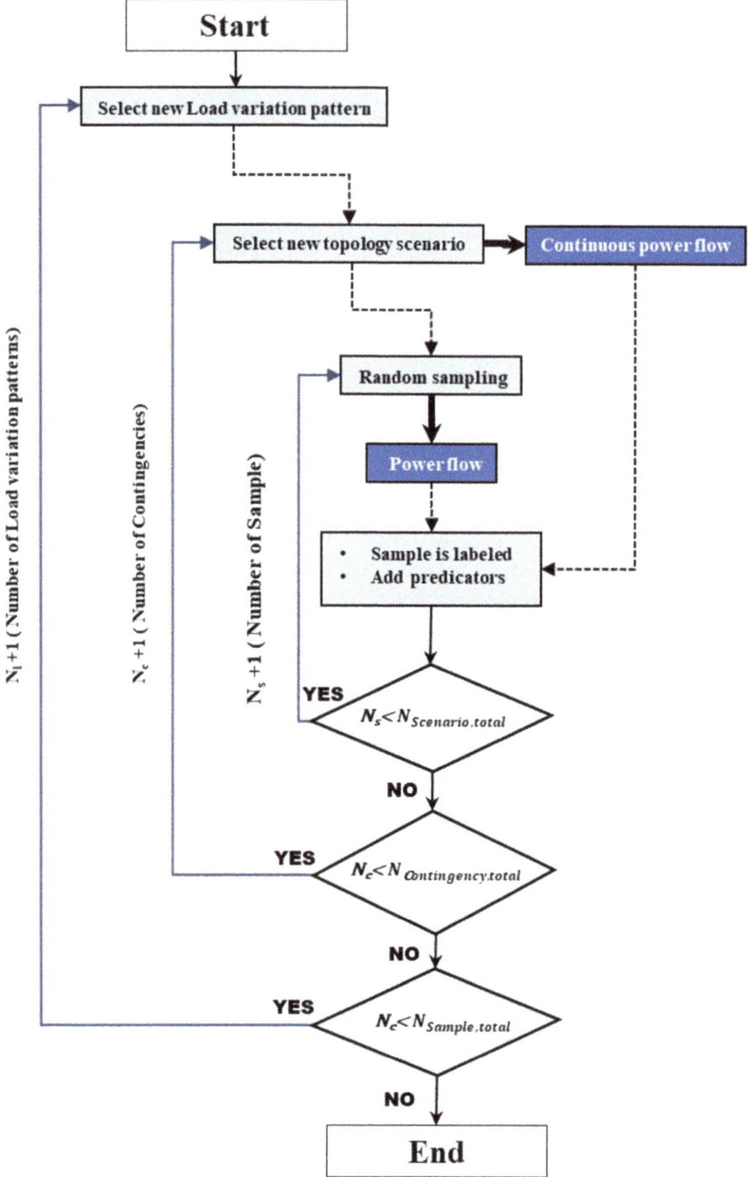

Fig. 9.3 Proposed approach for generating samples

2. **Topology Scenarios:** To contemplate both normal and contingency situations, single transmission lineouts are considered as a contingency. In contingencies that power flow never converge is neglected. Hence, the number of total topology scenarios and the overall number of OCs can be calculated as:

$$N_{scenario,total} = \left(N_{line-out,total} - N_{line-out,deverged}\right) + 1 \tag{9.7}$$

$$N_{OC,total} = N_{scenario,total} \times N_{load\ pattern} \tag{9.8}$$

3. **Random Sampling: For determined OCs in the last steps, CPF is conducted to calculate VSM. Furthermore, to obtain a more extensive database, a random sampling approach is applied for each OC. Final load and generation variation can be represented as:**

$$P_{L,i}^{j} = \left(1 + \lambda K_{L,i}\left(1 + k_j\right)\right) P_{L0,i}\ 1 < i < N_{Area}, 1 < j < N_{random} \tag{9.9}$$

$$Q_{L,i}^{j} = \left(1 + \lambda K_{L,i}\left(1 + k_j\right)\right) Q_{L0,i}\ 1 < i < N_{Area}, 1 < j < N_{random} \tag{9.10}$$

$$P_{G,i}^{j} = \left(1 + \lambda K_{G}^{j}\right) P_{G0,i}\ 1 < j < N_{random} \tag{9.11}$$

$$K_{G}^{j} = \frac{\sum\limits_{i \in \{Area\}} K_{L,i}\left(1 + k_j\right) P_{L0,i}}{\sum\limits_{n \in \{PV\}} P_{G0,n}} \tag{9.12}$$

where $P_{L,i}^{j}$ and $Q_{L,i}^{j}$ are the load active power and reactive power in each O. Also, $P_{G,i}^{j}$ is the generated active power, respectively. $K_{L,\ i}$ is indicate the load variation coefficient for each area, and k_j is a random number. K_{G}^{j} is a factor that determines the rate of increase in the production in generators. Finally, for each generated sample, the input variable X is constructed by combining various power system variables such as line-injected power or voltage of bus that are obtained from power flow calculation. Also, the target variable is labeled as "*Stable*" or "*Unstable*" according to WECC criteria for corresponding OC mentioned in Sect 9.2.

9.3.1.2 Importance Sampling

Importance sampling (IM) is a useful technique to shrink the database without the risk of losing information [44, 49]. Also, this technique reduces training time and improves accuracy. In the proposed framework, IM is employed to remove extremely heavy or light-loaded operating points and keep samples within the vicinity of VSM. To discriminate samples more comprehensively, the boundary for each topology scenario is calculated distinctly as:

$$P_{L,\ min}^{i} < P_{L,total}^{i} < P_{L,\ max}^{i}\ 1 < i < N_{scenario,total} \tag{9.13}$$

$$P^i_{L,max} = (1 + L_{max})VSM_{i,max} \; 1 < i < N_{scenario,total} \tag{9.14}$$

$$P^i_{L,min} = (1 + L_{min})VSM_{i,min} \; 1 < i < N_{scenario,total} \tag{9.15}$$

where L_{min} and L_{max} is criteria for the minimum and maximum limit of the boundary, respectively. Further, $VSM_{i,\ max}$ is the highest, and $VSM_{i,\ min}$ is the lowest stability margin for the determined topology scenario.

9.3.2 Training the Machine Learning Technique

In order to build an optimal model, the ML technique must be chosen carefully with regard to state of the problem. The online voltage stability assessment is a classification problem using supervised learning to determine whether system status is stable or unstable. As discussed previously, SVM, DT, and ensemble methods are capable techniques to apply in the online voltage stability assessment. Even though the proposed framework was adopted to utilize various techniques to train data, particularly, the ensemble method is employed to implement the ML model in the voltage stability assessment with respect to the impressive performance [44–46]. Furthermore, dimensionality reduction approaches such as feature selection [50, 51] and feature extraction [22, 27, 52, 53] are augmented to model to enhance training efficiency. Eventually, the implemented model can be optimized using hyperparameter tuning to obtain the online voltage stability assessment's most satisfactory performance.

9.3.2.1 Ensemble Method

The ensemble methods aim to combine multiple classification models with a given learning algorithm to improve performance over a single classifier. Various classification models can be utilized to compose the ensemble model. But DT is the base model to combine and construct a set of classifiers generally. Learning algorithms are applied to ensemble methods categorized into two major groups, the first known as averaging methods such as bagging trees and random forest. In these methods, several classifiers are combined independently, and the final result is obtained through averaging single classifiers' predictions. The second group is called boosting. The boosting methods compose classifiers sequentially to combine the outputs from the weak learner and create a powerful ensemble learner that eventually enhances the base learner's performance. Bagged tree and AdaBoost are considered as primary techniques utilized in the proposed framework. Details of which are given in the following subsections.

Bagged Tree

Bootstrap aggregating, often abbreviated as "*bagging*" is a type of ensemble learning algorithm employed to reduce the error generated by small fluctuations in the dataset and defined as the variance. To bag a weak learner such as a DT on a dataset, multiple bootstrapped subsamples of the dataset are generated. A DT is trained on each bootstrapped subsample. After subsamples are trained using DT, a voting algorithm is applied to aggregate DTs and form the most efficient classifier. Besides, independent DTs can be trained using parallel processing that improves speed significantly. As shown in Fig. 9.4, bootstrapping conducted random selection with a replacement, which means a sample can appear multiple times for each subset. Therefore, some of the observations can be excluded from the subset called "*out of bag*" samples. Usually, the voting approach is applied using equal weights although in online applications, weights can be updated instead of training a new model. Random forest (RF) is a bagging technique that combines the random decision trees with the bagging algorithm to achieve high-classification accuracy. As mentioned, the bagging technique reduces variance and can avoid overfitting issues. Thus, this technique is an appropriate choice for applying higher dimensionality data. Also, it can maintain performance for missing data.

AdaBoost

Boosting methods work in the same way as bagging methods: Combine a group of aggregated models to obtain an ensemble model with better performance. However, unlike bagging, which aims at reducing variance and avoiding overfitting, the boosting algorithm's purpose is to reduce bias. Boosting, like bagging, can be used for r both regression and classification applications. Adaptive boosting (AdaBoost) and gradient boosting are well known as boosting algorithms. AdaBoost involves training a sequence of weak learners on repeatedly modified versions of the

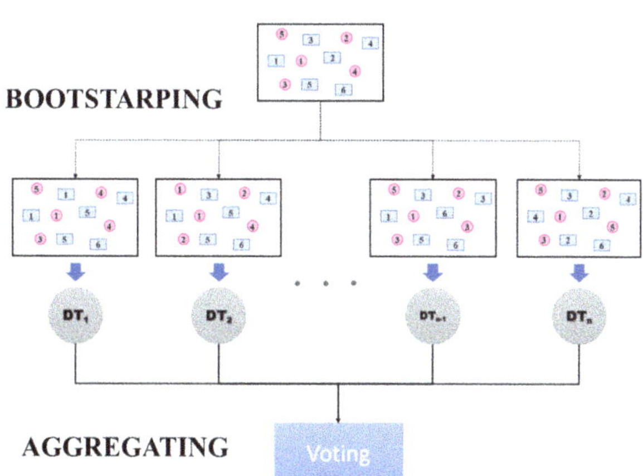

Fig. 9.4 Bootstrap-aggregating algorithms for training multiple DT

data. Further, all DTs' predictions are combined using a weighted voting approach to produce the final prediction. Eventually, an ensemble learner with lower bias and even better variance is obtained. The main procedure of the AdaBoost algorithm considering a DT as the base classifiers can be represented in major steps as follows [44]:

- Step 1) Input data:

1-. Load training set with *N* samples.
2-. Determination of the boosting iteration *T* and base DT specification.
3-. Training samples weight set equally as

$$w_n = \frac{1}{N} \tag{9.16}$$

- Step 2) Boosting approach:

1-. Start first iteration: $t = 1$.
2-. Fit DT using the initial training set.
3-. Calculate the learning error as

$$\varepsilon^t = \sum_{n=1}^{N} w_n^t \mathrm{I}(DT^t(X_n) \neq y_n) \tag{9.17}$$

where ε^t indicates the weighted classification error, and I is a logical function that returns *1* when the corresponding condition was satisfied. Otherwise, it's 0.

4-. Calculate the weight for the trained DT

$$\alpha^t = \frac{1}{2}\ log\left(\frac{1-\varepsilon^t}{\varepsilon^t}\right) \tag{9.18}$$

5-. Check iteration $t < T$.
6-. Update the weight of samples.

$$w_n^{t+1} = \frac{w_n^t}{Z_t} \exp\left(\theta_n \alpha^t\right) \tag{9.19}$$

where θ_n is an indicator function that is $\theta_n = -1$ when $DT^t(X_n) = y_n$ to decrease correctly classified samples weights and $\theta_n = 1$ when $DT^t(X_n) \neq y_n$ to increase correctly misclassified samples weight.

7-. Go to next iteration $t = t + 1$.

- Step 3) Build the final model:

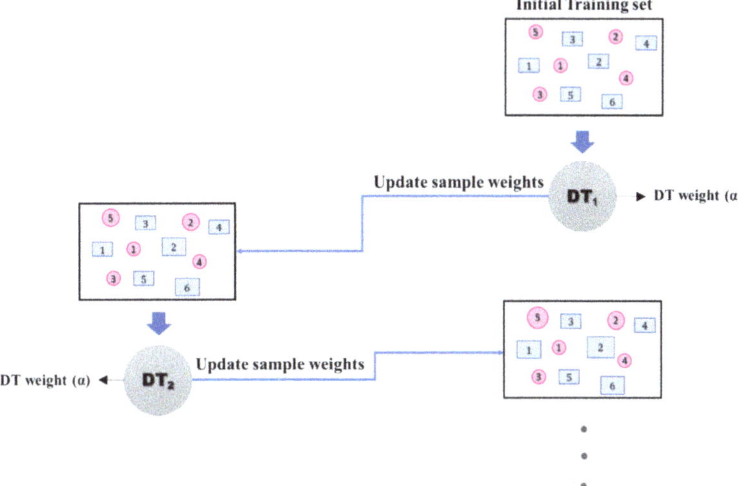

Fig. 9.5 Booting Algorithm for update weights

After the boosting iterations are completed, based on calculated weights, an ensemble model is obtained. Finally, the AdaBoost model computes prediction for new data using:

$$DT^{ensemble}(X_{new}) = \sum_{t=1}^{T} \alpha^t DT^t(X_{new}) \tag{9.20}$$

As shown in Fig. 9.5, when a DT misclassifies a sample in the sequence. Corresponding weight is increased so that the next DT take more efforts to predict inconvenient sample correctly. Weak models that are only slightly better than random guessing is often employed for boosting to reduce the bias error, such as small decision trees. Another important motivation to use weak classifiers with low variance for boosting is that fitting these models requires less computational effort. Indeed, as the boosting algorithm can't be done in parallel (unlike bagging), it could become time-consuming to obtain the final model using complex learners. Another drawback of using AdaBoost is the risk of overfitting according to higher variance.

9.3.2.2 Dimensionality Reduction

Fitting a predictive model to the dataset with numerous features is computationally expensive. Also, inattentive hyperparameter tuning increases the risk of overfitting. Dimensionality reduction is a useful technique that can be employed in the ML model to transform training data from a high-dimensional space into a low-dimensional space so that the low-dimensional representation keeps important essential information of the original data. According to the power system, it is a large-scale and complex system, and the power system's state can be represented by various features and measurements, so the voltage stability assessment is considered a high-dimensional space problem. Besides, utilizing PMU wide-area measurements in the voltage stability assessment requires online frameworks with desirable speed and accuracy. Consequently, augmenting dimensionality reduction techniques to ML-based framework is crucial to reduce training data size, speed up the calculation, improve accuracy, and avoid overfitting. Dimensionality reduction approaches can also be divided into feature selection and feature extraction.

Feature Selection

Feature selection is the process of identifying and selecting relevant features to the target variable. Generally, feature selection employed an evaluation process that scores the features. Further, a filtering criterion is applied to pick the most relevant feature for the ML technique input predictors. There are various methods for scoring features, and these methods are divided into three main approaches: knowledge-based, ML-based, and statistical-based. In the knowledge-based approach, practical experiences and technical knowledge are employed to indicate the most important features associated with targets. In the ML-based approach, after the model is trained, the relative rank represented by the model is utilized to arrange features concerning the target values' predictive ability. Also, the statistical-based methods select variables regardless of the model and prior knowledge. These methods use a statistical test such as correlation to determine the most interesting features. Univariate feature selection is a statistical-based method that scores each feature individually to evaluate the feature's relevance with the response variable. In the proposed framework, all three approaches are applied as a preprocessing module to choose optimal features set.

Feature Extraction

Unlike feature selection, feature extraction rebuilds whole new features intended to be informative as original features. Brand-new features have reduced the dimensionality of input data and also remove redundant features. Thus, training ML techniques using extracted features improve the classification performance. Extracted features can't be as interpretable as original features, so they can't be applied to corresponding preventive control. Besides, training ML techniques using feature extraction diminish the robustness of the model for noisy or missing data. There are various feature extraction algorithms such as principal component analysis (PCA), linear discriminant analysis (LDA), and autoencoder. The proposed framework uses PCA as a feature extraction method to reconstruct new and enhanced predictors.

The principal component analysis is defined as a linear transformation that maps data to a lower-dimensional space to maximize the variance of the data. Low-dimensional space is a new coordinated system in which each axis represents one of the principal components. The covariance matrix of the data is calculated to perform PCA, and then eigenvalues and corresponding eigenvectors of this covariance matrix are obtained. The eigenvectors that correspond to the largest eigenvalues can be interpreted as the principal components. Finally, these eigenvectors are applied to obtain new features set with a large fraction of the variance. Transformation of original data to low-dimensional space is performed as [54, 55]

$$t_i^k = X_i\, W^k \; k = 1, 2, \ldots L \; i = 1, 2, \ldots N \tag{9.21}$$

where L is the number of extracted space dimensional, and N is the number of input data dimensional, respectively. X is input features data that is subtracted the mean of each feature. W is vectors of coefficients to map data, and t is the principal component scores representing extracted features.

9.3.2.3 Hyperparameter's Tuning

A hyperparameter is a parameter of the ML model whose value is used to control the learning process such as the number of trees in the bagged tree technique. Hyperparameter tuning is an approach that finds the optimal ML model with respect to input data and the state of the problem. In the proposed framework, hyperparameter tuning is utilized to optimize ML model's performance using grid search. Grid search is a simple approach to find the best values for hyperparameters. It works by searching exhaustively through a determined boundary for hyperparameters.

9.3.3 Performance Evaluation

Performance evaluation of the implemented ML model is crucial to determine the model's efficiency in the online voltage stability assessment using wide-area measurements. However, accuracy and CPU time are employed as primary metrics in performance evaluation. They can't represent comprehensive detail of training performance. When training data are distributed unbalanced, the accuracy is inadequate to demonstrate classification performance. The training set for voltage stability assessment may be imbalanced according to the number of stable operating points in the power system that is much more than the number of unstable points. Therefore, a new assessment approach is necessary to be employed in the ML model evaluation. The confusion matrix useful tool is employed to demonstrate the performance of classification.

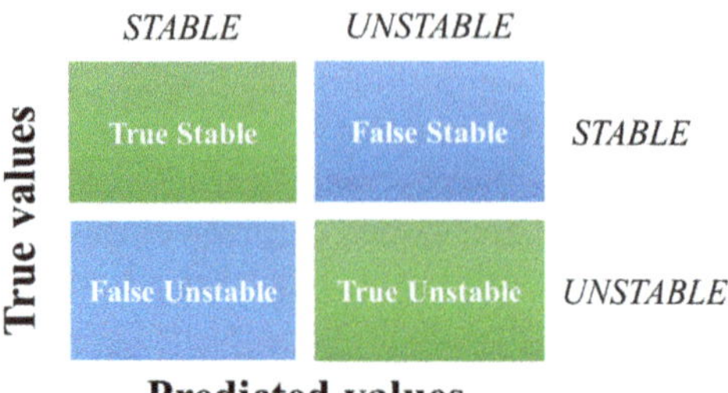

Fig. 9.6 Confusion matrix for the voltage stability assessment problem

As illustrated in Fig. 9.6, confusion matrix for voltage stability assessment is 2*2 table, in which table rows represent true labels, and columns represent predicted labels. Each cell of the table demonstrates a portion of data based on the corresponding labels. The confusion matrix describes the classification performance visually. Also, it can be employed to calculate several measures as [44, 56]

$$Accuracy(Accu) = \frac{N_{TrueStable} + N_{TrueUnstable}}{N_{Stable} + N_{Unstable}} \tag{9.22}$$

$$Misdetection(Mis) = \frac{N_{FalseStable}}{N_{Stable} + N_{Unstable}} \tag{9.23}$$

$$Fasle\ alarm(fal) = \frac{N_{FalseUnstable}}{N_{Stable} + N_{Unstable}} \tag{9.24}$$

$$Reliability(Rel) = \frac{N_{TrueUnstable}}{N_{TrueUnstable} + N_{FalseStable}} \tag{9.25}$$

$$Security(Sec) = \frac{N_{TrueStable}}{N_{TrueStable} + N_{FalseUnstable}} \tag{9.26}$$

$$G\ mean = \sqrt{\frac{N_{TrueStable}}{N_{TrueStable} + N_{FalseUnstable}} \cdot \frac{N_{TrueUnstable}}{N_{TrueUnstable} + N_{FalseStable}}} \tag{9.27}$$

Misdetection (Mis) represents the ratio of misclassified unstable cases, while false-alarm (Fal) indicates the proportion of stable samples predicted unstable. Besides, security and reliability stand for the proportion of secure cases that are correctly identified and the proportion of insecure cases correctly identified. G-mean (G) acts as a geometric mean of the security and reliability that evaluates learning's overall performance. Subsequently, K-Fold cross-validation is applied in the performance evaluation process to assess the effectiveness model with respect to an anonymous dataset. K-Fold cross-validation is a validation procedure that utilizes random splitting data to guarantee every sample of the original dataset has the chance of involving in training and test set.

9.3.3.1 PMU Uncertainly

As PMU measurements can be infected with noise or even missing, the ML model's performance should evaluate in noisy and missing data conditions to determine the robustness and efficiency. Therefore, two scenarios are considered for noisy data, in which the first scenario involves noise to all input data, and in the second scenario, noise is only added to the test data. Further, to produce synthetic noisy data, total vector error (TVE) criteria are employed. According to the IEEE standard [57], the TVE criteria for PMU data uncertainly should be under 1%. The TVE for the voltage measurements is represented in (9.28). For the missing data scenario, some PMUs are picked randomly, and then corresponding measurements in the testing samples for selected PMU are set to NaN.

$$TVE = \sqrt{\frac{\left|V_{real,\text{measured}} - V_{real,\text{ideal}}\right|^2 + \left|V_{imag,\text{measured}} - \mathrm{V}_{imag,\text{ideal}}\right|^2}{V_{real,\text{ideal}}{}^2 + V_{imag,\text{ideal}}{}^2}} \tag{9.28}$$

9.3.4 *Online Application*

In the last stage, after the developed model was approved in terms of accuracy, speed, and robustness using the performance evaluation process, the final model can be employed in online voltage stability assessment based on wide-area PMU measurements.

9.4 Simulations and Results

The proposed framework has been tested on two different-size case study. The first case is the new England 39-bus test system [58], and the second one is a bigger grid, the IEEE 118-bus test system [59]. In the case studies, the simulation programs including database generation, model training, and testing are performed within MATLAB® environment. Furthermore, all the experiments are executed on a system with Intel® Core™ i7 2.6 GHz processor and 16 GB of RAM.

9.4.1 IEEE 39-Bus Test System

9.4.1.1 Database Generation

This system involves 46 lines and 10 generators. In order to generate the database, the grid is divided into three areas based on the topographical location of each bus. Further, for each area, different confidences are picked to construct 20 load variation patterns. The buses included in each area, along with some instances of the load variation pattern, are presented in Table 9.1. Besides, in the topology scenario's definition, line 13 outage is excluded from contingencies according to power flow that can't converge in this scenario, and 45 + 1 topology scenarios are considered. The database is eventually generated using the proposed sampling approach, which involves 8155 stable samples and 7492 unstable. Figure 9.7 illustrates the distribution of the samples based on voltage magnitude.

Table 9.1 Existing buses in each area along sample coefficients

Area 1 buses	Area 2 buses	Area 3 buses
1–13	14–26	27–39
K area 1	**K area 2**	K area 3
0	0	0
0	0	0.2
0	0.2	0
0.2	0	0

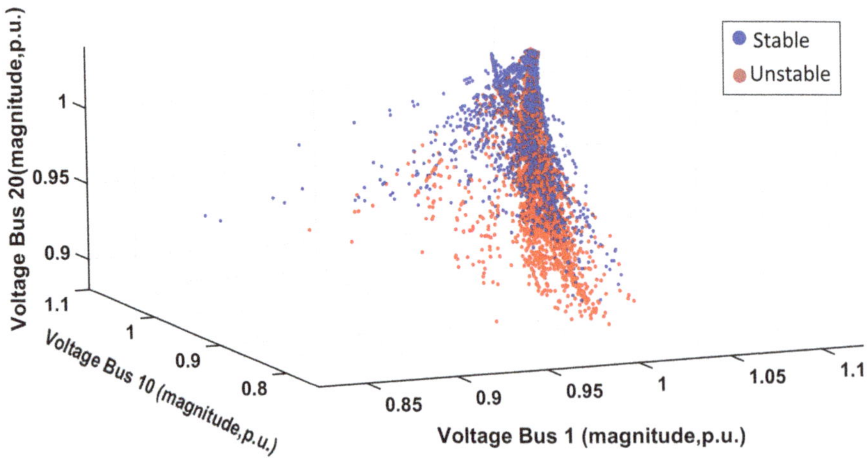

Fig. 9.7 Dataset sample distribution based on the voltage magnitude

9.4.1.2 Simple ML Model Training

In order to illustrate the effectiveness of the proposed framework, a simple ML model is essentially employed. This model employed the bagged tree techniques for training and voltage magnitude of buses as predictors. The bagged tree model involves 30 DTs with a depth of 15,646. The dataset was randomly split into two subsets, where the training set includes 80% of the data and the remaining 20% allocated to the test set. After training the ML model using the training set, the testing set's classification accuracy is evaluated as 80.70%. The predicted label's distribution for the testing set samples is shown in Fig. 9.8.

9.4.1.3 Optimal Dataset for Training

To represent the state of each sample, four combinations of the predictors are chosen based on the power system primary variables as follows:

$\boldsymbol{V_m}$: Represent the voltage magnitude of buses

$\boldsymbol{V_{m\text{-}\delta}}$: Represent the voltage magnitude and phase angle of buses

$\boldsymbol{P_{active\text{-}reactive}}$: Represent the injected active and reactive power of lines

$\boldsymbol{V_{m\text{-}\delta}\text{-}P_{active\text{-}reactive}}$: Represent the voltage magnitude and phase angle of buses and the injected active and reactive power of lines together

The importance sampling approach is applied to improve the classification performance by removing irrelevant samples. The dataset size is reduced by 41%, and the new dataset involves 4252 stable samples and 5260 unstable samples. Also, random selecting is conducted to construct a dataset with the same size as the reduced dataset and compare it with it. Four combinations of the predictors along three types of samples set are utilized to form 12 different training sets presented in

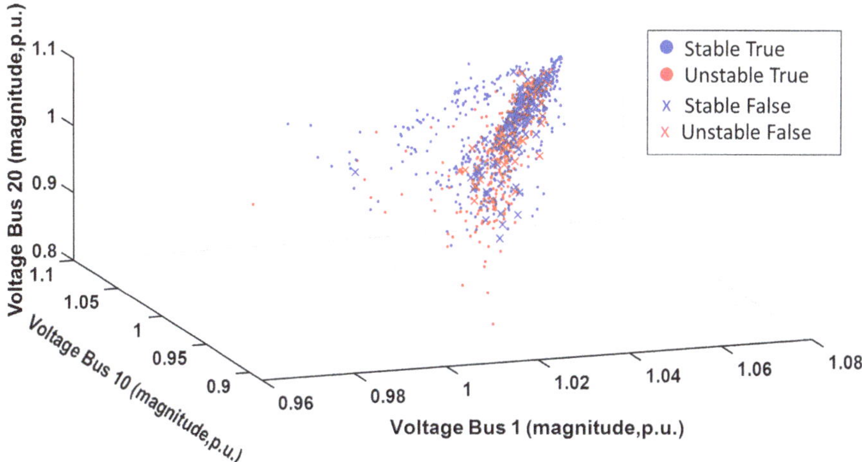

Fig. 9.8 Distribution of the predicted labels

Table 9.2 The number of samples and predicators for each dataset

Dataset	V_m		$V_{m\text{-}\delta}$		$P_{active\text{-}reactive}$		$P_{active\text{-}reactive}$	
	Predictor	Sample	Predictor	Sample	Predictor	Sample	Predictor	Sample
Original dataset	39	15,647	78	15,647	92	15,647	178	15,647
Importance sampling	39	9123	78	9123	92	9123	178	9123
Random selecting	39	9123	78	9123	92	9123	178	9123

Table 9.2. The performance of training various datasets by the bagged tree in the previous section is demonstrated in Table 9.3. Table 9.3 indicates that a larger set of predictors may give better performance. Also, reducing the dataset with the importance sampling approach will decrease computational time without significantly improving accuracy.

9.4.1.4 ML Techniques Performance

As previously discussed, DT and SVM are powerful ML techniques employed in the voltage stability assessment frequently. Besides, the bagged tree and the AdaBoost are presented as ensemble methods that combine several single models to improve classification performance. To compare these techniques and choose the finest ML technique, four different models are employed. The specification of these ML models is summarized in Table 9.4. Table 9.5 demonstrates the performance of training models using the original dataset. The superiority bagged tree technique is obvious in terms of accuracy and computational time. The DT is a fast technique but suffers a lack of classification accuracy. Further, SVM presents the desired accuracy, but it is heavily computational and may not be appropriate in the online voltage stability assessment. AdaBoost, due to its higher variance overfitted, the performance of classification is corrupted.

9.4.1.5 Dimensionally Reduction

In the proposed framework, three different approaches for feature selection techniques are conducted as follows:

Knowledge-base: In this approach, buses with the most voltage deviation and lines with the most loading are chosen as critical elements of the power system. Further, the most important input features are selected based on critical element variables.

ML-based: The bagged tree model trained in the previous section is employed to score features concerning the predictor importance that is evaluated by the bagged tree.

Statistical-base: In this approach, the relevance of the features with the target variable is evaluated using the chi-square test.

Figure 9.9 represents feature scores that are evaluated using the ML-based and statistical-based approaches. Also, PCA is employed to rebuild new input features with higher variance. As shown in Fig. 9.10, the distribution of samples using principal components has a greater distinction than the distribution based on selected variables. For a better comparison of the different methods, the new feature set's size is considered equal to $\frac{2}{3}$ of the number of original features. Therefore, 25,50,60 and 110 features are picked up for the V_m, $Vm\text{-}\delta$, $P_{active\text{-}reactive}$, and $Vm\text{-}\delta,\text{-}P_{active\text{-}reactive}$ predictors set, respectively. The performance of various dimension reduction

Table 9.3 Classification accuracies and training CPU time for each dataset

Dataset	V_m		$V_{m\text{-}\delta}$		$P_{active\text{-}reactive}$		$P_{active\text{-}reactive}$	
	Accuracy (%)	Time (s)	Accuracy (%)	Time (s)	Accuracy (%)	Time (s)	Accuracy (%)	Time (s)
Original dataset	80.70	65	85.36	59.75	95.16	64.53	94.87	80.84
Importance sampling	80	31.35	85	40.03	95.5	30.79	95	49
Random selecting	78	39.46	82	38.79	92	43.28	91	52.85

Table 9.4 Specification of employed ML models

ML technique	Specification		
DT	**Algorithm**	**Max number of splits**	**Split criterion**
	Cart	100	Gdi
SVM	**Kernel**	**Solver**	
	Radial basis function	Iterative single data algorithm	
AdaBoost	**Learning rate**	**Number of trees**	**Max number of splits**
	0.1	30	30
Bagged tree	**Learning rate**	**Number of trees**	**Max number of splits**
	0.1	30	15,646

methods is demonstrated in Table 9.6. From the results shown in Table 9.6, it is observed that the training speed of all proposed methods is improved according to input predictors reduction. Further, there is a minor downgrade in classification accuracy for knowledge-based and statistical-based methods. The results show that PCA proposes the best dimensionally reduction method, which enhances speed and accuracy simultaneously.

9.4.1.6 Hyperparameters Tuning

All trained models in the previous section are optimized using hyperparameters tuning. The tuning approach is applied with the grid search to find the best parameters for the model. Figure 9.11 illustrates the hyperparameters searching process for the bagged tree with the V_m predictors that involve three primary parameters: the number of DT, the max depth in each DT, and the number of principal components as the input features. Furthermore, a trade-off between speed and accuracy is required to pick optimum values with respect to the problem state. The chosen optimal values for the bagged three parameters are indicated with stars in Fig. 9.11.

9.4.1.7 Performance Evaluation

As mentioned in Sect. 9.3.3, the confusion matrix is an effective tool to demonstrate the classification performance. The confusion matrix for the DT and the bagged tree, which are trained using the *Vm* predictors set, is shown in Fig. 9.11. Further, calculated measures are based on the confusion matrix presented in Table 9.7. The table clearly shows that the bagged tree has better classification performance than the DT, especially in classification reliability, which means the bagged tree can separate stable samples from unstable samples more accurately (Fig. 9.12).

In addition to confusion matrix-based metrics, the robustness of the proposed approach is also evaluated. For the considered simulation, all buses of the power grid are assumed to be equipped with PMU. Therefore, the Vm-δ predictors set is considered as the input features. Let p be the percentage of the PMUs with noisy

Table 9.5 Classification accuracy and training CPU time for each technique

	V_m		$V_{m\text{-}\delta}$		$P_{active\text{-}reactive}$		$P_{active\text{-}reactive}$	
ML technique	Accuracy (%)	Time (s)	Accuracy (%)	Time (s)	Accuracy (%)	Time (s)	Accuracy (%)	Time (s)
DT	69	5.26	78.06	9.28	83.40	11	83.33	21.31
SVM	80.53	368.6	83.91	404.85	89.39	331.34	89.88	494.56
AdaBoost	69.27	105.57	77.08	219.53	84.37	230.70	87.63	497.82
Bagged tree	80.70	65.10	85.35	59.75	95.16	64.53	96.02	80.84

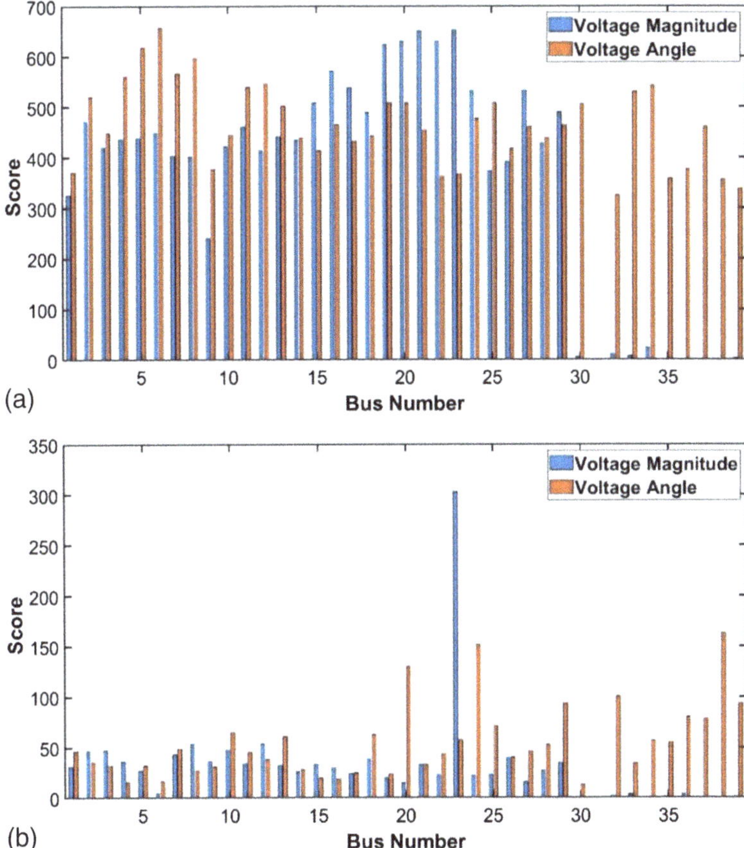

Fig. 9.9 Scores for the voltage-related features. (**a**) Univariate feature selection. (**b**) Predictor importance for bagged tree

data. So synthetic noisy data can be generated based on the PMU TVE criterion and the number of PMUs that are infected with noise. Classification accuracy using the bagged tree for the first scenario in which the noise included in the whole dataset is demonstrated in Table 9.8. Further, results for the second scenario that involves noisy data for the testing set, presented in Table 9.8 and Table 9.9, indicates that adding noise to the whole dataset can improve the robustness of the training ML model. Also, PCA is vulnerable to noise due to the high variance of the new extracted space. It is worth noting that the DT has shown the most robustness against noisy data according to its simple approach for learning.

Training performance in the presence of PMU missing data for given missing PMU ratio is shown in Table 9.10. It can be seen from the table that a high rate of missing data can ruin classification performance.

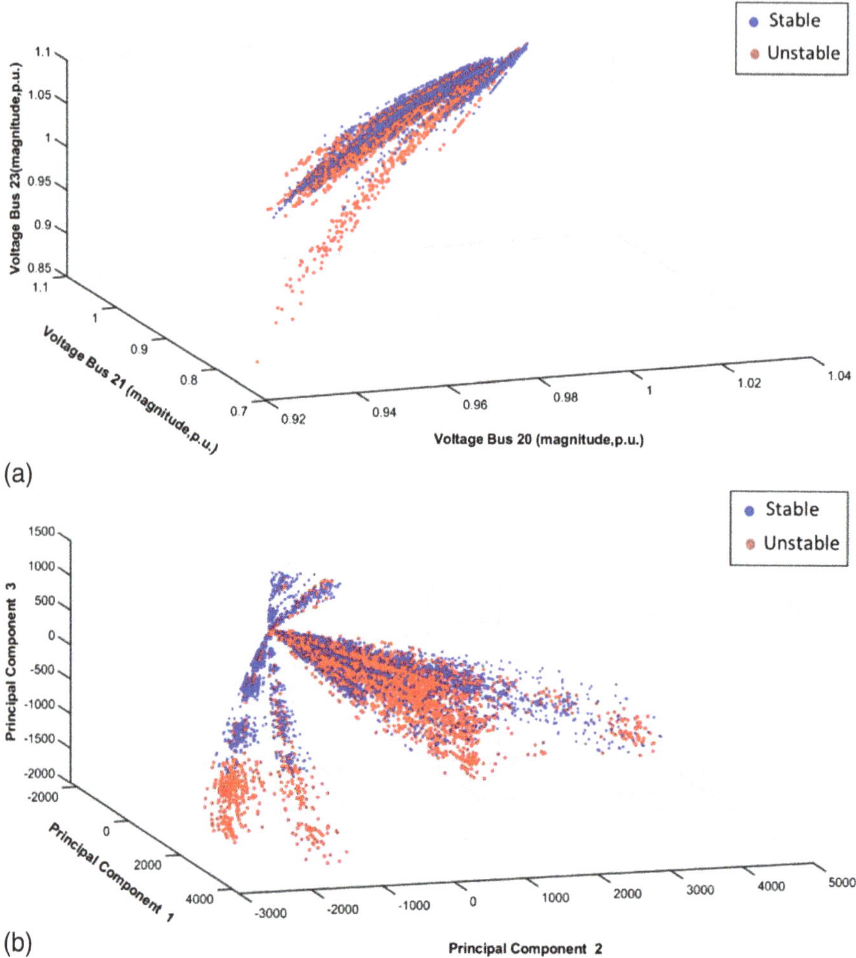

Fig. 9.10 Dataset samples distribution based on (a) selected voltage magnitude and (b) principal component

9.4.1.8 Online Application

In order to demonstrate the effectiveness of the proposed method in online applications, a synthesized 24-hour load curve is considered. The load curve involves 100 district points to represent various electric energy demands during the day. The trained model in previous sections is conducted to online voltage stability assessment based on the daily load curve. As shown in Fig. 9.13, the trained model has been capable to accurately detect instability during peak hours.

Table 9.6 Classification accuracy and training CPU time for each method

ML technique	V_m		$V_{m\text{-}\delta}$		$P_{active\text{-}reactive}$		$P_{active\text{-}reactive}$	
	Accuracy (%)	Time (s)	Accuracy (%)	Time (s)	Accuracy (%)	Time (s)	Accuracy (%)	Time (s)
None	80.70	65.1	85.35	59.75	95.15	64.53	94.87	80.84
Knowledge-base	79.03	48.43	84.69	53.81	94.74	45.37	94.31	60.21
Statistical-base	79.65	52.54	83.83	56.1	93.72	49.62	92.30	64.54
Bagged tree-base	80.95	50.18	86.54	55.79	95.29	46.28	95.52	62.46
PCA	93.12	31.14	94.82	38.84	96.12	36.19	96.02	47.88

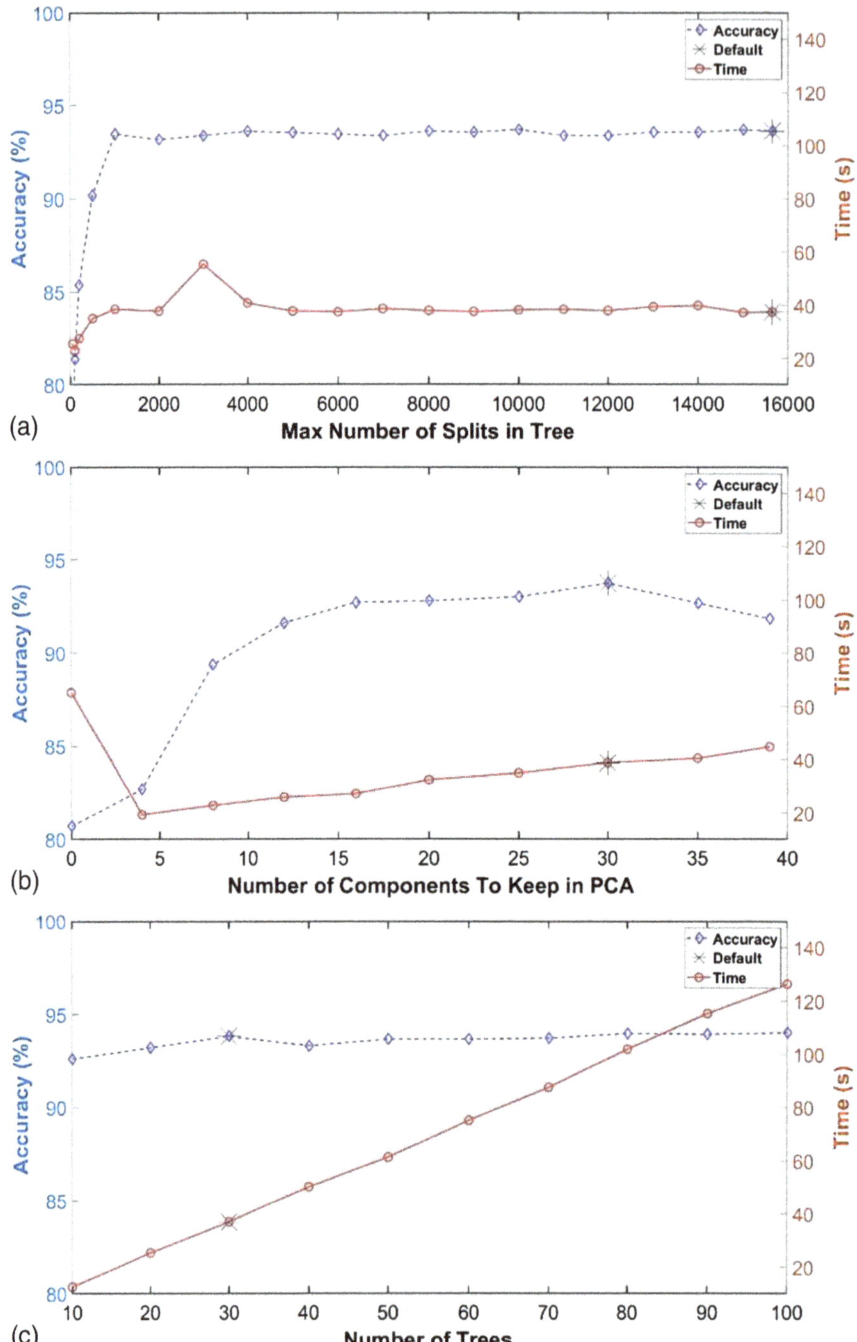

Fig. 9.11 Time and accuracy for corresponding parameter values. (**a**) Max number of splits. (**b**) Number of PCA components. (**c**) Number of trees

Table 9.7 Evaluation measures for each dataset

ML technique	V_m		$V_{m\text{-}\delta}$		$P_{active\text{-}reactive}$		$P_{active\text{-}reactive}$	
	DT	Bagged tree	DT	Bagged tree	DT	Bagged tree	DT	Bagged tree
Accuracy (%)	68.99	80.70	78.06	85.36	83.40	95.16	83.33	94.87
Reliability (%)	57.92	80.31	74.53	84.82	81.57	94.00	82.12	94.03
Security (%)	81.06	81.11	81.90	85.93	85.40	96.38	85.80	95.82
G-mean (%)	68.52	80.71	78.13	85.37	83.46	95.20	83.94	94.91

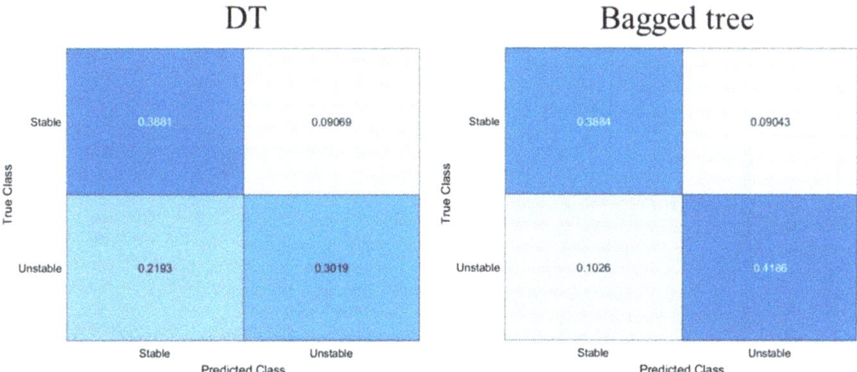

Fig. 9.12 Confusion matrix. (Left) decision tree. (Right) bagged tree

Table 9.8 Accuracy for given noisy PMU ratio (p) – scenario 1

ML technique	Accuracy (%)							
	$p = 0\%$		$p = 10\%$		$p = 50\%$		$p = 90\%$	
	Base	With PCA	Base	With PCA	Base	With PCA	Base	With PCA
Tree	78.06	77.40	78.01	77.78	77.70	76.38	76.65	76.46
SVM	83.91	88.24	81.47	86.48	78.80	85.91	76.40	85.06
AdaBoost	77.08	80.76	77.31	77.45	76.68	76.79	76.35	77.30
Bagged tree	85.35	94.82	85	91.77	83.70	88.28	81.84	86.59

9.4.2 118-Bus Test System

The IEEE 118-bus test case is employed to demonstrate the effectiveness of the proposed framework for a larger power system. The simulation details for this case are summarized in Table 9.12. As shown in Fig. 9.14, the distribution of stable and unstable samples is imbalanced, which means the number of stable OPs is significantly greater than unstable in contract with the case 39-bus. The performance of

Table 9.9 Accuracy for given noisy PMU ratio (p) – scenario 2

ML technique	Accuracy (%)							
	$p = 0\%$		$p = 10\%$		$p = 50\%$		$p = 90\%$	
	Base	With PCA	Base	With PCA	Base	With PCA	Base	With PCA
Tree	78.06	77.40	72.66	65.87	73.69	59.77	72.95	56.95
SVM	83.91	88.24	62.36	79.13	47.82	47.89	47.88	47.74
AdaBoost	77.08	80.76	76.95	76.67	75.59	69.53	74.23	67.72
Bagged tree	85.35	94.82	84.46	75.64	82.31	64.69	79.12	59.03

Table 9.10 Accuracy for given missing PMU ratio (p)

ML technique	Accuracy (%)			
	$p = 0\%$	$p = 10\%$	$p = 25\%$	$p = 50\%$
Tree	78.06	74.33	74.33	66.03
SVM	83.91	52.12	52.12	52.12
AdaBoost	77.08	72.40	72.40	59.76
Bagged tree	85.35	72.29	72.29	58.76

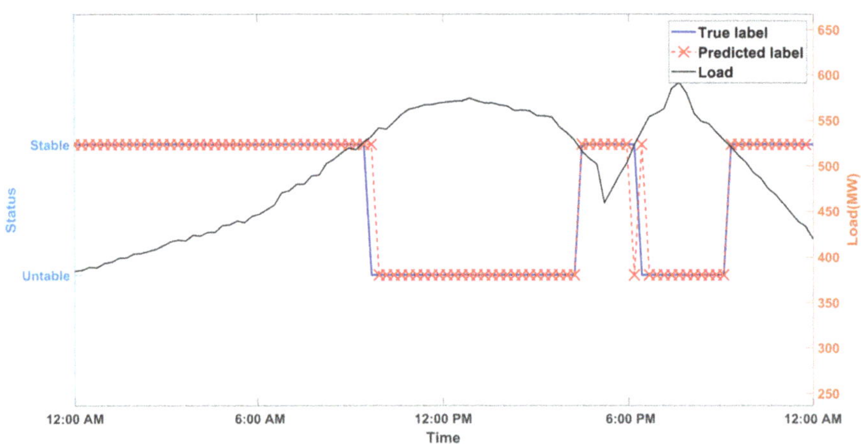

Fig. 9.13 Online voltage stability assessment based on the daily load curve

classification using the bagged tree for the *Vm* predictors set is demonstrated in Table 9.12. From the Table 9.11, it is observed that despite better accuracy than the previous case, it has a lower G-mean due to the unbalanced distribution of samples.

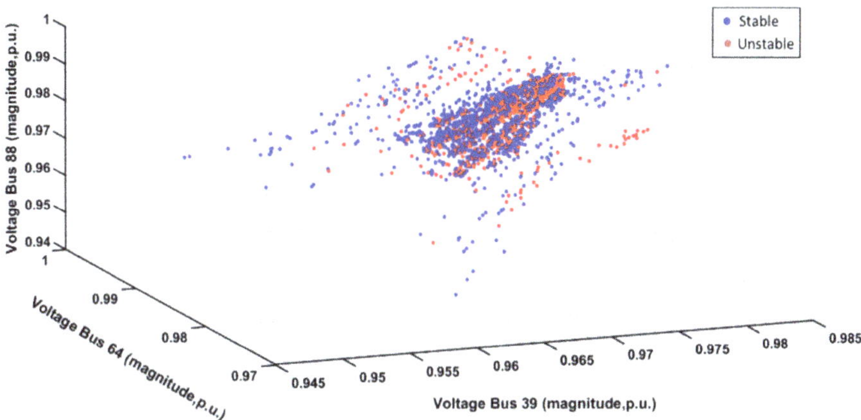

Fig. 9.14 Dataset sample distribution based on the voltage magnitude

Table 9.11 Given details generate 118-bus system database

Specification			
Powe system	**Bus**	**Line**	**Generator**
	118	186	54
Load variation	**Area**	**Load variation pattern**	
	3	12	
Topology scenario	**Contingency**	**Divergent contingencies**	
	186 lineout +1 normal condition	7–9–113-116-133-134-176-177-183-184	
Dataset	**Total samples**	**Stable samples**	**Unstable samples**
	62,135	45,978	16,157

Table 9.12 Evaluation measures for given predictors set

ML technique	V_m		$V_{m\text{-}\delta}$		$P_{active\text{-}reactive}$		$P_{active\text{-}reactive}$	
	Base	With PCA	Base	With PCA	Base	With PCA	Base	With PCA
Accuracy	82.74	91.30	84.81	95.36	93.78	96.50	93.31	96.90
Reliability	60.14	78.91	64.25	89.04	85.89	91.66	86.64	92.63
Security	91.76	96.26	92.48	97.89	97.04	98.44	96.76	98.62
G-mean	74.30	87.15	77.08	93.36	91.30	94.99	90.50	95.56

9.5 Conclusion

A framework for the development of the ML application in the online voltage stability assessment is presented. The framework provides several options to implement in the optimal model. These options can be utilized in the database generation like importance sampling or the training stage, i.e., feature selection, feature extraction, and hyperparameters tuning. The proposed framework was tested on the IEEE 39-bus and IEEE 118-bus test systems. The results of case studies demonstrate that the bagged tree model with PCA presents the best classification performance in terms of accuracy and speed the robustness of the framework in the presence of PMU uncertainty is evaluated. So it indicates that a high rate of noisy or missing data can ruin the voltage stability assessment using the ML model.

Relevant future works aspire to augment the online weighted majority voting in the bagged tree technique. Weighted voting enables online model updating. Therefore, the bagged tree can be adapted to unforeseen changes in system conditions. Also, robust feature extraction approaches can be employed to ensure desirable classification performance in the presence of noise and missing data.

References

1. Begovic, M., *Voltage collapse mitigation.* IEEE Power System Relaying Committee, Working Group K12, 1997. IEEE Publ. No. 93, THO596-7PWR
2. V. Ajjarapu, B. Lee, Bibliography on voltage stability. IEEE Trans. Power Syst. **13**(1), 115–125 (1998)
3. Z. Bo et al., An analysis of previous blackouts in the world: Lessons for China′s power industry. Renew. Sust. Energ. Rev. **42**, 1151–1163 (2015)
4. M.Z. El-Sadek, Preventive measures for voltage collapses and voltage failures in the Egyptian power system. Electr. Power Syst. Res. **44**(3), 203–211 (1998)
5. O.P. Veloza, F. Santamaria, Analysis of major blackouts from 2003 to 2015: Classification of incidents and review of main causes. Electr. J. **29**(7), 42–49 (2016)
6. P. Kessel, H. Glavitsch, Estimating the voltage stability of a power system. IEEE Transactions on Power Delivery **1**(3), 346–354 (1986)
7. A. Tiranuchit, R.J. Thomas, A posturing strategy against voltage instabilities in electric power systems. IEEE Trans. Power Syst. **3**(1), 87–93 (1988)
8. C.W. Taylor, *Power System Voltage Stability*. (McGraw-Hill, 1994)
9. J. Hongjie, Y. Xiaodan, Y. Yixin, An improved voltage stability index and its application. Int. J. Electr. Power Energy Syst. **27**(8), 567–574 (2005)
10. R. Tiwari, K.R. Niazi, V. Gupta, Line collapse proximity index for prediction of voltage collapse in power systems. Int. J. Electr. Power Energy Syst. **41**, 105–111 (2012)
11. V. Ajjarapu, C. Christy, The continuation power flow: A tool for steady state voltage stability analysis. IEEE Trans. Power Syst. **7**(1), 416–423 (1992)
12. C.-Y. Lee, S.-H. Tsai, Y.-K. Wu, A new approach to the assessment of steady-state voltage stability margins using the P– Q– V curve. Int J Elec Power Energ Syst **32**, 1091–1098 (2010)
13. O.A. Alimi, K. Ouahada, A.M. Abu-Mahfouz, A review of machine learning approaches to power system security and stability. IEEE Access **8**, 113512–113531 (2020)
14. M. Amroune, Machine learning techniques applied to on-line voltage stability assessment: A review. Archives of Computational Methods in Engineering (2019)

15. B. Jeyasurya, Artificial neural networks for power system steady-state voltage instability evaluation. Electr. Power Syst. Res. **29**(2), 85–90 (1994)
16. A.A. El-Keib, X. Ma, Application of artificial neural networks in voltage stability assessment. IEEE Trans. Power Syst. **10**(4), 1890–1896 (1995)
17. L.D. Arya, L.S. Titare, D.P. Kothari, Determination of probabilistic risk of voltage collapse using radial basis function (RBF) network. Electr. Power Syst. Res. **76**, 426–434 (2006)
18. D. Devaraj, J.P. Roselyn, R.U. Rani, Artificial neural network model for voltage security based contingency ranking. Appl. Soft Comput. **7**(3), 722–727 (2007)
19. D.Q. Zhou, U.D. Annakkage, A.D. Rajapakse, Online monitoring of voltage stability margin using an artificial neural network. IEEE Trans. Power Syst. **25**(3), 1566–1574 (2010)
20. D. Devaraj, J. Preetha Roselyn, On-line voltage stability assessment using radial basis function network model with reduced input features. Int. J. Electr. Power Energy Syst. **33**(9), 1550–1555 (2011)
21. M. Moghavvemi, S. Yang, ANN application techniques for power system stability estimation. Electric Power Components and Systems **28**, 167–177 (2000)
22. S. Hashemi, M.R. Aghamohammadi, Wavelet based feature extraction of voltage profile for online voltage stability assessment using RBF neural network. Int. J. Electr. Power Energy Syst. **49**, 86–94 (2013)
23. S. Rajan, S. Kumar, D. Mathew, Online static security assessment module using artificial neural networks. Power Systems, IEEE Transactions on **28**, 4328–4335 (2013)
24. A. Bahmanyar, F. Karami, Power system voltage stability monitoring using artificial neural networks with a reduced set of inputs. Int. J. Electr. Power Energy Syst. **58**, 246–256 (2014)
25. S. Ashraf et al., Voltage stability monitoring of power systems using reduced network and artificial neural network. Int. J. Electr. Power Energy Syst. **87**, 43–51 (2017)
26. W.M.L.-L., J.M. Villa-Acevedo, D.G. Colomé, *Voltage Stability Margin Index Estimation Using a Hybrid Kernel Extreme Learning Machine Approach.* Energies, 2020. 13 (857)
27. A. Ghaghishpour, A. Koochaki, An intelligent method for online voltage stability margin assessment using optimized ANFIS and associated rules technique. ISA Trans. **102**, 91–104 (2020)
28. B. Gharehpetian, Power system on-line static security assessment by using multi-class support vector machines. J. Appl. Sci. **12**, 8 (2008)
29. H. Mohammadi et al., Voltage stability assessment using multi-objective biogeography-based subset selection. Int. J. Electr. Power Energy Syst. **103**, 525–536 (2018)
30. K. Sundaram, S. Swarup, Classification of static security status using multi-class support vector machines. Journal of Engineering Research **9**, 21–30 (2012)
31. M.V, S. and B. C.K, Fast assessment of voltage stability margin of a power system. Journal of Electrical Systems, 2014. 10: p. 305–316
32. M.V. Suganyadevi, C.K. Babulal, Support vector regression model for the prediction of Loadability margin of a power system. Appl. Soft Comput. **24**, 304–315 (2014)
33. K.S. Sajan, V. Kumar, B. Tyagi, Genetic algorithm based support vector machine for on-line voltage stability monitoring. Int. J. Electr. Power Energy Syst. **73**, 200–208 (2015)
34. H. Yang et al., PMU-based voltage stability prediction using least square support vector machine with online learning. Electr. Power Syst. Res. **160**, 234–242 (2018)
35. L. Wehenkel, M. Pavella, Decision tree approach to power systems security assessment. Int. J. Electr. Power Energy Syst. **15**(1), 13–36 (1993)
36. R. Diao et al., Decision tree-based online voltage security assessment using PMU measurements. IEEE Trans. Power Syst. **24**(2), 832–839 (2009)
37. R.F. Nuqui, et al. Fast on-line voltage security monitoring using synchronized phasor measurements and decision trees. In 2001 *IEEE Power Engineering Society Winter Meeting. Conference Proceedings (Cat. No.01CH37194).* (2001)
38. Z. Li, W. Wu, Phasor measurements-aided decision trees for power system security assessment. In *2009 Second International Conference on Information and Computing Science.* (2009)

39. C. Zheng, V. Malbasa, M. Kezunovic, Regression tree for stability margin prediction using synchrophasor measurements. IEEE Trans. Power Syst. **28**(2), 1978–1987 (2013)
40. H. Mohammadi, M. Dehghani, PMU based voltage security assessment of power systems exploiting principal component analysis and decision trees. Int. J. Electr. Power Energy Syst. **64**, 655–663 (2015)
41. V. Krishnan, J.D. McCalley, Progressive entropy based contingency grouping for deriving decision trees for multiple contingencies. Int. J. Electr. Power Energy Syst. **45**(1), 35–41 (2013)
42. X. Meng et al., Construction of decision tree based on C4.5 algorithm for online voltage stability assessment. Int. J. Electr. Power Energy Syst. **118**, 105793 (2020)
43. X.Z. Meng, P. Zhang, Decision tree for online voltage stability margin assessment using C4.5 and relief-F algorithms. Energies **13**, 3824 (2020)
44. M. Beiraghi, A.M. Ranjbar, Online voltage security assessment based on wide-area measurements. IEEE Transactions on Power Delivery **28**(2), 989–997 (2013)
45. H. Su, T. Liu, Enhanced-online-random-Forest model for static voltage stability assessment using wide area measurements. IEEE Trans. Power Syst. **33**(6), 6696–6704 (2018)
46. J.D. Pinzón, D.G. Colomé, Real-time multi-state classification of short-term voltage stability based on multivariate time series machine learning. Int. J. Electr. Power Energy Syst. **108**, 402–414 (2019)
47. K.D. Dharmapala et al., Machine learning based real-time monitoring of long-term voltage stability using voltage stability indices. IEEE Access **8**, 222544–222555 (2020)
48. A. Abed, WECC voltage stability criteria, load shedding strategy, and reactive power reserve monitor methodology. IEEE Power Eng. Soc. **1**, 191–197 (1999)
49. V. Krishnan et al., Efficient database generation for decision tree based power system security assessment. IEEE Trans. Power Syst. **26**(4), 2319–2327 (2011)
50. C.A. Jensen, M.A. El-Sharkawi, R.J. Marks, Power system security assessment using neural networks: Feature selection using fisher discrimination. IEEE Trans. Power Syst. **16**(4), 757–763 (2001)
51. K. Verma, K.R. Niazi, Supervised learning approach to online contingency screening and ranking in power systems. Int. J. Electr. Power Energy Syst. **38**, 97–104 (2012)
52. M. Sun, I. Konstantelos, G. Strbac, A deep learning-based feature extraction framework for system security assessment. IEEE Transactions on Smart Grid **10**(5), 5007–5020 (2019)
53. S. Varshney, L. Srivastava, M. Pandit, ANN based integrated security assessment of power system using parallel computing. Int. J. Electr. Power Energy Syst. **42**(1), 49–59 (2012)
54. W.J. Ridgman, *Principles of Multivariate Analysis*. By W. J. Krzanowski. 563 pages. (Clarendon Press, Oxford, 1988). (Price £65.00 (hard covers). ISBN 0 19 852211 8. The Journal of Agricultural Science, 1989. 112(1): p. 141–142)
55. J.E. Jackson, *A User's Guide to Principal Components*. (Wiley, 2003)
56. L. Zhu et al., Imbalance learning machine-based power system short-term voltage stability assessment. IEEE Transactions on Industrial Informatics **13**(5), 2533–2543 (2017)
57. *IEEE Standard for Synchrophasors for Power Systems*. IEEE Std C37.118–2005 (Revision of IEEE Std 1344–1995), 2006: p. 1–65
58. A. Pai, *Energy Function Analysis for Power System Stability*. (Springer, New York, 1989)
59. Gan, R.Z.a.D., *MATPOWER: A MATLAB Power System Simulation Package*. [Online]. www.pserc.cornell.edu/matpower/

Chapter 10
Evaluation and Classification of Cascading Failure Occurrence Potential Due to Line Outage

Morteza Abedi, Mohammad Reza Aghamohammadi, and Mohammad Taghi Ameli

10.1 Introduction

The modern power systems have developed a continuous topology to increase reliability and economic performance. Increasing the dimensions of the power system on the one hand and the need for continuous monitoring of the network for secure operation on the other hand has caused the issue of assessing the security of the power system to face challenges. In such a condition, considering economic factors and the fast growth of consumers in a power system that should be supplied, the power systems are operated close to their allowed operating limit [1]. Since a power system consists of a large number of transmission lines, the location of the lines in the power system and their multiplicity compared to other equipment such as transformers and generators make the lines more vulnerable. Therefore, the sudden loss of one or more transmission lines can lead to violations of network operation limits and endanger the stability margins of the power system and affect the performance of other equipment and the power system in general [2].

Cascading failure usually occurs as a result of an initial failure like outage of a transmission line due to heavy loading caused by a short-circuit fault. In this situation, cascading failure in the power system may lead to uncontrolled separation and islanding and eventually collapse and blackout on each island [3]. In the blackout event on 14 August 2003, the cascading failure had an important role. Due to cascading failure, the power system is separated into independent islands and finally leads to collapse and blackout in each island [4]. Therefore, if the potential of cascading failure and the resulting blackout due to the outage of the line can be

M. Abedi · M. R. Aghamohammadi (✉) · M. T. Ameli
Faculty of Electrical Engineering, Shahid Beheshti University, Tehran, Iran
e-mail: mor_abedi@sbu.ac.ir; M_Aghamohammadi@sbu.ac.ir; m_ameli@sbu.ac.ir

M. Nazari-Heris et al. (eds.), *Application of Machine Learning and Deep Learning Methods to Power System Problems*, Power Systems,
https://doi.org/10.1007/978-3-030-77696-1_10

evaluated and predicted, respectively, a set of preventive actions can be taken to reduce the vulnerability of the power system and prevent cascading failure caused by line outages in the power systems.

The development of power systems has increased the number of operating variables of the power system. Extracting information of operating variables provides knowledge for the PSCC that can make fast and effective decisions in threatening situations and prevent failures [5]. When a specific line is disconnected, a limited number of operating variables of the power system would be effective. Also, using the information on all operating variables of the power systems for evaluating the potential of cascading failure resulting from line outage is not cost-effective, because receiving information of all operating variables by the PSCC requires a large number of measurement and communication devices, and storing this large volume of information increases the computational costs and the cost of providing devices. Thus, to evaluate the potential of cascading failure resulting from the line outage requires reducing the dimensions of the operating variables and identifying a limited number of dominant variables with rich information regarding the operation conditions.

In recent scientific studies, identification and extraction of DOVs in various contexts of power systems, including evaluating the security of the power system in terms of voltage stability [6], rotor fault detection [7], effective fault location [8], short-term price and load prediction [9], and transient stability analysis [10], have been investigated. However, extracting information from variables and identifying DOVs for evaluating the potential of cascading failure due to a line outage in power systems have not been studied. In [11], two variance reduction methods based on meta-modeling have been presented for quick approximation of cascading failure probability. In [12], factors affecting the probability of power system equipment failure as a function of system load level have been studied and used in a simple model to assess the probable risk of cascading failure. In [13], considering the effect of the hidden fault on a protection system, the probability of cascading failure is evaluated. In [14], an improved OPA model has been presented for cascading failure in which the issues regarding line outages, simulation of cascading failure, and the probability distribution of outage severity have been studied. In [15], the total number of line outages has been predicted in cascading failure by using a branching process of Galton-Watson. In [16], the graph theory has been used to analyze the interaction among the devices undergoing failure to specify the general propagation pattern of the cascading failure. In [17], the Random Chemistry search algorithm has been used to present a model for estimating the risk of large cascading outages as a result of multiple failures. In [18], two different branch process models have been used to simulate the probability distribution of load shedding using OPA and TRELSS in cascading blackout. In [19], a tool has been presented to evaluate the risk of cascading failure for measuring the power system reliability. The proposed method looks for the initial N-x event that has the maximum impact on cascading failure. In [20], a new random model of cascading failure has been presented. In this model, AC power flow and decentralized transformation have been used to analyze the vulnerability of the power system, and the random probability distribution has been used to model load shedding relays. Also, the effect of wind turbine penetration of cascading failure has been studied. In [21], an approach based on sample-induced semi-analytic has been proposed to determine the effect of the failure of power system devices on the blackout risk during cascading failure and identifying the relationship between the failure of devices and blackout risk. The Markov sequence model has been used to develop an accurate relationship between the failure of devices and blackout risk. In [22], an approach based on data mining has been presented for predicting the vulnerability after cascading failure in a power system in real time. In this method, the singular value decomposition and post-events data are used to determine the vulnerable areas. The method employed in [23] can predict the blackout probability online by analyzing the transmission line event along with the probabilistic framework using SVM as the learning tool. In [24], the propagation and development of transmission lines' outage are estimated, and the parameters of the probabilistic branch process model resulting from cascading failure are obtained. Then, the branch process model is used to predict the total number of outages for one initial outage. In [25], a new intelligent machine method has been presented for protection relays based on SVM, SCADA, and communications among protection relays, which is generally called intelligent protection relays. By making intelligent decisions about the trip and its time, and using SCADA information, intelligent relays can reduce cascading failure and global blackouts to a great extent. In [26], using the decision tree and calculating the brittleness index at each instant, the blackout of the power system during a cascading failure is predicted.

Many of the reviewed studies are evaluated based on the offline data obtained from the system model. The studies that have employed the online model cannot be used for online evaluation of the potential of cascading failure and prediction of the resulting blackout severity using the information before the failure occurrence. The reviewed studies look to find the propagation pattern of the cascading failure, and none of them have evaluated the potential of cascading failure before the occurrence of the initial failure. Also, the introduced indices represent the power system vulnerability against cascading failure and do not estimate the blackout severity using the information before the occurrence of the cascading failure. To this end, a proper intelligent scheme (like DT) should be presented to predict the blackout severity if there is a potential of cascading failure using the information before the occurrence of the cascading failure.

In this study, a new approach is presented for early prediction of CFOP and the resulting blackout severity using the DT technique and the DOVs for each line of the system. To this end, a three-step intelligent method using three DTs predicting the CSPDT and DOVs is presented. In the proposed method, the DOVs are identified using a method based on entropy, and mutual information theory between the operating variables and their information is given to the CSPDTs instantaneously using WAMS/PMU as an input vector. At each specific operating point and in real time, using the information of the DOVs before line outage and using the result of CSPDT performance, the CFOP resulting from line outage and the blackout severity are estimated. The proposed scheme determines the CFOP and predicts and classifies the blackout severity resulting from cascading failure by the CSPDTs in three levels.

Therefore, in the presented scheme, the CFOP in the power system before line outage is evaluated instantaneously and online. Also, if the CFOP is existed, the blackout severity resulting from the outage of each line is also predicted. Therefore, the PSCC can take preventive actions to reduce the potential. To construct CSPDT, the C4.5 algorithm, which is more accurate and robust against noisy data [27, 28] is used.

The rest of this paper is organized as follows. Section 10.2 presents the general framework of the proposed scheme. Section 10.3 introduces the employed DT. Section 10.4 describes the identification of DOVs for a specific line #L. Section 10.5 presents the three-step scheme completely. In Section 10.6, the proposed scheme is applied to sample networks, and the results are analyzed and validated. Finally, the paper is concluded in Sect. 10.7.

10.2 Overview of the Proposed Method

The occurrence of cascading failure is a complicated process in which cascading outages weaken the power system and increase the potential of small or large blackouts. Evaluating the potential of cascading failure helps the operator to decide what preventive actions to take.

In the proposed scheme, operational information of the power system before line outage is used to evaluate and predict the potential of cascading failure and blackout severity, respectively. Also, the occurrence of cascading failure due to outage of line #L, at the operating point (i), is detected by the outage of at least two lines after the outage of #L line with more than 5% load shedding.

In the proposed scheme, the line that its outage results in cascading failure and blackout in the power system is known as the critical line. In other words, the outage of the critical line changes some of the operating variables, resulting in cascading outage of other devices and cascading blackout. Therefore, the operating point at which the #L line is disconnected from the power system due to a fault occurrence and causes cascading failure is known as a critical scenario.

Figure 10.1 shows the conceptual structure of the proposed three-step predictor scheme. First, the proposed scheme is designed, and then it is implemented in PSCC. The design of the proposed scheme includes identifying DOVs among all monitorable operating variables of the power system and evaluating the potential of cascading failure and predicting the severity of the resulting blackout through training CSPDTs. Information of DOVs before the failure of #L line is collected continuously and instantaneously using WAMS/PMU and given to the CSPDTs. To functionalize the correlation of potential of cascading failure and the blackout severity with DOVs, the intelligent decision tree method is used.

The three-step DT is responsible for evaluating the potential and predicting the magnitude and severity of the blackout; the blackout magnitude is defined as the percentage of the total load shedding to the total initial load of the power system. To discriminate the severity of the system's cascading failure in terms of the magnitude

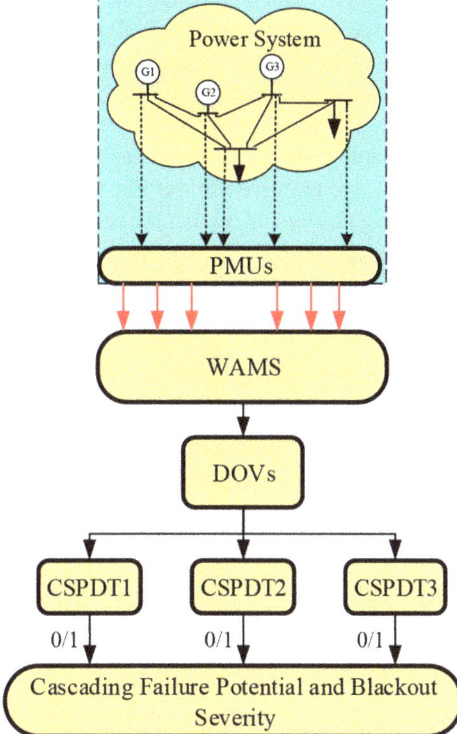

Fig. 10.1 Conceptual structure of the proposed scheme

Table 10.1 Output of CSPDTs for predicting severity of blackout

System vulnerability	CSPDT1	CSPDT2	CSPDT3	Severity of predicted blackout
Non-critical	0	0	0	B.S < 5%
Critical	1	0	0	5% ≤ B.S < 20%
Severely critical	1	1	0	20% ≤ B.S < 40%
Extremely critical	1	1	1	40% ≤ B.S ≤ 100%

B.S=Blackout severity

of blackouts, three decision trees CSPDT1, CSPDT2, and CSPDT3 are used to predict the specific magnitude of blackouts.

Considering the potential of cascading failure and blackout severity, the system vulnerability is divided into four states of non-critical, critical, severely critical, and extremely critical regarding the reasonable combination of the CSPDTs' outputs. Table 10.1 represents the reasonable combinations of the CSPDTs' outputs for deciding about the blackout severity resulting from cascading failure. For each vulnerability state of the power system, the combination of CSPDT outputs should be as shown in Table 10.1. According to Table 10.1, CSPDT1 is responsible for evaluating the potential for cascading failure so that its zero output indicates that there is no potential for cascading failure to occur due to specified L # line outage.

Conversely, output 1 indicates the potential of cascading failure occurrence due to the outage of the specified L # line and the specific magnitude of a blackout. CSPDT2 and CSPDT3 are responsible to predict the blackout severity resulting from cascading failure due to outage of #L line. In this study, the blackout severity conceptual is the same as the percentage of load shedding in the power system. For implementing the proposed scheme, the identified DOVs and the trained CSPDTs for each line are given to the PSCC. Thus, the PSCC can evaluate the potential of cascading failure and predict the resulting blackout severity for each line and in each operating point of the power system.

10.3 Cascading Failure Severity Predictor DT

In this study, DT is used to evaluate the potential of cascading failure and predict the blackout severity using DOVs of the power system. To train and build the DT, various algorithms like CHAID [29], CART [30], and C4.5 are used. In this paper, the C4.5 algorithm is employed for the proposed CSPDT.

10.3.1 C4.5 Algorithm [31]

In the C4.5 DT, the statistical values called entropy and information gain are used to determine how much the feature can split the training samples based on their classification.

10.3.1.1 Entropy

The entropy determines the purity of a set of examples. If the set Z includes positive and negative examples of a concept or objective, the entropy of Z regarding this Boolean class is defined as follows:

$$Entropy(Z) = -p_{\oplus} \log_2 p_{\oplus} - p_{\ominus} \log_2 p_{\ominus} \tag{10.1}$$

where $p_{\oplus}$ is the ratio of positive examples to total examples and $p_{\ominus}$ is the ratio of negative examples to total examples.

10.3.1.2 Information Gain

The information gain of a feature is the reduction of entropy resulting from splitting the examples using this feature. In other words, the information gain of Z and A which shown as Gain (Z, A) for a feature like A with respect to the set of examples Z is defined as follows:

$$Gain(Z,A) = Entropy(Z) - \sum_{v \in Values(A)} \frac{|Z_v|}{|Z|} Entropy(Z_v). \tag{10.2}$$

10.3.2 Considering Continuous Attributes

Algorithm C5 also includes features with continuous values. For a continuous feature such as A, a Boolean class, such as Ac, is defined so that Ac is true if A < C and otherwise false.

10.3.2.1 Gain Ratio

Gain ratio demonstrates the uniformity and extensiveness of the data split by a feature. It is defined as in Eq. (10.3):

$$Gain\,Ratio(Z,A) = \frac{Gain(Z,A)}{Split\,Information(Z,A)} \tag{10.3}$$

The split information of Z and A is known as SI and defined by Eq. (10.4):

$$Split\,Information(Z,A) = -\sum_{i=1}^{c} \frac{|Z_i|}{|Z|} \log_2 \frac{|Z_i|}{|Z|} \tag{10.4}$$

In Eq. (10.4), C is the number of subsets of Z.

10.4 Dominant Operating Variables

The outage potential of a line followed by cascading failure depends on the operating condition of the power system. The operating condition of the power systems can be described using the operation information before line outage like the voltage of buses, power flow of lines, loads, and generation of the generators.

In real power systems, there is a large volume of operational information on each operating point. However, considering the criticality of a specific line #L, all operational information does not significantly depend on the line condition. In other words, among all operating variables, only a limited number of the variables known as DOVs are associated with line criticality and depend on it significantly. In fact, the DOVs associated with the outage of each line are the minimum number of operating variables that can be used to evaluate the criticality of a specific line #L.

Therefore, dimension reduction methods are used to identify the DOVs among a large number of operating variables. These methods are usually used to process data and select the dominant features. The main purpose of reducing the data dimension and the number of features to the minimum possible value is to increase the data classification accuracy [32]. The feature selection and identification methods reduce the dimensions by selecting a subset of features that minimize a specific cost function. Feature identification is used in many applications of expert and intelligent systems like data mining, learning machine, image processing, anomaly detection, and bioinformatics [33]. Also, the feature identification process is known as the variable identification, feature reduction, or detection of a subset of variables.

The MIM method is one of the DOV variable dimension reduction and DOV identification methods, which is defined based on the mutual information theory. In this method, the relationship and dependency among all variables are studied [34]. Then, the operating variables with rich information are identified, and the operating variables with low or same information are neglected.

As mentioned before, for evaluating the line criticality, the identified DOVs should be used. To this end, various operating points with maximum coverage of the operating space should be developed in the operational space of the power system, such that the line status is evaluated as critical for some operating points and non-critical for the others.

10.4.1 DOVs' Identification Algorithm

In power systems, the system status in a specific operating point is defined by a set of operating variables {VAR} such as bus voltage, power flow, load, and generation of the generators. At each operating point, various behavioral characteristics can be attributed to the power system using a causal relationship between the operating variables and the behavioral characteristics. However, considering the behavioral characteristic B, all sets of operating variables do not have an equal contribution in the formation of the behavioral characteristic. Some variables are strongly correlated with the behavioral characteristic, while others have a weak correlation. For example, the overloading after an outage of #L line is considered as a behavioral characteristic that is associated with a number of operating variables before a line outage. Therefore, for a set of operating variables of the power system {VAR}, considering a specific behavioral characteristic of the power system B, a subset of the variables $\{VAR_D\}$ with a strong correlation with B is known as the DOVs associated with B. It should be mentioned that for each set of operating variables {VAR}, there are various DOVs considering various behavioral characteristics.

In this study, the total set of operating variables before line outage is considered as {VAR}. The line (#L) outage potential for initiating of cascading failure in the power system is known as the behavioral characteristic B of the power system. Considering the behavioral characteristic B, the goal is to identify the subsets of operating variables {VAR} associated with B. To identify the DOVs compatible with {VAR_D} and associated with the behavioral characteristic B among all operating variables of the set {VAR}, the mutual information theory and entropy can be used as follows.

If an operating variable in N operating points is considered as a set like X = {x_1, x_2,. . . . ., x_N}, such that x_i represents the values that the operating variable can adopt at different operating points, a random variable x_i with probability $p(x_i)$ is defined for each operating variable. For the dataset X, an entropy function H(x) is obtained using Eq. (10.5), which can be used to obtain useful information about the operating variable X:

$$H(x) = -\sum_{x \in X} p(x). \log_2 p(x) \tag{10.5}$$

such that H(X) is the entropy function and known as the positive Shannon function [35]. Considering two independent random variables (X, Y), the bivariate entropy function is presented as in Eq. (10.6):

$$H(X, Y) = -\sum_{x, y} p(x, y). \log_2 p(x, y) \tag{10.6}$$

where p(x,y) is the bivariate probability function, and the value of H(X,Y) represents the information hidden in variable X to describe the status of variable Y [35]. If the value of H(X,Y) is calculated for Y = y_0, the bivariate entropy is known as the conditional entropy of variable X for Y = y_0, and it is described using Eq. (10.7). This conditional entropy represents the information of variable X for a specific value of Y:

$$H(X|y_0) = -\sum_{x} P(x|y_0). \log_2 P(x|y_0) = H(X, Y) - H(Y) \tag{10.7}$$

where $P(x| y_0)$ $p(x| y_0)$ is the random bivariate conditional probability function and H $(X|y_o)$ is the conditional entropy of variable X for Y = y_0 [35]. Using Eqs. (10.6) and (10.7), (10.8) is obtained as the mutual information between two variables of X and Y. Mutual information between two variables shows the amount of common information between the two variables:

$$I(X; Y) = H(\mathrm{X}) - \mathrm{H}(\mathrm{X}|\mathrm{Y}) = \sum_{x}\sum_{y} P(x, y). \log \frac{P(x, y)}{p(x).p(y)} \tag{10.8}$$

In other words, mutual information represents the information of a variable about the other variable [35]. In this study, the variable X is considered as an operating

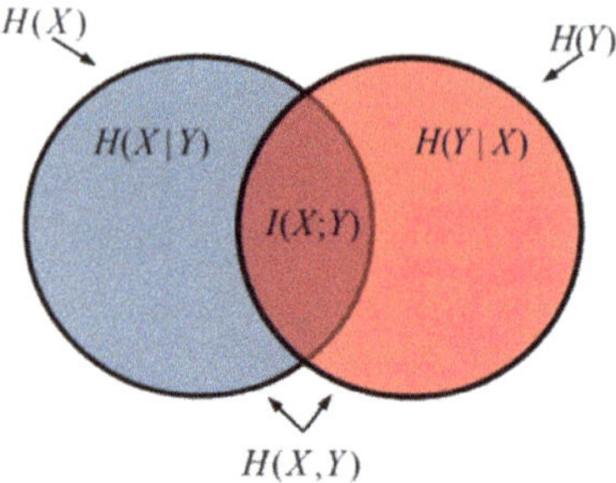

Fig. 10.2 Venn diagram for the mutual information

variable of the set {VAR}, and variable Y is considered as the behavioral characteristic B of the power system. Figure 10.2 shows the Venn diagram of Eq. (10.8).

To employ the mutual information theory method in a power system for identifying the DOVs associated with critical condition of each line, several operating points should be selected in the operating space of the power system. At each operating point i, the information of operating variables of the set {VAR} is unique. Thus, at each operating point i, considering the critical condition of #L line, for the set of operating variables $\{VAR^i\}$, there is a behavioral characteristic B^i that describes the #L line being either critical or non-critical. Therefore, for N operating points, there are N sets of operating variables {VAR} that constitute a data matrix along with a matrix including N behavioral characteristics. The set of operating variables includes an N*m data matrix, where m is the total number of operating variables of the power system. N behavioral characteristics constitute an N*1 vector.

To identify the DOVs regarding the outage of each line, the MIM [36] method based on information theory is presented. The mathematical formulation of MIM is described by Eq. (10.9). In Eq. (10.9), X and Y are the operating variable of {VAR} and the behavioral characteristic B of the potential of cascading failure, respectively. In this method, the mutual information between all operating variables of {VAR} and the behavioral characteristic B is calculated. Then, the results of mutual information calculation are sorted maximum to minimum. In Eq. (10.9), X_k is an operating variable, k is the number of operating variables, and $VAR_K(X_k)$ is the set of identifying k variables with the highest mutual information. $I(X_k;\ Y)$ is the mutual information between X_k and Y, $H(X_k)$ and H(Y) are single variable entropy functions, and $H(X_k;\ Y)$ is a bivariate entropy function:

$$VAR_k(X_K) = \arg\max I(X_k; Y) = H(X_k) + H(Y) - H(X_k, Y) \qquad (10.9)$$

Considering Eq. (10.9), the number of operating variable k can be changed. In other words, Eq. (10.9) can identify a set of k dominant variables among all operating variables considering the various values of k. Figure 10.3 shows the structural relationship between sets of operating variables with the corresponding behavioral characteristics for N different operating points.

Thus, the result of MIM is examined by the learning machine for different values of k. A set of operating variables that has the maximum accuracy in evaluating the

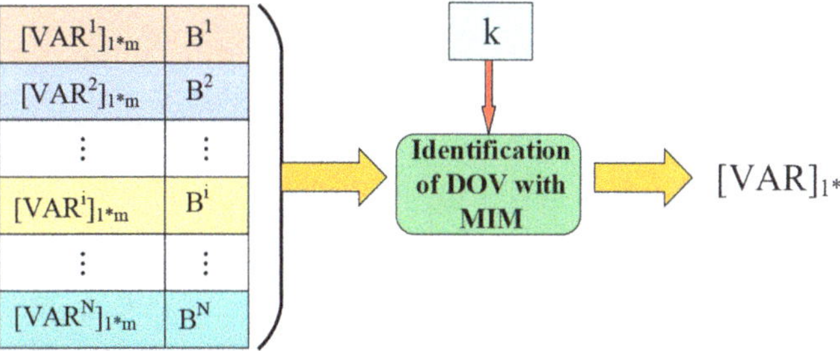

Fig. 10.3 relationship between operating variables with corresponding behavioral characteristics

potential of cascading failure and predicting the blackout severity is introduced as the set of dominant operating variables known as DOVs. The k parameter corresponding to the set of DOVs is the number of dominant operating variables and represented by D. Thus, using the proposed MIM method, the set of dominant operating variables $\{VAR_D\}$ called DOVs and dimension of D < <m, which is strongly correlated with behavioral characteristic B, is identified.

10.5 The Proposed Method

Using the proposed MIM method and various disturbance and perturbation scenarios, a set of unique DOVs are detected for each line of the system. Also, using the proposed CSPDTs and the detected DOVs, a relationship is created between the system operating condition and CFOP evaluation and prediction of blackout severity. Therefore, the main advantage of CSPDTs is the evaluation of CFOP and blackout prediction for unobserved disturbances. This capability of the CSPDTs is based on interpolation capability, which is obtained considering various disturbance and perturbation scenarios and covering the whole disturbance space of the power system. Figure 10.4 shows the general structure of the proposed method for evaluating the CFOP and blackout severity prediction using DOVs, CSPDT1, CSPDT2, and CSPDT3.

Since the aim of the proposed scheme is to evaluate the CFOP in the power system before the first failure, the measured information of the identified DOVs is given to the CSPDTs by the WAMS at each instant. The instantaneous information of the DOVs for evaluating the CFOP and blackout severity prediction resulting from outage of line #L as the vector $VAR_D^L(t)$ with D members is used as input at time t to the CSPDTs as given in Eq. (10.10):

$$VAR_D(t) = [\mathrm{var}_1(t), \mathrm{var}_1(t), \mathrm{var}_1(t), \ldots, \mathrm{var}_D(t)] \tag{10.10}$$

Asynchronous data of WAMS and temporary loss of some WAMS data might affect CFOP evaluation. In this study, asynchronous or lost data of $VAR_D^L(t)$ can be

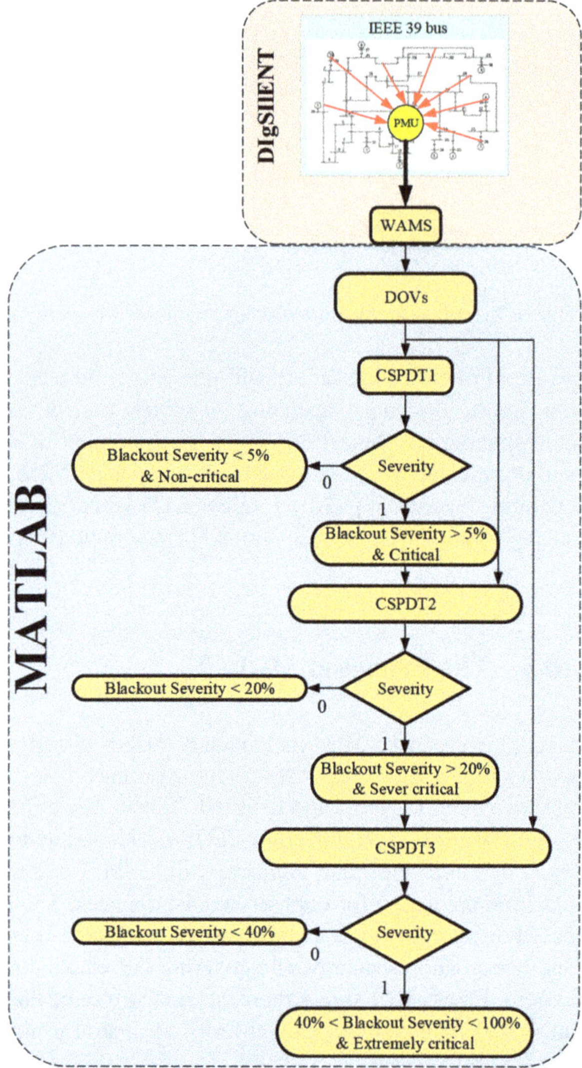

Fig. 10.4 general structure of the proposed scheme

replaced with the measured DOV of the previous sample (t-Δt). Δt is the time step of data transmission by PMU and WAMS to the CSPDTs.

According to Fig. 10.4, the blackout severity procedure is classified in three levels, and each level is associated to a unique CSPDT. At each time t, by presenting the vector $VAR_D^L(t)$ to each CSPDT simultaneously, each CSPDT returns 0 or 1 at the output; by combining the outputs of all CSPDTs, the CFOP and blackout severity resulting from outage of line #L are predicted.

The zero output for CSPDT1 indicates that if the line #L in the power system has experienced outage as a result of a failure, the predicted blackout severity is smaller than 5% of the power system loading level. Also, the cascading failure is not caused by the result of outage of line #L; in other words, the CFOP does not exist at the operating point. This state of the power system is called non-critical in which the outputs of CSPDT2 and CSPDT3 are zero. Output 1 for CSPDT1 indicates that if the line #L experiences outage as a result of a failure, the predicted blackout severity would be more than 5% of the power system loading level. Also, the cascading failure is caused by the result of outage of line #L; in other words, there is the potential of cascading failure at the operating point.

The zero output for CSPDT2 indicates that if the line #L experiences outage as a result of failure, the predicted blackout severity would be 5% to 20% of the power system loading level. Also, cascading failure is caused by the result of outage of line #L; in other words, there is the potential of cascading failure at the operating point. This state is called critical in which the outputs of CSPDT1 and CSPDT3 are 1 and 0, respectively. The output of the CSDPT2 being 1 indicates that if line #L of the power system experiences outage as a result of a failure, the predicted blackout severity would be more than 20% of the power system loading. Also, the cascading failure is the result of outage of line #L, indicating that there is the CFOP at the operating point.

The zero output for CSPDT3 being zero indicates that in line #L which experiences outage as a result of failure, the predicted blackout severity would be 20% to 40% of the power system loading level. Also, the cascading failure is the result of outage of line #L, indicating that there is the potential of cascading failure at the operating point. This state is called severe critical in which the outputs of CSPDT1 and CSPTS2 are one. The output of CSPDT3 being one indicates that if the line #L experiences outage as a result of a failure, the predicted blackout severity would be 40% to 100% of the power system loading level. Also, the cascading failure is the result of outage of line #L, indicating that there is the potential of cascading failure at the operating point. This state is called extremely critical in which the outputs of CSPDT1 and CSPDT2 are one.

After evaluating the potential and blackout severity prediction in the power system considering the output of the designed CSPDTs, to reduce the system stress and CFOP, regarding the vulnerability of the system and impact priority, preventive actions might be taken. Among these actions, disconnecting the load/generation [37], emergency line disconnection [38], and island control [39] can be mentioned. Also, to prevent and stop excess loading of the line, which is the most effective mechanism for power system outage, proper load removal can be adopted as a preventive action [40]. Also, to prevent undesired performance of the distance relay resulting from power fluctuations, zone 3 of the distance relay can be blocked [41].

10.6 Simulation Studies

In this section, the proposed scheme is applied to lines 17–18 and 21–22 of the IEEE39-bus system, and the results are studied. In Fig. 10.5, the IEEE39-bus power system is shown. This network has 10 generators, 34 transmission lines, and 19 loads [42].

Table 10.2 shows various monitorable operating variables of the IEEE39-bus power system, where the total number of the operating variables considering the characteristics of the IEEE39-bus power system is 405. In the first and third columns of Table 10.2, the type of the operating variable and its unit are given. In the second and fourth columns, the complete name of the operating variables is given. Among these variables, a number of variables like the voltage magnitude of the generators, speed of the generators, and the rotor phase angle of the slack generator are always constant in all operating points of the power system. Therefore, considering the

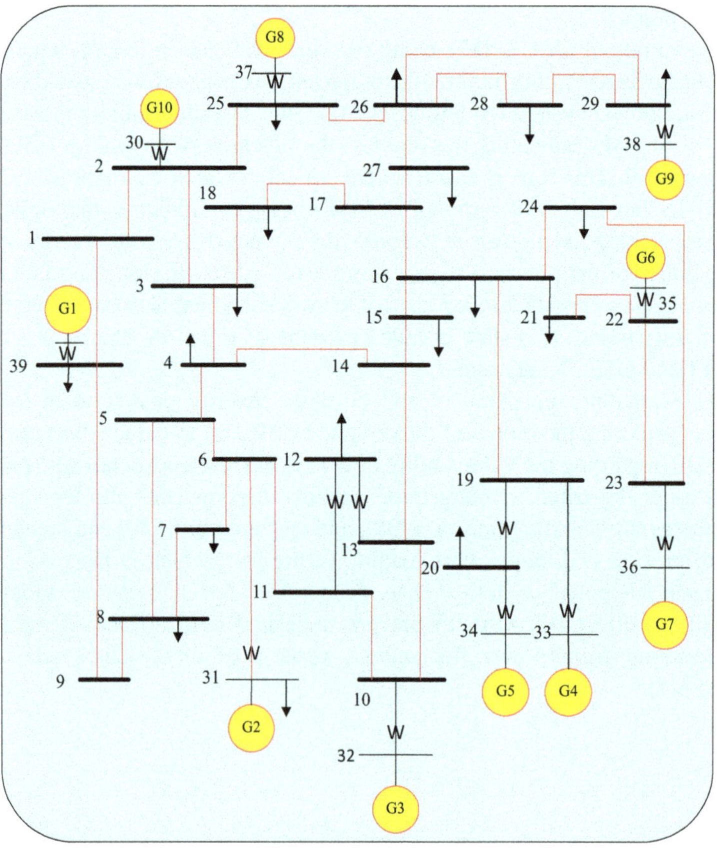

Fig. 10.5 IEEE39-bus power system

Table 10.2 IEEE39-bus power system monitorable operating variables

Var.	Var. name	Var.	Var. name
PLine(p.u)	Line active power	V_{phase}(deg)	Bus voltage angle
QLine(p.u)	Line reactive power	δ_{Phase}(deg)	Rotor angle(estimated)
Line %	Line loading	PG(p.u)	Gen. active power
PLoad(p.u)	Load active power	QG(p.u)	Gen. reactive power
QLoad(p.u)	Load reactive power	Gen %	Gen. loading
V(p.u)	Bus voltage	Tr %	Trans. loading

characteristics of the power system, 383 operating variables are not constant, which describe the operating point of the power system.

10.6.1 Preparing Scenarios

The modeling and simulation used to implement the cascading failure and the proposed scheme are based on time-domain calculations. Determining the violation of the devices from the operating limit using the time-domain method requires a precise dynamic modeling of the power system devices. Also, the dynamic model of the power system devices has an effective role in the formation of the cascading failure. Thus, dynamic modeling of the power system is essential for simulating the cascading failure. In the studied power system, sixth order of the generator, O.S protection, generator primary actuator, generator voltage control system, low-frequency generator protection, line distance protection, load shedding under frequency/voltage protection and tap changer of the transformers connected to a load, frequency and voltage load removal relays, and overcurrent relays are modeled.

In this study, DIgSILENT Power Factory 15.1 is used for modeling of the IEEE39-bus power system. The training scenarios resulting from the outage of line #L are provided through offline simulation using this software. Also, MATLAB is used to identify the DOVs and train the CSPDTs; the proposed three-step scheme is also implemented in MATLAB. All three CSPDTs are trained offline using MATLAB. Therefore, all scenarios which prepared offline by DIgSILENT are provided to MATLAB software in order to train CSPDTs.

When the training of CSPDTs is finished, the identified DOVs and the trained CSPDTs are provided to the PSCC for online evaluating the potential of cascading failure and predicting the blackout severity caused by the outage of #L line.

Considering the above models, 770 operating points are studied for developing various training scenarios. All scenarios are formed considering the basic load of 6000 MW at 11 load levels (0.7–0.75-0.8-0.85-0.9-0.95-1-1.05-1.1-1.15-1.2). Such that 70 operating points are considered at each load level, and the operating points are created such that the generators, lines, and the transformers are at maximum loading at different load and generation levels especially high loading levels. Thus, the power system is always under stress. Also, to predict the blackout severity in the prepared training scenarios, the load shedding is determined after the occurrence of each scenario.

10.6.2 Evaluating the Criticality of the Lines

The proposed scheme is used to evaluate the potential of cascading failure and predicting blackout severity for the outage of lines 17–18 and 21–22 as a result of a three-phase short circuit, and the DOVs are identified for each line.

10.6.2.1 Evaluating Criticality of Line 17–18

748 operating points are created for the outage of lines 17–18, and 383 operating variables are sampled at each operating point, according to Table 10.2 before the outage of lines 17–18. The specifications of the operating points are given in Table 10.3 for identifying the DOVs and providing the test and training scenarios.

A three-phase short circuit fault is applied to the lines 17–18 at all of the 748 operating points. Thus, 266 critical scenarios and 482 non-critical scenarios are obtained. Among the 748 scenarios, 523 scenarios (80% of the all scenarios) are randomly selected as training scenarios and are provided to the MIM method. Among the 523 selected scenarios, the ratio of the scenarios that result in cascading failure to the ones that do not result in cascading failure is 30% to 70%.

The MIM method identifies a set of operating variables based on the mathematical relationships of mutual information theory and entropy and a value of K. Then, the identified operating variables and the training scenarios are used to train the three-step trees of evaluating potential of cascading failure and predicting blackout. After training the CSPDTs, each CSPDT is validated using test scenarios to determine the accuracy and performance of the CSPDTs.

10.6.2.2 Identifying the DOVs and Training the CSPDTs for Lines 17–18

The MIM method identifies a set of operating variables with K members based on mutual information theory and considering the value of K. In this study, the value of

Table 10.3 Number of different scenarios created as a result of outage of lines 17–18

Line	Fault type	Operating point	Operating point		Operating point	
			Critical scenarios	Non- critical scenarios	Training scenarios	Testing scenarios
17–18	3φ	748	266	482	523	225

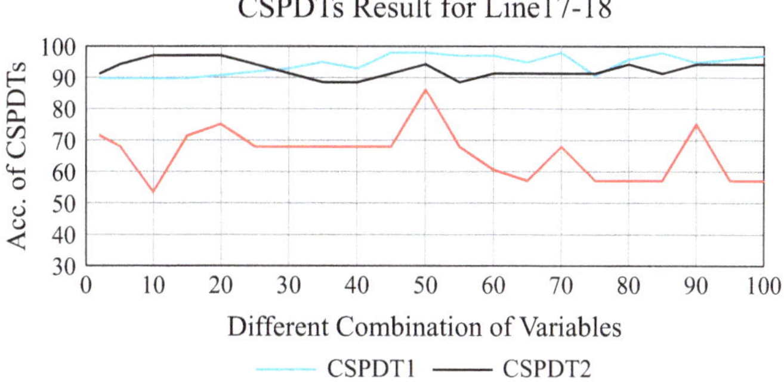

Fig. 10.6 The accuracy of CSPDTs for different Ks due to outage of lines 17–18

Table 10.4 The best accuracy of CSPDTs due to outage of lines 17–18

CSPDT1		CSPDT2		CSPDT3		
Acc. %	Num. of test scenarios	Acc. %	Num. of test scenarios	Acc. %	Num. of test scenarios	Num. of DOVs
97.959	225	94.285	225	86.142	225	50

K is selected from the set {2, 5, 10, 20, 25, 30, 35, 40, 45, 50, 55, 60, 65, 70, 75, 80, 85, 90, 95, 100}. Figure 10.6 shows the accuracy of CSPDTs for different K and, consequently, the different number of identified operating variables to evaluate the potential of cascading failure and to predict blackout. Therefore, the sets of operating variables with different numbers of members are identified using MIM. CSPDTs are trained using the identified operating variables and the provided training scenarios. Finally, their accuracy and performance are evaluated using the test scenarios. A set of operating variables identified by MIM for which the CSPDTs have maximum accuracy in evaluating the potential of cascading failure and predicting blackout severity due to outage of lines 17–18 is identified as the set of DOVs.

In Fig. 10.6, a diagram is represented for the evaluation accuracy of each CSPDT. According to Fig. 10.6, the maximum accuracy of CSPDTs with the minimum number of operating variables can be identified. Table 10.4 shows the best accuracy of CSPDTs in evaluating the potential of cascading failure and predicting blackout severity for test scenarios due to outage of lines 17–18.

Considering the results of Table 10.4, it is seen that the maximum accuracy in evaluating the potential of cascading failure and predicting blackout severity is obtained for 50 operating variables. The combination of these operating variables identified by the MIM method is according to Table 10.5. The numbers after the operating variable's name are the bus number.

Based on Fig. 10.6 and Table 10.5, the combination of 50 operating variables identified using MIM has the maximum accuracy in evaluating the potential of the cascading failure and predicting blackout severity for the outage of lines 17–18. Thus, these identified operating variables are called DOVs of lines 17–18. For online evaluation of the power system vulnerability regarding the outage of lines 17–18, it

Table 10.5 Dominant operating variables for lines 17–18

Number	Operating Var	Num.	Operating Var
1	Qline_19–16	26	Qline_27–17
2	Qline_13–10	27	Qline_6–5
3	Qline_26–25	28	Pline_15–14
4	Qline_24–23	29	Qline_2–1
5	Qline_3–2	30	Qline_39–1
6	QG_38	31	Qline_18–17
7	Qline_5–4	32	Pline_18–17
8	Qline_14–13	33	Qline_22–21
9	QG_37	34	Line_27–26 (%)
10	QG_33	35	QG_39
11	Qline_11–10	36	V_{phase} _28
12	QG_32	37	Line_7–6 (%)
13	Pline_23–22	38	PG_36 (%)
14	Qline_29–28	39	Qline_27–26
15	Qline_8–7	40	Pline_3–2
16	Pline_27–17	41	Qline_16–15
17	Qline_23–22	42	Tr_25–37 (%)
18	QG_31	43	Tr_22–35 (%)
19	Line_3–2 (%)	44	Line_25–2 (%)
20	Pline_24–16	45	Qline_15–14
21	Pline_9–8	46	Tr_ 23–36 (%)
22	V_{phase}_29	47	PG_38 (%)
23	Pline_39–9	48	Tr_6–31 (%)
24	Qline_8–5	49	Tr_2–30 (%)
25	Qline_14–4	50	V_{phase} _23

is sufficient to provide the DOVs of Table 10.5 along with the trained CSPDTs to the PSCC. In PSCC, the information of DOVs at each instant and each operating point is received by WAMS and provides them as input to the trained CSPDTs. Then, considering the output of each CSPDT, the potential of cascading failure and blackout severity for the outage of lines 17–18 is determined with high accuracy.

10.6.2.3 Evaluating the Criticality of Lines 21–22

The proposed scheme is applied to lines 21–22 similar to lines 17–18. The specifications of the operating points are given in Table 10.6 for identifying the DOVs and providing the test and training scenarios.

According to Table 10.6, a three-phase short circuit fault is applied to the lines 21–22 at all of the 751 operating points. Thus, 580 critical scenarios and 171 non-critical scenarios are obtained. Among the 751 scenarios, 526 scenarios (80% of the all scenarios) are randomly selected as training scenarios and are provided to the MIM method.

Table 10.6 Number of different scenarios created as a result of outage of lines 21–22

Line	Fault type	Operating point	Operation point		Operating point	
21–22	3φ	751	Non-critical scenarios	Critical scenarios	Testing scenarios	Training scenarios
			171	580	225	526

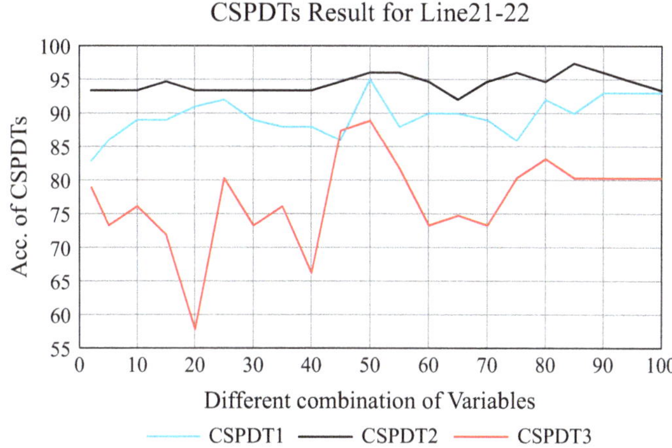

Fig. 10.7 The accuracy of CSPDTs for different K's due to outage of lines 21–22

Table 10.7 The best accuracy of CSPDTs due to outage of lines 21–22

CSPDT1		CSPDT2		CSPDT3		
Acc. %	Num. of test scenarios	Acc. %	Num. of test scenarios	Acc. %	Num. of test scenarios	Num. of DOVs
94.949	225	96.334	225	88.872	225	50

10.6.2.4 Identifying the DOVs and Training the CSPDTs for Lines 21–22

Figure 10.7 shows the accuracy of CSPDTs for different K and, consequently, the different number of identified operating variables to evaluate the potential of cascading failure and to predict blackouts. Similar to lines 17–18, for lines 21–22, the sets of operating variables with different numbers of members are identified using MIM. CSPDTs are trained using the identified operating variables and the provided training scenarios. Finally, their accuracy and performance are evaluated using the test scenarios. A set of operating variables for which the CSPDTs have maximum accuracy in evaluating the potential of cascading failure and predicting blackout severity resulting from the outage of lines 21–22 is identified as the set of DOVs.

According to Fig. 10.7, the maximum accuracy of CSPDTs can be identified considering the minimum number of operating variables. Table 10.7 represents the

Table 10.8 Dominant operating variables for lines 21–22

Num.	Operating var.	Num.	Operating var.
1	QG_30	26	Pline_23–22
2	QG_32	27	Qline_15–14
3	QG_31	28	Qline_8–5
4	Qline_6–5	29	Qline_14–4
5	QG_33	30	Qline_27–26
6	Qline_14–13	31	V_{phase}_29
7	Qline_13–10	32	Qline_7–6
8	Qline_22–21	33	Pline_15–14
9	Qline_19–16	34	Pline_9–8
10	Qline_24–23	35	Qline_16–15
11	Qline_26–25	36	Pline_39–9
12	Qline_5–4	37	Tr_12–13(%)
13	Qline_11–10	38	V_{phase}_22
14	Qline_18–3	39	Pline_18–3
15	QG_37	40	Pline_26–25
16	Pline_27–17	41	load_24
17	Pline_18–17	42	Line_22–21(%)
18	QG_36	43	Qline_27–17
19	QG_35	44	Tr_2–30 (%)
20	Qline_8–7	45	Line_27–26 (%)
21	Qline_3–2	46	Tr_23–36 (%)
22	QG_39	47	Qline_18–17
23	Pline_24–16	48	Tr_6–31 (%)
24	Pline_22–21	49	Tr_20–34 (%)
25	QG_38	50	Tr_25–37 (%)

best performance accuracy of the CSPDTs in evaluating the potential of cascading failure and predicting blackout severity for test scenarios due to outage of lines 21–22.

According to the results of Table 10.7 and Fig. 10.7, it is seen that the maximum accuracy in evaluating the potential of cascading failure and predicting the blackout severity is obtained for 50 operating variables. The combination of these operating variables identified by the MIM method is according to Table 10.8.

The combination of 50 operating variables identified using MIM according to Table 10.8 provides the maximum accuracy in evaluating the potential of cascading failure and predicting the blackout severity. Thus, the identified operating variables according to Table 10.8 are called the DOVs of lines 21–22. Considering the results, PSCC can determine the potential of cascading failure and blackout severity for lines 17–18 and 21–22, independently at each operating point of the system using the related DOVs and the trained CSPDTs.

10.6.3 Performance Validation

The performance of the proposed scheme is validated for four different scenarios. Specifications of the four scenarios are given in Table 10.9. None of the above scenarios are used to identify the operating variables and train the CSPDTs. In Table 10.9, the amount and percentage of load shedding for the four scenarios are determined. Based on the proposed scheme, information of the DOVs of each line is received from the WAMS of the power system before the fault occurrence and provide to the trained CSPDTs of each line. According to the output of the CSPDTs, the potential of cascading failure and blackout severity of each scenario is predicted and represented in Table 10.10. To validate the proposed scheme, the output of each CSPDT should be examined considering the percentage of load shedding as blackout severity. According to the results in Table 10.10, the proposed three-step scheme has performed well in evaluating the potential of cascading failure and predicting blackout severity for four different scenarios. According to the percentage of load shedding for scenarios 1–4 represented in Table 10.9, it is expected that the system vulnerability should be non-critical, critical, extremely critical, and severely critical, respectively. According to Table 10.10, it is seen that the expected results are achieved correctly. The numbers after the operating variable's name are the bus number.

Table 10.9 Specification of fault scenarios for performance validation

Fault scenario	Load level (MW)	Cascading failure status	Fault type	Fault location		Load shedding (MW)	Load shedding/load level (%)
				Line	Line %		
1	6214	Non-critical	3 phase	17–18	20	205	3.29
2	5557	Critical	3 phase	21–22	50	598	10.76
3	7089	Extremely critical	1 phase	17–18	70	5500	77.58
4	7124	Sever critical	Resistive fault 10 Ω	21–22	50	2841	39.87

Table 10.10 Result of performance validation

Fault scenario	Actual system vulnerability	CSPDT1 output	CSPDT2 output	CSPDT3 output	Prediction of system vulnerability
1	Non-critical	0	0	0	Non-critical
2	Critical	1	0	0	Critical
3	Extremely critical	1	1	1	Extremely critical
4	Severely critical	1	1	0	Sever critical

10.6.4 Global Performance

In the proposed scheme, three CSPDTs should operate based on a logical and synchronous process to evaluate the potential of cascading failure and predict the resulting blackout severity. Therefore, the illogical performance of the three CSPDTs concerning each other might be considered as incorrect performance. Table 10.11 represents the eight possible combinations of the CSPDTs' outputs, and their logical performance can be interpreted as follows:

1. Combinations 1–4 are logical and the decision made by CSPDTs as the initial decision is verified compared to the logical decision.
2. Combinations 5–6 are illogical and the decision made by CSPDTs as the initial decision is verified compared to the logical decision.
3. Combinations 7–8 are illogical and the decision made by CSPDTs as the initial decision is not verified, and it is modified and verified by the logical decision.

10.7 Conclusion

In this study, the DOVs and DT technique are used to present a three-step predictor scheme for evaluating the potential of cascading failure and predicting the blackout severity resulting from the outage of each line of the power system. This intelligent scheme includes three CSPDTs that the logical combination of their outputs makes

Table 10.11 Logical combinations of CSPDTs outputs

State num.	CSPDT1	CSPDT2	CSPDT3	Logic/ illogical	Malfunction	Initial decision	Logical decision
1	0	0	0	Logical	–	Non-critical	Non-critical
2	1	0	0	Logical		Critical	Critical
3	1	1	0	Logical	–	Severely critical	Severely critical
4	1	1	1	Logical	–	Extremely critical	Extremely critical
5	1	0	1	Illogical	CSPDT2 or CSPDT3	Extremely critical or critical	Extremely critical or critical
6	0	1	1	Illogical	CSPDT1	Extremely critical	Extremely critical
7	0	1	0	Illogical	CSPDT1 or CSPDT2	Critical or severely critical	Non-critical or severely critical
8	0	0	1	Illogical	CSPDT3	Extremely critical	Non-critical

the proposed scheme able to evaluate the potential of cascading failure and predict the blackout severity is different sizes.

In the proposed scheme, the DOVs are identified using MIM based on mutual information theory and entropy to evaluate the potential of cascading failure and predict blackout severity. Then, the CSPDTs are trained using DOVs and various disturbance scenarios. According to the results, the CSPDTs can evaluate the potential of cascading failure and predict the resulting blackout severity with high accuracy. The DOVs and trained CSPDTs of each line are provided to the PSCC to evaluate the potential of cascading failure. In PSCC, the DOVs information is prepared online at each instant by the WAMS and provide to the trained CSPDTs of each line. The zero output of all CSPDTs at each operating point of the system describes the secure operation of the system without CFOP.

The proposed scheme can be used as a warning system to enable emergency actions and preventive control as a prediction of the severe blackout. In such a case, PSCC can determine the potential of cascading failure and the resulting blackout severity at each operating point, knowing the DOVs and the corresponding CSPDTs of each line. Then, the PSCC can transfer the operating point of the power system from critical to non-critical, if necessary, so that the power system always operates in the secure area without cascading failure occurrence.

10.8 Suggestion for Future Research

In this study, the power system vulnerability after evaluating CFOP is examined by predicting the blackout severity. The blackout severity index is used as a static vulnerability measure of the power system. In future studies, in addition to the blackout severity prediction as a measure of the power system vulnerability, other indices like frequency and rotor angle can be used to predict the dynamic stability of the power system.

References

1. S. Henry et al., New trends for the assessment of power system security under uncertainty. IEEE Power Systems Conference and Exposition **3**, 1380–1385 (2004)
2. D. Ajendra, *Vulnerability Analysis and Fault Location in Power Systems Using Complex Network Theory*. PhD thesis, (College of Science, Engineering and Health RMIT University, 2011)
3. L. Zhou, *Multi-Agent System Based Special Protection and Emergency Control Scheme against Cascading Events in Power System*. PhD thesis, (The Faculty of Engineering, Science and Medicine, Aalborg University, 2013)
4. NERC. *Protection System Response to Power Swings*. http://www.nerc.com (2013)
5. A. Abedi et al., Review of major approaches to analyze vulnerability in power system Reliability Engineering & System Safety. Reliability Engineering & System Safety **183**, 153–172 (2019)

6. D. Seyed Javan et al., Information extraction from effective descriptor variables in reconstruction of power system security region by considering correlation between loads. International Transaction Electrical Energy System **13**, 145–181 (2017)
7. R. Casimir, E. Boutleux, G. Clerc, A. Yahoui, The use of features selection and nearest neighbors rule for faults diagnostic in induction motors. Eng. Appl. Artif. Intell. **19**(2), 169–177 (2006)
8. Chang H and Viet L.N (2017) Statistical feature extraction for fault locations in nonintrusive fault detection of low voltage distribution systems. Energies 10(5):611
9. H. Muhammad Faisal et al., Prediction of building energy consumption using enhance convolutional neural network. Web, Artificial Intelligence and Network Applications **927**, 1157–1168 (2019)
10. Saunders C.S et al. (2019) Feature extraction-based real-time transient stability analysis. Techno Econ Smart Grids Sustain Energy 4(15)
11. F. Cadini et al., Estimation of rare event probabilities in power transmission networks subject to cascading failures. Reliability Engineering & System Safety **158**, 9–20 (2017)
12. P. Henneaux, Probability of failure of overloaded lines in cascading failures. Int. J. Electr. Power Energy Syst. **73**, 141–148 (2015)
13. S.N. Ashida et al., Risk assessment of dynamic system cascading collapse for determining the sensitive transmission lines and severity of total loading conditions. Reliability Engineering & System Safety **157**, 113–128 (2017)
14. M. Shengwei et al., An improved OPA model and blackout risk assessment. IEEE Trans. Power Syst. **24**(2), 814–823 (2009)
15. H. Ren et al., Using transmission line outage data to estimate cascading failure propagation in an electric power system. IEEE Transactions on Circuits and systems **55**(9), 927–931 (2008)
16. J. Qi et al., An interaction model for simulation and mitigation of cascading failures. IEEE Trans. Power Systems **30**(2), 804–819 (2015)
17. P. Rezaei et al., Estimating cascading failure risk with random chemistry. IEEE Trans. Power Syst. **30**(5), 2726–2735 (2015)
18. J. Kim et al., Estimating propagation and distribution of load shed in simulations of cascading blackouts. IEEE Syst. J. **6**(3), 548–557 (2012)
19. L. Che et al., Identification of cascading failure initiated by hidden multiple-branch contingency. IEEE Trans. Reliab. **68**(1), 149–160 (2019)
20. M.H. Athari, Z. Wang, Stochastic cascading failure model with uncertain generation using unscented transform. IEEE Transactions on Sustainable Energy **11**(2), 1067–1077 (2019)
21. J. Guo et al., Quantifying the influence of component failure probability on cascading blackout risk. IEEE Trans. Power Syst. **33**(5), 5671–5681 (2018)
22. J. Cepeda et al., Data-mining-based approach for predicting the power system post-contingency dynamic vulnerability status. International Transactions on Electrical Energy Systems **25**(10), 2515–2546 (2015)
23. S. Gupta et al., Support-vector-machine-based proactive Cascade prediction in smart grid using probabilistic framework. IEEE Trans. Ind. Electron. **62**(4), 2478–2486 (2015)
24. I. Dobson, Estimating the propagation and extent of cascading line outages from utility data with a branching process. IEEE Trans. Power Syst. **27**(4), 2146–2155 (2012)
25. Y. Zhang et al., Mitigating blackouts via smart relays: A machine learning approach. Proc. IEEE **99**(1), 94–118 (2011)
26. M.R. Salimian, M.R. Aghamohammadi, A three stages decision tree-based intelligent blackout predictor for power systems using brittleness indices. IEEE Transactions on Smart Grid **9**(5), 5123–5131 (2018)
27. M. Abedi, M.R. Aghamohammadi, DT based intelligent predictor for out of step condition of generator by using PMU data. Electr. Power Energy Syst. **99**, 95–106 (2018)
28. N. Senroy et al., Decision tree assisted controlled islanding. IEEE trans. Power System **21**, 1790–1797 (2006)

29. G.V. Kass, An exploratory technique for investigating large quantities of categorical data. Application Statistic **29**(2), 119–127 (1980)
30. L. Breiman et al., *Classification and Regression Trees* (Wadsworth International Group, Belmont, CA, 1984)
31. K. Max, K. Johnson, *Applied Predictive Modeling* (Springer, New York, 2013)
32. H. Liu et al., Feature selection with dynamic mutual information. Elsevier, Pattern Recognition **42**(7), 1330–1339 (2009)
33. B. Remeseiro, V. Bolon-Canedo, A review of feature selection methods in medical applications. Comput. Biol. Med. **112**, 1–9 (2019)
34. T. Cover, Thomas, *Elements of Information Theory* (Wiley, New York, 2006)
35. M. Bennasar et al., Feature selection using joint mutual information maximization. Expert Syst. Appl. **42**, 8520–8532 (2015)
36. B. Zhang et al., Mutual information maximization-based collaborative data collection with calibration constraint. IEEE Access **7**, 21188–21200 (2019)
37. M. Khaji, M.R. Aghamohammadi, Online emergency damping controller to suppress power system inter-area oscillation using load-generation tripping. Electrical Power System Research **140**, 806–820 (2016)
38. Khaji M and Aghamohammadi M.R (2017) Emergency transmission line switching to suppress power system inter-area oscillation. Int. J. Electr. Power Energy Syst. 87:52–64
39. M.R. Salimian, M.R. Aghamohammadi, Intelligent out of step predictor for inter area oscillations using speed-acceleration criterion as a time matching for controlled islanding. IEEE Transactions on Smart Grid **9**(4), 2488–2497 (2016)
40. M. Majidi, M.R. Aghamohammadi, New design of intelligent load shedding algorithm based on critical line overloads to reduce network cascading failure risks. Turkish Journal of Electrical Engineering & Computer Sciences **22**(6), 1395–1409 (2014)
41. M.R. Aghamohammadi et al., A new approach for mitigating blackout risk by blocking minimum critical distance relays. Int. J. Elect. Power Energy Syst. **75**, 162–172 (2016)
42. Z. Shuangxi et al., *Power System Voltage Stability and its Control* (China Electric Power Press, Beijing, 2003)

Chapter 11
LSTM-Assisted Heating Energy Demand Management in Residential Buildings

Amin Mansour-Saatloo, Arash Moradzadeh, Sahar Zakeri, and Behnam Mohammadi-Ivatloo

11.1 Introduction

The proliferation of urbanization in recent years requires a great deal of multiple energy sources to supply citizens. However, lack of energy sources and limitation of fossil fuel-based sources usage owing to climate change and global warming push researchers to find optimal energy management methods to handle this issue [1, 2]. Load prediction is one of the promising solutions, which can assist buildings' energy systems to operate in an optimal and reasonable way. In the case of residential buildings, thermal energy is the foremost used energy. Heating load prediction is the premise for heating operation and dispatching [3].

According to research, commercial and residential buildings today have the highest daily energy consumption in people's lives. Therefore, intelligent control of building energy consumption can be considered as the most important tool in energy saving and management. Many solutions have been introduced in various studies for energy management of buildings, each of which has its own advantages and disadvantages. Meanwhile, forecasting the heating load of a building can be considered as one of the most basic management solutions in the energy of buildings, which can have better advantages than other methods [3, 4].

A. Mansour-Saatloo · A. Moradzadeh · S. Zakeri
Faculty of Electrical and Computer Engineering, University of Tabriz, Tabriz, Iran
e-mail: amin_mnsr97@ms.tabrizu.ac.ir; arash.moradzadeh@tabrizu.ac.ir; s_zakeri@tabrizu.ac.ir

B. Mohammadi-Ivatloo (✉)
Faculty of Electrical and Computer Engineering, University of Tabriz, Tabriz, Iran

Department of Energy Technology, Aalborg University, Aalborg, Denmark
e-mail: bmohammadi@tabrizu.ac.ir

M. Nazari-Heris et al. (eds.), *Application of Machine Learning and Deep Learning Methods to Power System Problems*, Power Systems,
https://doi.org/10.1007/978-3-030-77696-1_11

Prediction of heat load has a complicated procedure that requires dealing with a nonlinear optimization problem. So far, many scholars were studied heat load prediction deploying various methods. Some literature was deployed regression analysis algorithms, e.g., a nonlinear autoregressive exogenous model (NARX) with external input was used in [5] to commercial buildings heat load prediction. In [6], a real-time prediction based on nonlinear predictive control besides advanced machine learning methods was developed to heating and cooling load prediction. The performance of the linear and nonlinear autoregressive model with exogenous inputs was discussed in [7] to different load prediction. In addition, in [8, 9] autoregressive moving average (ARMA) and in [10, 11] autoregressive integrated moving average (ARMIA) scheme were deployed to load prediction. In [12], seasonal autoregressive integrated moving average (SARIMA) was deployed to heat load prediction. However, owing to multiple dimensions of heat load characteristics, regression-based methods cannot be further developed. In addition, regression methods require time series and cannot utilize long-term historical data.

Realization of the data-driven modeling and prediction with neural network and machine learning development cause to resolve the aforementioned problems. An online sequential extreme learning machine (OSELM) was developed in [13] to heating and cooling load prediction. Multilayer perceptron (MLP) and support vector machine (SVM) performance for heat load prediction were analyzed in [14]. Nine different extreme machine learning methods were developed for heating load forecasting in [15]. A dynamic neural network was deployed in [16] to heating load prediction of an apartment building. Various machine learning techniques including SVM, regression tree, etc. were applied in [17] to residential and commercial buildings load prediction. One of the deep learning applications called deep neural network (DNN) has been employed in [18] to predict the heating and cooling loads of a residential building. The proposed scheme in this study is trained based on the structural features of the building as network input to predict the heating and cooling loads as the network output. In [19], the firefly searching algorithm to optimize SVM parameters was used in the case of heating load prediction based on the SVM algorithm. In a similar context, the least-square SVM (LSSVM) was used to multiple load prediction in [20]. Support vector regression (SVR), DNN, and extreme gradient boosting were utilized to heating load prediction in [21]. In [22], a temporal conventional neural network (CNN) was deployed to heating load prediction. In [23], strand-based long short-term memory (LSTM) was used to heating load prediction. A review of some valuable literature related to heating load forecasting was done. Table 11.1 lists the literature in question by utilized method, evaluation metric, data type, and purpose.

Some of the above reviewed studies suffer from the inability of the proposed method to analyze continuous data and others from the lack of correlation between input data. In this paper, prediction of heating load in building is performed using the LSTM method as a suitable method for time series data analysis. The LSTM predicts the heating load by applying on data related to basic factors in the structure of the building.

Table 11.1 Review of studies related to building heating load predicting

Refs.	Year	Data type	Method	Evaluation metric	Propose
[5]	2017	Real	NARX	Mean squared normalized error	Estimating heat demand of commercial buildings
[6]	2014	Real	Adaptive Gaussian process, hidden Markov model, episode discovery and semi-Markov model	Root mean square error (RMSE)	Estimating local weather condition to reduce the heating and cooling energy consumption
[7]	2014	Real	Linear ARX, NARX	Mean absolute percentage error (MAPE)	Estimating of heating, cooling and load forecasting
[8]	2003	Real	ARMA	Relative error	Estimating electrical load
[9]	2013		ARMA, ARIMA, autoregressive ANN	Root mean square error (RMSE), mean bias error (MBE)	Comparison of different methods in forecasting of inflow of Dez dam reservoir
[11]	2018	Real	ARMIA	Mean absolute percentage error (MAPE)	Estimating electrical load of residential and industrial buildings
[12]	2016	Real	SARMIA	Adapted MAPE (AMAPE)	Estimating heat demand
[13]	2018	Simulated	OSELM	Mean absolute error (MAE)	Estimating energy consumption and energy-efficient building design
[14]	2018	Real	MLP, SVM	Root mean square error (RMSE)	Estimating heat demand
[15]	2016	Real	Multi-step ahead predictive model based on extreme leaning machine	Root means square error (RMSE)	Estimating heat demand
[16]	2016	Real	Dynamic neural network (DNN)	Root mean square error (RMSE), the mean absolute percentage error (MAPE), and the absolute fraction of variation (R2)	Estimating heating and cooling demand of apartment buildings
[17]	2016	Real	SVM, regression tree, feed forward neural network, multiple linear regression	Root mean square error (RMSE)	Estimating of heat demand of residential and commercial buildings
[18]	2020	Simulated	DNN		

(continued)

Table 11.1 (continued)

Refs.	Year	Data type	Method	Evaluation metric	Propose
				Variance accounted for (VAF)	Estimating energy consumption and energy-efficient building design
[19]	2016	Real	Optimized SVM	RMSE, Pearson correlation coefficient, coefficient of determination	Estimating heat demand
[20]	2020	Real	LSSVM	MAPE, RMSE	Estimating of electricity, heating, cooling, and gas demand of industrial park
[21]	2019	Real	SVR, DNN, XGBoost	RMSE	Estimating heat demand
[22]	2020	Real	TCN	MAE, MAPE, RMSE	Estimating heat demand of residential buildings
[23]	2020	Real	Strand-based LSTM	Mean absolute percentage error (MAPE)	Heating operation management and dispatching

LSTM falls into the recurrent neural network (RNN) and shows excellent performance in different applications. Indeed, the LSTM is the improved version of the RNN that covers disadvantages of it. In [24], LSTM was utilized to wind energy prediction. Extreme Gradient Boost (XGBoost) and LSTM as shallow learning and deep learning tools were compared in [25] in the building short-term (i.e., 1 h ahead) load prediction area. Long-term energy consumption was predicted using LSTM in [26]. In [27, 28], photovoltaic energy production was predicted using LSTM. In [29], the LSTM network was trained using pinball loss instead of mean square error (MSE) for individual consumers load prediction. In addition to the power system applications, LSTM was used in medical diagnosis [30, 31], air pollution prediction [32], human action recognition [33], etc.

According to the reviewed literature and to the best of the author's knowledge, there is no application of the LSTM in the residential buildings heating load prediction. To fill this research gap, this chapter focuses on the heating load prediction of residential buildings using LSTM as a deep learning technique. As it is known, heating consumption of buildings is mainly depending on weather and environmental conditions. In doing so, this chapter aims to conduct the prediction based on weather and environmental conditions. Environmental and weather characteristics along with buildings structure characteristics are the input of the LSTM and heating value is the target.

The rest of this chapter is organized as follows: Section 11.2 describes the detail of the utilized dataset. Section 11.3 provides the LSTM mathematical model.

Section 11.4 discusses the obtained results from the simulation. Finally, the last section concludes the chapter.

11.2 Dataset

The utilized dataset in this study contains 768 samples with 8 attributes and 1 decision variable for each sample. z_1, z_2, . . ., z_8 are the attributes, which refer to eight different aspects, i.e., relative compactness, surface area, wall area, roof area, overall height, orientation, glazing area, and glazing area distribution, that aim to predict the heat value w as a decision variable. All the simulations were conducted at Ecotect software for 12 residential buildings by Tsanas and Xifara [34]. Each building was made up using 18 elementary cubes ($3.5 \times 3.5 \times 3.5$) with identical materials for all buildings. It was assumed that buildings are located in Greece, Athens, and 60% humidity, 0.6 clothing, 0.3 m/s air speed, and 300 lx lightning level were considered as environmental factors. Moreover, infiltration of 0.5 for air change with the sensitivity of 0.25 air changer per hour was applied during simulations. Table 11.1 reviews the features and response of the utilized dataset.

11.3 LSTM Mathematical Model

Today, deep learning plays an increasing role in most parts of people's daily lives. It has been able to make significant progress in scientific and industrial applications such as diagnosing a variety of diseases in medicine, face recognition and fingerprinting in security issues, fault detection in power systems, and energy forecasting issues around the world. Deep learning has a variety of techniques, each of which can play a role in different applications [35, 36]. Among the various types of deep learning techniques, the LSTM is used as a powerful tool with a very high ability to process time series data. The LSTM is the improved algorithm of the RNN procedure, which can overcome RNN's drawbacks such as the vanishing gradient problem with keeping memory in their channels [37]. Adding and removing information are possible in the LSTM neural network by using cell state through the following gates that make the LSTM superior to the RNN [23]. The mathematical structure of LSTM is formulated as follows [23, 37]:

$$i_t = \sigma(W_{ii}z_t + b_{ii} + W_{hi}h_{t-1} + b_{hi}) \tag{11.1}$$

$$f_t = \sigma\left(W_{if}z_t + b_{if} + W_{hf}h_{t-1} + b_{hf}\right) \tag{11.2}$$

$$g_t = \tanh\left(W_{ig}z_t + b_{ig} + W_{hg}h_{t-1} + b_{hg}\right) \tag{11.3}$$

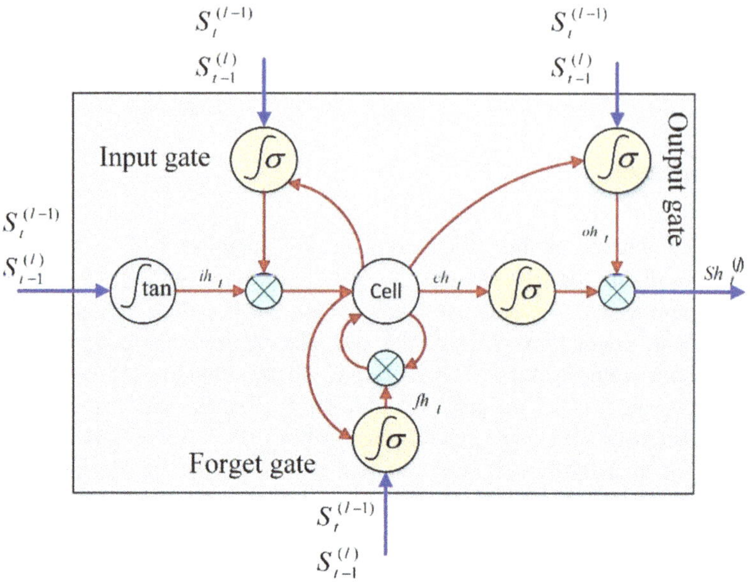

Fig. 11.1 Main block diagram corresponding to the one unit of LSTM

$$o_t = \sigma(W_{io}z_t + b_{io} + W_{ho}h_{t-1} + b_{ho}) \tag{11.4}$$

$$c_t = f_t * c_{t-1} + i_t * g_t \tag{11.5}$$

$$h_t = o_t * \tanh(c_t) \tag{11.6}$$

$$\sigma(z) = \frac{1}{1 + e^{-z}} \tag{11.7}$$

where, i_t, f_t, g_t, $o_t \in \mathbb{R}^h$ are the input gate, forget gate, cell gate, and output gate, respectively; c_t, $h_t \in \mathbb{R}^h$ are the cell state and hidden state; $z_t \in \mathbb{R}^d$ is the input vector; σ is the sigmoid function that is given in Eq. (11.7); W_{ii}, W_{if}, W_{ig}, $W_{io} \in \mathbb{R}^{h \times d}$ are the trainable parameters of input vector for input, forget, cell, and output gates, respectively, as well as W_{hi}, W_{hf}, W_{hg}, $W_{ho} \in \mathbb{R}^{h \times h}$ for output vector; b_{ii}, b_{hi}, b_{if}, b_{hf}, b_{ig}, b_{hg}, b_{io}, $b_{ho} \in \mathbb{R}^h$ are trainable biases; and * is the scalar product and *tanh* is the active function. Figure 11.1 shows the LSTM block diagram based on the equations provided.

11.4 Simulation and Numerical Results

In this chapter, predicting the heating load of a building by the LSTM is done by considering the influential factors in the structure of the building as input data and the amount of heating load related to each condition as a target. When the network is

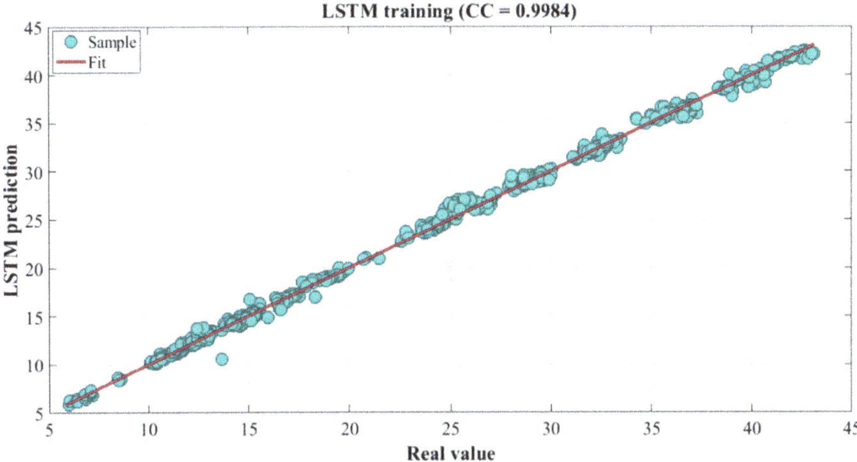

Fig. 11.2 Regression results for heating load prediction by LSTM in training stage

trained in the amount of heating load used for each building condition, it will be able to determine the amount of consumption for each structure. In the training phase, the network learns data-related behavioral patterns so that it can predict system behavior for the future. Thus, when the LSTM is trained based on some basic parameters affecting the structure of the building and the heating load associated with them, it will be possible to predict the heating load of other buildings based on structural parameters. Accordingly, an ideal approach to managing heating energy consumption in buildings is presented to consumers. Once designed, the LSTM is trained with 70% of the data, and the remaining 30% of the data is used for network test. Figure 11.2 shows the heating load prediction results for the training data. This figure provides a correlation coefficient (CC) for LSTM in predicting the heating load by training data. It can be seen that the prediction was done with high accuracy (CC = 0.9984) and the network was able to train well the behavioral patterns of the building heating load in various conditions. Figure 11.3 evaluates the error values related to the forecast results for the training stage in mean squared error (MSE) and mean absolute error (MAE) formats.

The network learns data-related behavioral patterns during the training stage and after completing this stage can be saved as a black box to be used for future predictions. The heating load prediction results for the test data are presented in Fig. 11.4. It can be seen that when the network is trained with high accuracy, it is able to provide a favorable prediction for test data. The evaluation of the network error values for predicting test data by MSE and RMSE indicators is shown in Fig. 11.5. After training and testing the network and based on the accuracy and error values obtained, a prediction error in the form of a histogram can be obtained. Figure 11.6 shows the test error in the form of a histogram.

From the results presented in Fig. 11.4, it is clear that the LSTM is able to predict the heating load of the building for test data with high accuracy (0.9980) with

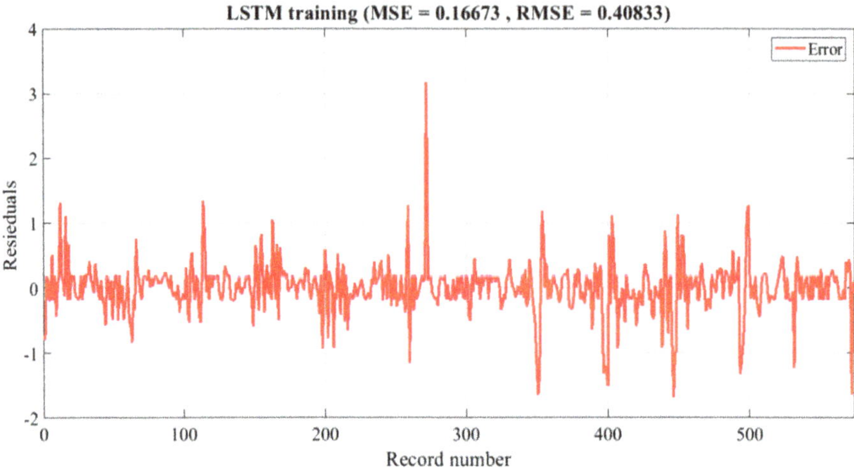

Fig. 11.3 Error values for training stage of heating load prediction by LSTM in the forms of MSE and RMSE

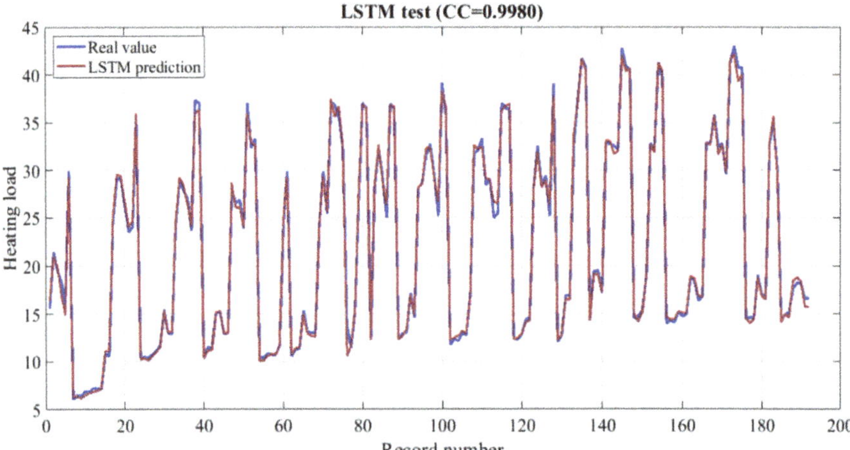

Fig. 11.4 Heating load prediction by LSTM for test data

previous appropriate training. Also, the error values presented in Figs. 11.5 and 11.6 indicate the proper performance of the LSTM network during the test stage. According to the results presented in the training and test of the LSTM network to predict the heating load of the building, it is now possible to predict the amount of heating load by considering the new input data for the LSTM network. So, it is understandable that deep learning techniques in addition to the ability in other sciences and industrial applications can be considered as a powerful tool in the energy management of buildings or energy networks. It should be noted that the

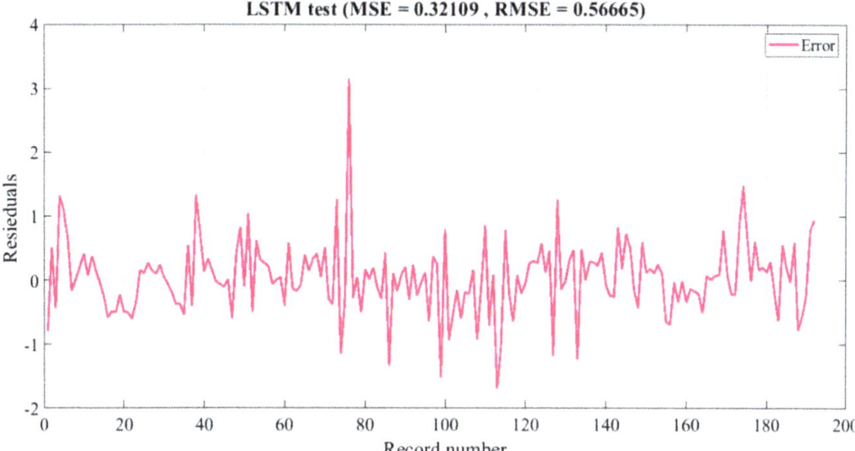

Fig. 11.5 LSTM error values for predicting heating load by test data in the forms of MSE and RMSE

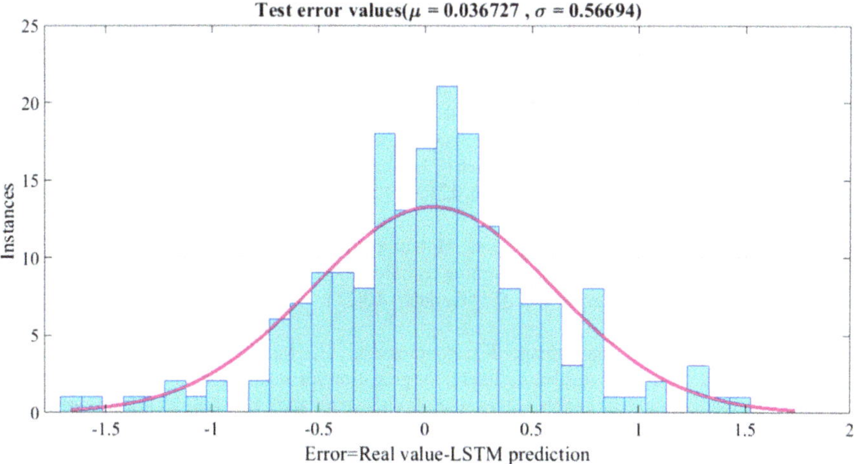

Fig. 11.6 LSTM error values for predicting heating load by test data in the form of histogram

proposed method can provide acceptable performance for real-world data, especially time-series data.

11.5 Conclusion

Due to the amount of energy consumption by commercial and residential buildings, energy management in these buildings has created many challenges. In this paper, energy management in a residential building was performed by predicting the heating load of the building. To make an accurate prediction, one of the deep learning techniques called LSTM was introduced and employed. To use the proposed technique, data related to a simulated building in Ecotect software were utilized. The basic parameters in the structure of the building were considered as the LSTM input variables to predict the heating load of the building as an output variable. After training and test the network, the results were analyzed using statistical evaluation indicators. The results showed the ability of the LSTM method with accuracy CC = 99.84% and CC = 99.80% for the training and test stages of heating load forecasting of a residential building, respectively.

References

1. A. Mansour-Saatloo, A. Moradzadeh, B. Mohammadi-Ivatloo, A. Ahmadian, A. Elkamel, Machine learning based PEVs load extraction and analysis. Electronics (Switzerland) **9**(7), 1–15 (2020). https://doi.org/10.3390/electronics9071150
2. M.Z. Oskouei, B. Mohammadi-Ivatloo, M. Abapour, A. Ahmadian, M.J. Piran, A novel economic structure to improve the energy label in smart residential buildings under energy efficiency programs. J. Clean. Prod. **260**, 121059 (2020). https://doi.org/10.1016/j.jclepro.2020.121059
3. A. Moradzadeh, A. Mansour-Saatloo, B. Mohammadi-Ivatloo, A. Anvari-Moghaddam, Performance evaluation of two machine learning techniques in heating and cooling loads forecasting of residential buildings. Appl. Sci. (Switzerland) **10**(11), 3829 (2020). https://doi.org/10.3390/app10113829
4. A. Moradzadeh, O. Sadeghian, K. Pourhossein, B. Mohammadi-Ivatloo, A. Anvari-Moghaddam, Improving residential load disaggregation for sustainable development of energy via principal component analysis. Sustainability (Switzerland) **12**(8), 3158 (2020). https://doi.org/10.3390/SU12083158
5. A. Sandberg, F. Wallin, H. Li, M. Azaza, An analyze of Long-term Hourly District heat demand forecasting of a commercial building using neural networks. Energy Procedia **105**, 3784–3790 (2017). https://doi.org/10.1016/j.egypro.2017.03.884
6. B. Dong, K.P. Lam, A real-time model predictive control for building heating and cooling systems based on the occupancy behavior pattern detection and local weather forecasting. Building Simulation **7**(1), 89–106 (2014). https://doi.org/10.1007/s12273-013-0142-7
7. K.M. Powell, A. Sriprasad, W.J. Cole, T.F. Edgar, Heating, cooling, and electrical load forecasting for a large-scale district energy system. Energy **74**, 877–885 (2014)
8. S.-J. Huang, K.-R. Shih, Short-term load forecasting via ARMA model identification including non-Gaussian process considerations. IEEE Trans. Power Syst. **18**(2), 673–679 (2003)
9. M. Valipour, M.E. Banihabib, S.M.R. Behbahani, Comparison of the ARMA, ARIMA, and the autoregressive artificial neural network models in forecasting the monthly inflow of Dez dam reservoir. J. Hydrol. **476**, 433–441 (2013)
10. J. Contreras, R. Espinola, F.J. Nogales, A.J. Conejo, ARIMA models to predict next-day electricity prices. IEEE Power Engg. Rev. **22**(9), 57 (2002)

11. D. Alberg, M. Last, Short-term load forecasting in smart meters with sliding window-based ARIMA algorithms. Vietnam J. Comput. Sci. **5**(3–4), 241–249 (2018)
12. T. Fang, R. Lahdelma, Evaluation of a multiple linear regression model and SARIMA model in forecasting heat demand for district heating system. Appl. Energy **179**, 544–552 (2016)
13. S. Kumar, S.K. Pal, R.P. Singh, A novel method based on extreme learning machine to predict heating and cooling load through design and structural attributes. Energ. Buildings **176**, 275–286 (2018). https://doi.org/10.1016/j.enbuild.2018.06.056
14. M. Dahl, A. Brun, O.S. Kirsebom, G.B. Andresen, Improving short-term heat load forecasts with calendar and holiday data. Energies **11**(7), 1678 (2018). https://doi.org/10.3390/en11071678
15. S. Sajjadi et al., Extreme learning machine for prediction of heat load in district heating systems. Energ. Buildings **122**, 222–227 (2016)
16. S. Sholahudin, H. Han, Simplified dynamic neural network model to predict heating load of a building using Taguchi method. Energy **115**, 1672–1678 (2016)
17. S. Idowu, S. Saguna, C. Åhlund, O. Schelén, Applied machine learning: Forecasting heat load in district heating system. Energ. Buildings **133**, 478–488 (2016). https://doi.org/10.1016/j.enbuild.2016.09.068
18. S.S. Roy, P. Samui, I. Nagtode, H. Jain, V. Shivaramakrishnan, B. Mohammadi-ivatloo, Forecasting heating and cooling loads of buildings: A comparative performance analysis. J. Ambient. Intell. Humaniz. Comput. **11**(3), 1253–1264 (2020). https://doi.org/10.1007/s12652-019-01317-y
19. E.T. Al-Shammari et al., Prediction of heat load in district heating systems by support vector machine with firefly searching algorithm. Energy **95**, 266–273 (2016). https://doi.org/10.1016/j.energy.2015.11.079
20. Z. Tan et al., Combined electricity-heat-cooling-gas load forecasting model for integrated energy system based on multi-task learning and least square support vector machine. J. Clean. Prod. **248**, 119252 (2020)
21. P. Xue, Y. Jiang, Z. Zhou, X. Chen, X. Fang, J. Liu, Multi-step ahead forecasting of heat load in district heating systems using machine learning algorithms. Energy **188**, 116085 (2019). https://doi.org/10.1016/j.energy.2019.116085
22. J. Song, G. Xue, X. Pan, Y. Ma, H. Li, Hourly heat load prediction model based on temporal convolutional neural network. IEEE Access **8**, 16726–16741 (2020). https://doi.org/10.1109/ACCESS.2020.2968536
23. J. Liu, X. Wang, Y. Zhao, B. Dong, K. Lu, R. Wang, Heating load forecasting for combined heat and power plants via Strand-based LSTM. IEEE Access **8**, 33360–33369 (2020). https://doi.org/10.1109/ACCESS.2020.2972303
24. F. Shahid, A. Zameer, A. Mehmood, M.A.Z. Raja, A novel wavenets long short term memory paradigm for wind power prediction. Appl. Energy **269**, 115098 (2020)
25. Z. Wang, T. Hong, M.A. Piette, Building thermal load prediction through shallow machine learning and deep learning. Appl. Energy **263**, 114683 (2020)
26. J.Q. Wang, Y. Du, J. Wang, LSTM based long-term energy consumption prediction with periodicity. Energy **197**, 117197 (2020)
27. K. Wang, X. Qi, H. Liu, Photovoltaic power forecasting based LSTM-Convolutional Network. Energy **189** (2019). https://doi.org/10.1016/j.energy.2019.116225
28. M. Gao, J. Li, F. Hong, D. Long, Day-ahead power forecasting in a large-scale photovoltaic plant based on weather classification using LSTM. Energy **187**, 115838 (2019)
29. Y. Wang, D. Gan, M. Sun, N. Zhang, Z. Lu, C. Kang, Probabilistic individual load forecasting using pinball loss guided LSTM. Appl. Energy **235**, 10–20 (2019). https://doi.org/10.1016/j.apenergy.2018.10.078
30. M. Rahman, I. Saha, D. Islam, R.J. Mukti, A deep learning approach based on convolutional LSTM for detecting diabetes. Comput. Biol. Chem., 107329 (2020)

31. İ. Kırbaş, A. Sözen, A.D. Tuncer, F.Ş. Kazancıoğlu, Comperative analysis and forecasting of COVID-19 cases in various European countries with ARIMA, NARNN and LSTM approaches. Chaos, Solitons & Fractals, 110015 (2020)
32. Y.-S. Chang, H.-T. Chiao, S. Abimannan, Y.-P. Huang, Y.-T. Tsai, K.-M. Lin, *An LSTM-Based Aggregated Model for Air Pollution Forecasting* (Atmospheric Pollution Research, New York, 2020)
33. Z. Zhang, Z. Lv, C. Gan, and Q. Zhu, *Human Action Recognition Using Convolutional LSTM and Fully-Connected LSTM with Different Attentions*. (Neurocomputing, 2020)
34. A. Tsanas, A. Xifara, Accurate quantitative estimation of energy performance of residential buildings using statistical machine learning tools. Energ. Buildings **49**, 560–567 (2012). https://doi.org/10.1016/j.enbuild.2012.03.003
35. A. Moradzadeh and K. Pourhossein, Short circuit location in transformer winding using deep learning of its frequency responses. In *Proceedings 2019 International Aegean Conference on Electrical Machines and Power Electronics, ACEMP 2019 and 2019 International Conference on Optimization of Electrical and Electronic Equipment, OPTIM 2019*, (2019), pp. 268–273, doi: https://doi.org/10.1109/ACEMP-OPTIM44294.2019.9007176
36. A. Shrestha, A. Mahmood, Review of deep learning algorithms and architectures. IEEE Access **7**, 53040–53065 (2019). https://doi.org/10.1109/ACCESS.2019.2912200
37. A. Moradzadeh, S. Zakeri, M. Shoaran, B. Mohammadi-Ivatloo, F. Mohamamdi, Short-term load forecasting of microgrid via hybrid support vector regression and long short-term memory algorithms. Sustainability **12**(17), 7076 (2020)

Chapter 12
Wind Speed Forecasting Using Innovative Regression Applications of Machine Learning Techniques

Arash Moradzadeh, Amin Mansour-Saatloo, Morteza Nazari-Heris, Behnam Mohammadi-Ivatloo, and Somayeh Asadi

12.1 Introduction

The increasing energy users and the widespread expansion of renewable energy sources (RESs) have significantly expanded the world's distribution systems. Consumers and especially users of sensitive loads tend to have access to a reliable and sustainable power supply [1]. Therefore, power producers need short-term planning for sustainable investment, production, and operation. On the other hand, the exorbitance use of fossil fuels, including oil, natural gas, and coal, paves the way for climate change and global warming by releasing a high amount of greenhouse gases, especially carbon dioxide [2, 3]. For a better future, it is needed to limit the usage of fossil fuels. One promising solution is utilizing renewable energy sources

A. Moradzadeh · A. Mansour-Saatloo
Faculty of Electrical and Computer Engineering, University of Tabriz, Tabriz, Iran
e-mail: arash.moradzadeh@tabrizu.ac.ir; amin_mnsr97@ms.tabrizu.ac.ir

M. Nazari-Heris
Faculty of Electrical and Computer Engineering, University of Tabriz, Tabriz, Iran

Department of Architectural Engineering, Pennsylvania State University,
State College, PA, USA
e-mail: mun369@psu.edu

B. Mohammadi-Ivatloo (✉)
Faculty of Electrical and Computer Engineering, University of Tabriz, Tabriz, Iran

Department of Energy Technology, Aalborg University, Aalborg, Denmark
e-mail: bmohammadi@tabrizu.ac.ir

S. Asadi
Department of Architectural Engineering, Pennsylvania State University,
State College, PA, USA

M. Nazari-Heris et al. (eds.), *Application of Machine Learning and Deep Learning Methods to Power System Problems*, Power Systems,
https://doi.org/10.1007/978-3-030-77696-1_12

(RESs) in the energy sector instead of fossil fuel-based plants. Wind, solar, hydro, biomass, tidal, and geothermal are the main RESs worldwide [4]. However, wind energy owing to its potentials in implementing and operating has a significant place among these alternatives. So, a viable and secure forecasting mechanism can assist the power system's operation more steadily and reliably [5, 6]. The wind energy is mostly depending on wind speed, so it is stochastic and volatile, which causes an adverse impact on the power system [7]. Wind speed is one of the weather variables with a stochastic nature and is complicated to forecast. Generally, from the forecasting time interval point of view, three different forecasting methods can be defined as short-term forecasting with less than 6 h intervals, midterm forecasting with 6 h to 1-day intervals, and long-term forecasting with more than 1 day [8, 9]. Since short-term forecasting is an appropriate method for scheduling and operation of the power system, this work is focused on this method.

So far, many scholars attempted to introduce models to forecast wind energy [10]. Some of them utilized physical methods, including atmosphere physical conditions such as mass, momentum, etc. by applying physic laws, e.g., in [11], Kalman filter, and in [12], flower pollination method were used to wind speed and wind power forecasting. Some other studies were applied statistical methods, e.g., a hybrid wavelet transform (WT) based on repeated auto-regressive integrated moving average (ARIMA) was deployed in [13] to wind speed forecasting, which in order to enhance the accuracy a new model, namely, RWT-ARIMA, was introduced. In a similar context, nested ARIMA was deployed in [14] to wind speed forecasting. In [15], auto-regressive moving average (ARMA) was applied to wind speed and wind direction forecasting. The ARMA based algorithm for wind speed forecasting was introduced in [16], in which variational mode decomposition was applied to decompose the wind speed to linear, nonlinear, and noise parts. In [17], wind speed has been forecasted via several solutions such as simple average strategy, traditional combination forecasting method (TCFM) with multi-objective grasshopper optimization algorithm (MOGOA), and optimized extreme learning machine (ELM). In this study, the results of comparisons emphasize the optimized ELM's better performance than other methods. In [18], for wind speed forecasting, the general regression neural network to deal with the system's nonlinearity and the ARIMA model was developed. In [19], statistical methods were applied to forecast wind power loss due to ice growth. However, statistical models provide a linear correlation, and they are not suitable for nonlinear relationships.

In recent years artificial intelligence and machine learning techniques such as artificial neural network (ANN) [20, 21], generalized neural network (GRNN) [22], fuzzy logic models [23], support vector machine (SVM) [24], multilayer perceptron (MLP) [25], group method of data handling (GMDH) [26], etc. were developed rapidly, in which nonlinear structures can be effectively captured. In doing so, many kinds of literature have been deployed machine learning techniques to forecast the wind speed. For example, in [27], support vector regression (SVR), and in [28], SVM, were applied to forecast the wind speed. A probabilistic extreme learning machine model to wind generation forecasting was introduced in [29], where bootstrap methods were used to consider the forecasting model's uncertainty.

Three different machine learning methods, i.e., SVR, ANN, and Gaussian process, were utilized in [29] to capture the correlation between numerical weather prediction and wind output. Extreme machine learning and deep learning methods were applied in [30]. To reach a more accurate model, generalized correntropy was used instead of the Euclidean norm of the mean square error. LSTM network was utilized in [31] to extract the temporal features of different nodes in deep learning from nearby wind forms wind speed and wind direction. A tree-based machine learning method was proposed in [32] for short-term wind speed forecasting. In [33], a short-term wind power forecasting using a hybrid model, decomposition and extreme learning machine was proposed. The authors in [34] have proposed a deep learning neural network for forecasting wind power based on the databases of supervisory control and data acquisition (SCADA) with a high frequency rate. In this reference, the proposed model's main objectives are mentioned as reduction of the computational efforts and costs in the forecasting process and attaining high accurate results for the forecasted power. In [35], a numerical weather forecasting model for wind speed error correction based on gated recurrent unit neural networks has been presented to obtain a short-term prediction of wind power. A mathematical modelling is for wind power prediction is proposed in [36], where genetic algorithm as heuristic optimization method is applied for dealing with the forecasting time delay. The authors in [37] have proposed a combined model for forecasting of the wind speed for different time horizons in various locations, where three various approaches including ANN, ARIMA, and the integration of these methods have been used for evaluating the results of the forecasting process. In this reference, numerical error evaluation techniques are used for prediction of the model accuracy.

According to the reviewed scholars, there is no focus on applying the MLP and GMDH methods of machine learning in the wind power forecasting. The current work aims to investigate the performance of these methods, i.e., MLP and GMDH, in the short-term wind power forecasting. MLP is a multi-input multi-output (MIMO) system with a backpropagation learning algorithm. The specific advantages of the nonlinear multilayer structure and fast convergence make the MLP favorable for the regression problems [38, 39]. GMDH is another data-driven method that benefits from subtractive clustering for data selection and is easy to implement in modeling complex systems [26].

The rest of the chapter is organized as follows: Section 12.2 describes the utilized data set. Section 12.3 provides methods and mathematical theories. The next section discusses the simulation results, and, finally, the last section concludes the chapter.

12.2 Data Set

The selected wind farm for this study is located in the Khaf area with a longitude of 34.567° north and 60.148° east. The utilized data includes air temperature, relative humidity, solar radiation, and wind speed at a 30-m height measured with a 10-min resolution for 15 months from July 2007 to September 2008. The target is

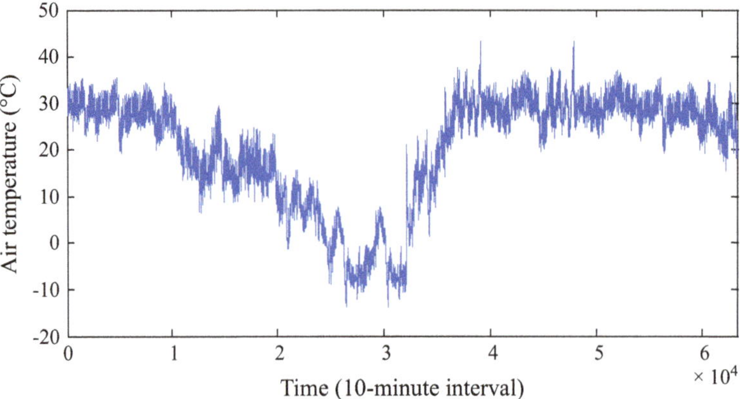

Fig. 12.1 Air temperature

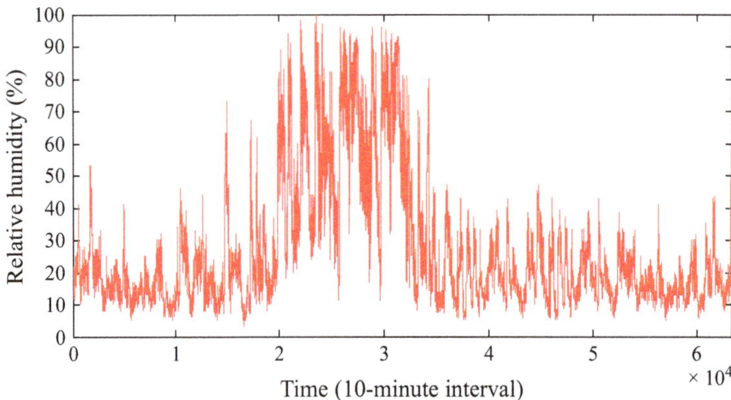

Fig. 12.2 Relative humidity

forecasting the wind speed, so the deployed machine learning techniques find the correlation between air temperature, humidity, solar radiation, and wind direction as inputs and wind speed as output. The air temperature for the studied location is shown in Fig. 12.1, which is based on a 10-min resolution. Figure 12.2 shows the utilized data set of the network for humidity. The solar radiation for the selected location is demonstrated in Fig. 12.3, which shows that the radiation has lower values at the second and third months considering the sun's location compared to the earth. Finally, the wind direction and wind speed, which are used for defining the power output of the wind turbines, are shown in Figs. 12.4 and 12.5, respectively. Because wind speed is based on many variables consisting of temperature and climate parameters such as air temperature and humidity, solar radiation, and wind direction are needed for performing the forecasting process.

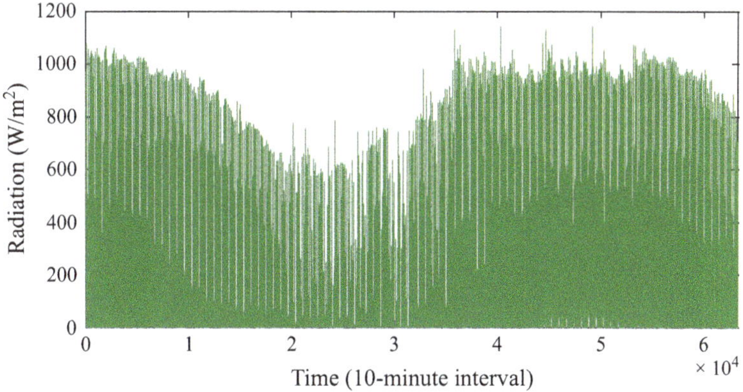

Fig. 12.3 Solar radiation

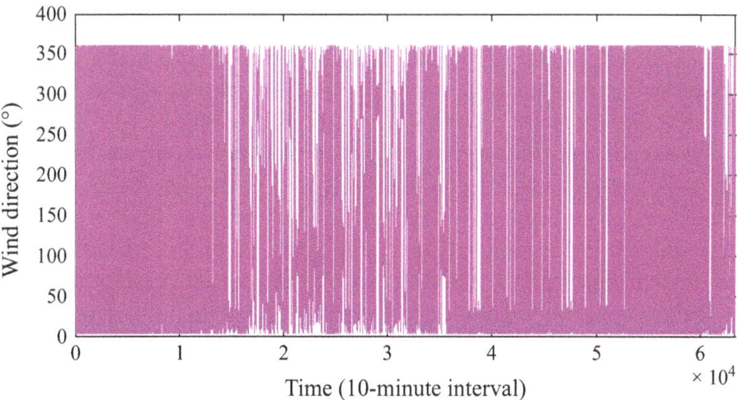

Fig. 12.4 Wind direction

12.3 Methods

Data mining was introduced as an emerging and growing technology [36]. Data mining has many techniques, each of which is used in some way to process data and a tool to discover hidden knowledge of large and complex data. Hidden knowledge in data can be considered as intrinsic patterns, features, correlations, and relationships between data. The most widely used data mining techniques are artificial neural networks and machine learning [37]. Artificial neural network (ANN) techniques, derived from human intelligence, have long been used as tools for pattern recognition, classification, feature extraction, and prediction applications. These techniques can predict future events by learning and supervising past behavior [38]. During the training process, the ANN techniques model the correlation between input variables, and then the prediction operation is performed based on

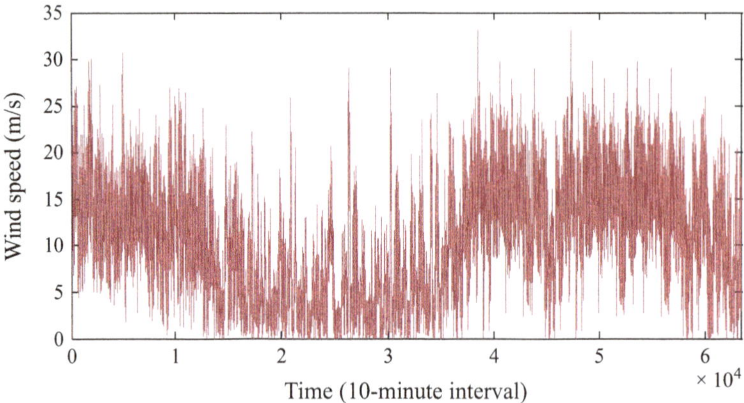

Fig. 12.5 Wind speed

the estimated model [39]. The ANNs have a variety of algorithms, one of the most widely used being multilayer perceptron (MLP) which is used in this paper for wind speed forecasting. Machine learning is one of the applications of data mining that aims to develop computer programs and artificial intelligence. Machine learning algorithms are able to identify data patterns, develop predictive models, and estimate in regression and classification modes. Learning techniques in machine learning algorithms are performed in three modes: supervised, semi-supervised, and unsupervised. High-dimensional data processing is one of the most important features of machine learning techniques [37]. Among all machine learning algorithms, the group method of data handling (GMDH) is selected as a powerful tool in big data processing in this paper for wind speed forecasting.

12.3.1 Multilayer Perceptron (MLP)

The MLP, as a neural network with a layer-to-layer and feed-forward structure, has a special position in solving regression problems and predicting continuous data. The MLP consists of one input layer, one output layer, and one or multiple hidden layers, which uses a supervised learning algorithm, namely, backpropagation. Each layer's neurons are fully connected to the previous and next layers' neurons through weighting elements. The number of layers and parameters related to the MLP structure is directly related to the problem under consideration. While the number of hidden layer neurons is mainly adjusted and selected by trial and error method. The training procedure in the MLP method is based on the minimization of the selected cost function and weight functions are defined during this procedure [25]. Each perceptron maps the inputs to the outputs using a nonlinear activation function and output signals generated via a nonlinear transfer function. The typical architecture of an MLP network is shown in Fig. 12.6 [40].

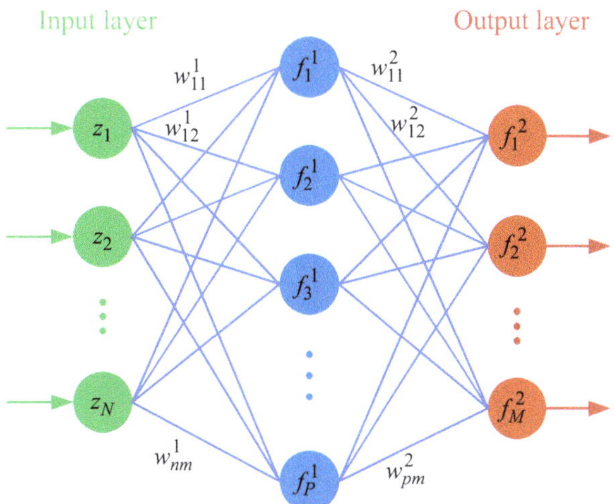

Fig. 12.6 Typical architecture of the MLP network

The input layer consists of N dimension and the hidden layer includes P neurons with n weight elements for each neuron. The matrix $I_{N \times P}$ denotes the input weight as Eq. (12.1), and the matrix $J_{P \times M}$ denotes the output weight as Eq. (12.2), where M is the number of output signals. $\mathrm{f}^1 = \left[f^1_1, f^1_2, \cdots, f^1_P\right]^{\mathrm{T}}$ and $\mathrm{f}^2 = \left[f^2_1, f^2_2, \cdots, f^2_P\right]^{\mathrm{T}}$ are the transfer function vectors of the hidden layer and output layer, respectively, as Eqs. (12.3)–(12.4); $\mathrm{z} = [z_1, z_2, \cdots, z_N]^{\mathrm{T}}$ is the input vector; $\mathrm{b}^1 = \left[b^1_1, b^1_2, \cdots, b^1_N\right]^{\mathrm{T}}$ $\mathrm{b}^2 = \left[b^2_1, b^2_2, \cdots, b^2_N\right]^{\mathrm{T}}$ are the bias vectors of the hidden layer and output layer, respectively; and $\mathrm{y}^2 = \left[y^2_1, y^2_2, \cdots, y^2_M\right]^{\mathrm{T}}$ is the output vector.

$$\mathrm{I}_{N \times P} = \begin{bmatrix} w^1_{11} & \cdots & w^1_{1P} \\ \vdots & \ddots & \vdots \\ w^1_{N1} & \ldots & w^1_{NP} \end{bmatrix} \tag{12.1}$$

$$\mathrm{J}_{P \times M} = \begin{bmatrix} w^2_{11} & \cdots & w^2_{1M} \\ \vdots & \ddots & \vdots \\ w^2_{P1} & \ldots & w^2_{PM} \end{bmatrix} \tag{12.2}$$

$$\mathrm{y}^1 = \mathrm{f}^1\left(\mathrm{b}^1 + \mathrm{I.z}\right) \tag{12.3}$$

$$\mathrm{y}^2 = \mathrm{f}^2\left(\mathrm{b}^2 + \mathrm{J.y}^1\right) \tag{12.4}$$

1.1. Group Method of Data Handling.

GMDH is a machine learning method that introduced by Alexey Grigorevich Ivakhnenko [41]. The GMDH is also known as a polynomial neural network and is a

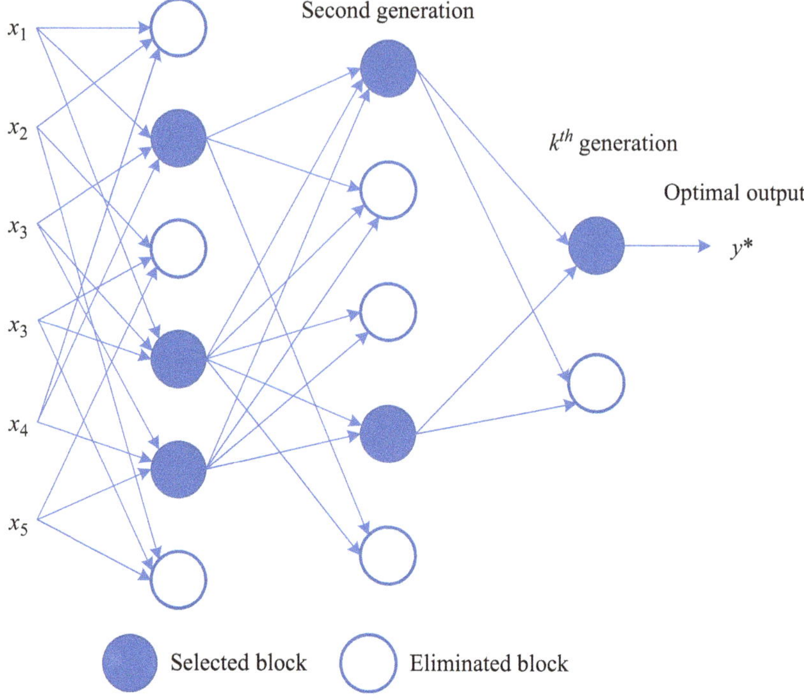

Fig. 12.7 The GMDH structure

nonlinear regression algorithm. The GMDH aims to find a precise function as much as possible that predicts the output of $\widehat{y}$ a given input vector x = $[x_1, x_2, \ldots, x_N]$ in complex systems based on natural selection to handle the system's complexity. The GMDH method uses the Kolmogorov-Gabor polynomial function as following [42]:

$$\widehat{y} = c_0 + \sum_{i=1}^{m} c_i x_i + \sum_{i=1}^{m} \sum_{j=1}^{m} c_{ij} x_i x_j + \sum_{i=1}^{m} \sum_{j=1}^{m} \sum_{g=1}^{m} c_{ijg} x_i x_j x_g + \ldots \tag{12.5}$$

where, c = $[c_0, c_i, c_{ij}, \ldots]$ is the vector of weights. To determine the model's structure and find the optimal model to generate the output, the GMDH algorithm uses the lower order of the Kolmogorov-Gabor polynomial function. The first order of the polynomial is given in Eq. (12.6), which is used in each generation, as shown in Fig. 12.7. The generation of models continues until the process shows over-fitting.

$$\widehat{y} = c_0 + c_1 x_1 + c_2 x_2 + \ldots c_N x_N \tag{12.6}$$

12.4 Simulation and Numerical Results

Forecasting wind speed based on the required data as input variables and influencing the wind speed curve formation requires tools and methods of processing big data. In this chapter, machine learning applications called GMDH and MLP have been selected to forecast wind speed. Using these methods requires a database as input and network formation. Given that wind speed is a parameter dependent on temperature and climatic conditions, accordingly in this chapter, the parameters of air temperature, relative humidity, solar radiation, and wind direction are considered as input variables to form the input data set. After forming the input data set and the design of the networks, each network is trained using 70% of the data. The rest of the data is used as test data to predict wind speed. In the training phase, each network analyzes the input data and, by determining the weights and applying the bias to the input data, predict the output value. After each of the training and test stages, each stage's results should be evaluated using acceptable evaluation metrics. It should be noted that the evaluation of results can be considered the most important part of research work. In this chapter, the results of the training and test stages to predict wind speed are evaluated using three statistical performance metrics such as correlation coefficient (R), mean square error (MSE), root mean square error (RMSE), and mean absolute error (MAE) for each method. Details of each of these indicators are provided in [8, 43]. Figure 12.8 shows each network's result in the training phase for predicting and recognizing patterns related to input data.

It can be seen that in the training phase, the GMDH model with 99.91% of R-value had better efficiency than the MLP method. Based on this, it is expected that this network will provide better results for the test stage and forecasting of wind energy than the MLP method. After training, test data is employed as network input to predict the wind speed. Each network provides forecasting the wind speed based on the done training. Figure 12.9 shows the results of wind speed prediction by the test data. Figure 12.10 presents the evaluation of wind speed forecast results using MSE and RMSE metrics for GMDH and MLP networks.

It can be seen that Fig. 12.10 shows the amount of prediction error values for each test sample (each input data for 10 min) and finally a mean value of errors is presented for the whole test step. It should be noted that the error values obtained for each model shown in Fig. 12.10 are in m/s unit. The results of the evaluations compare the accuracy and efficiency of each of the methods used. It can be seen that, as the training phase, in this phase, the GMDH method was able to provide better predictive results with lower error values. However, in this chapter, a comparative approach is presented to emphasize the effectiveness of the GMDH method in big data processing as well as to provide appropriate wind speed forecasts. In this approach, the MLP and GMDH methods' results are compared with the results of other similar studies. It should be noted that the comparison of the methods' performance should be done so that the data used for all methods are the same. This comparison is performed so that in each of these studies, the input variables selected for each method are the same. The compared methods used the same

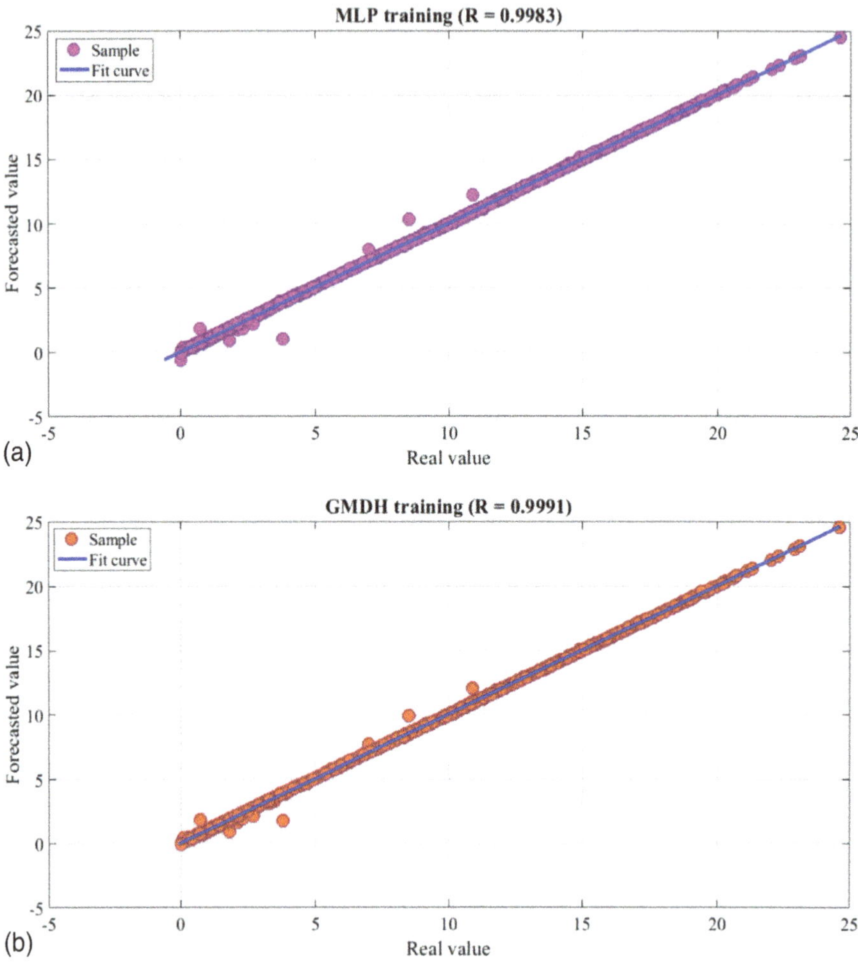

Fig. 12.8 The results of R value for each network in the training stage; (**a**) MLP (**b**) GMDH

meteorological data with the same time interval for sampling to predict wind speed. Table 12.1 makes this comparison.

After comparing the results and evaluating each method's performance, it is observed that the GMDH method can be used as an ideal method for processing large data due to its layer-to-layer structure. It was observed that the methods used in this chapter were able to predict wind speed at any point in time as well as in different weather conditions. Accurate wind speed forecasts can provide proper planning to wind power plants to have a proper performance for generating electricity through wind energy. Besides, other machine learning techniques can be used as an accurate tool to perform the types of predictions needed in power system issues.

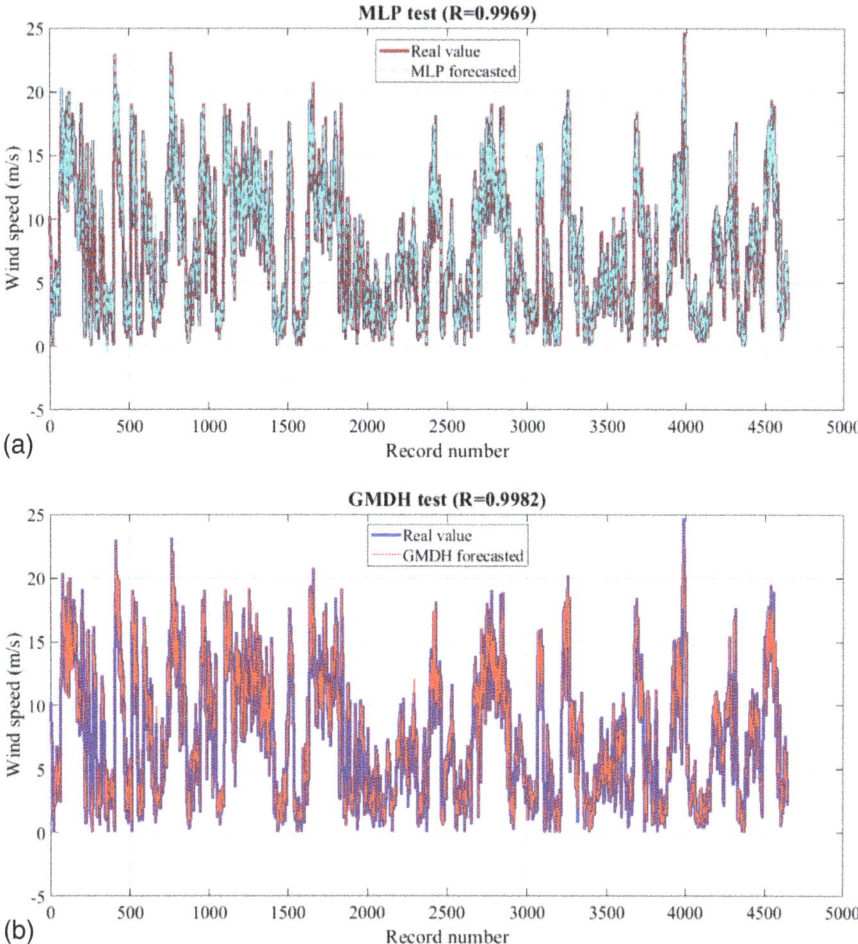

Fig. 12.9 Results of wind speed prediction in the test phase; (**a**) MLP (**b**) GMDH

12.5 Conclusion

Today, wind power plants are considered as one of the essential RESs around the world. Wind speed is one of the most effective factors in the production of electrical energy by wind turbines, the accurate prediction of which can be used as an impressive approach in the performance of wind power plants. In this chapter, two machine learning applications called the MLP and GMDH were employed to forecast the wind speed. To express the effectiveness of the suggested methods, data related to the Khaf region in Iran were utilized. Each of the MLP and GMDH techniques predicted the wind speed by training with input variables such as air temperature, humidity, solar radiation, and wind direction. The results of the predictions were evaluated with different performance evaluation metrics that it was

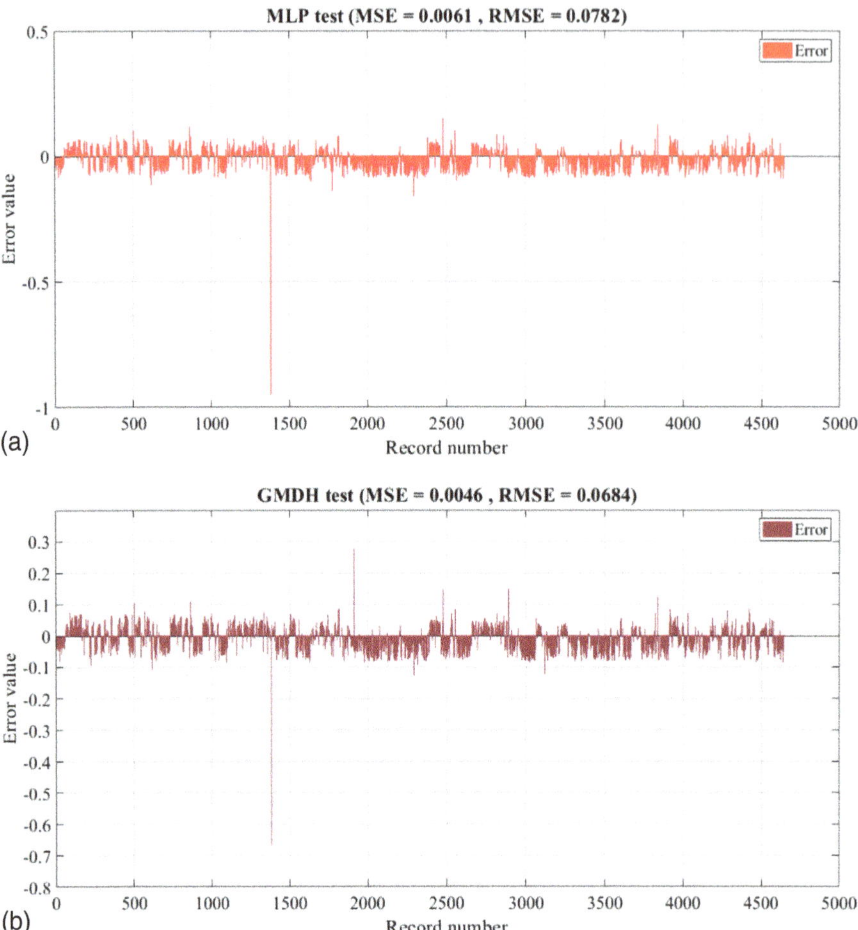

Fig. 12.10 Rate of MSE and RMSE errors to forecast wind speed; (**a**) MLP (**b**) GMDH

Table 12.1 Performance evaluation of various methods used to forecast wind speed

Method	R	MSE	RMSE	MAE
MLP in this chapter	0.9969	0.0061	0.0782	0.0445
GMDH in this chapter	0.9982	0.0046	0.0684	0.0439
WT-ARIMA [13]	–	–	0.4328	0.3369
RWT-ARIMA [13]	–	–	0.3204	0.2269
Nested ARIMA [14]	–	–	–	0.0446
TCFM-MOGOA [17]	–	–	0.4026	0.2046
Optimized ELM [17]	–	–	0.2531	0.1845

observed that the GMDH method with high accuracy (R = 99.82%) than the MLP (R = 99.62%) presented better performance. Finally, in a comparative approach with different methods used to forecast wind speed, the high capability of the GMDH technique was emphasized. It should be noted that the GMDH method can be used as a powerful tool to solve other power system problems that depend on big data.

Wind power forecasting based on wind speed predictions and the improvement of machine learning algorithms for time series data predictions can be considered as future work. In addition, using statistical regression to evaluate and analyze the sensitivity of each of the input variables in the forecast model can significantly improve the forecast results in future work.

References

1. O. Sadeghian, A. Moradzadeh, B. Mohammadi-Ivatloo, B. Mohammadi-Ivatloo, M. Abapour, F.P.G. Marquez, Generation units maintenance in combined heat and power integrated systems using the mixed integer quadratic programming approach. Energies **13**(11), 2840 (2020). https://doi.org/10.3390/en13112840
2. R. K. Pachauri et al., *Climate change 2014: synthesis report. Contribution of Working Groups I, II and III to the fifth assessment report of the Intergovernmental Panel on Climate Change.* (IPCC, 2014)
3. A. Moradzadeh, O. Sadeghian, K. Pourhossein, B. Mohammadi-Ivatloo, A. Anvari-Moghaddam, Improving residential load disaggregation for sustainable development of energy via principal component analysis. Sustainability (Switzerland) **12**(8), 3158 (2020). https://doi.org/10.3390/SU12083158
4. Ren21. 2019. Renewables 2019 Global Status Report. Paris: Ren21Secretariat. Accessed 16 Nov 2019
5. S. Madadi, B. Mohammadi-Ivatloo, S. Tohidi, Dynamic line rating forecasting based on integrated factorized Ornstein–Uhlenbeck processes. IEEE Trans Power Deliv **35**(2), 851–860 (2019)
6. S. Madadi, B. Mohammadi-Ivatloo, and S. Tohidi, *Probabilistic Real-Time Dynamic Line Rating Forecasting Based on Dynamic Stochastic General Equilibrium with Stochastic Volatility* (IEEE Transactions on Power Delivery, 2020)
7. Y. Hao, C. Tian, A novel two-stage forecasting model based on error factor and ensemble method for multi-step wind power forecasting. Appl. Energy **238**, 368–383 (2019)
8. D.B. Alencar, C.M. Affonso, R.C.L. Oliveira, C.R. Jose Filho, Hybrid approach combining SARIMA and neural networks for multi-step ahead wind speed forecasting in Brazil. IEEE Access **6**, 55986–55994 (2018)
9. A. Moradzadeh, S. Zakeri, M. Shoaran, B. Mohammadi-Ivatloo, F. Mohamamdi, Short-term load forecasting of microgrid via hybrid support vector regression and long short-term memory algorithms. Sustainability (Switzerland) **12**(17), 7076 (2020). https://doi.org/10.3390/su12177076
10. M. Lei, L. Shiyan, J. Chuanwen, L. Hongling, Z. Yan, A review on the forecasting of wind speed and generated power. Renew. Sust. Energ. Rev. **13**(4), 915–920 (2009)
11. F. Cassola, M. Burlando, Wind speed and wind energy forecast through Kalman filtering of numerical weather prediction model output. Appl. Energy **99**, 154–166 (2012)
12. W. Zhang, Z. Qu, K. Zhang, W. Mao, Y. Ma, X. Fan, A combined model based on CEEMDAN and modified flower pollination algorithm for wind speed forecasting. Energy Convers. Manag. **136**, 439–451 (2017)
13. S.N. Singh, A. Mohapatra, Repeated wavelet transform based ARIMA model for very short-term wind speed forecasting. Renew. Energy **136**, 758–768 (2019)
14. S.-K. Sim, P. Maass, P.G. Lind, Wind speed modeling by nested ARIMA processes. Energies **12**(1), 69 (2019)

15. E. Erdem, J. Shi, ARMA based approaches for forecasting the tuple of wind speed and direction. Appl. Energy **88**(4), 1405–1414 (2011)
16. S. Smyl, A hybrid method of exponential smoothing and recurrent neural networks for time series forecasting. Int. J. Forecast. **36**(1), 75–85 (2020)
17. P. Jiang, Z. Liu, Variable weights combined model based on multi-objective optimization for short-term wind speed forecasting. Appl. Soft Comput. **82**, 105587 (2019)
18. W. Zhao, Y.-M. Wei, Z. Su, One day ahead wind speed forecasting: A resampling-based approach. Appl. Energy **178**, 886–901 (2016)
19. S. Scher, J. Molinder, Machine learning-based prediction of icing-related wind power production loss. IEEE Access **7**, 129421–129429 (2019)
20. H. Li, J. Wang, R. Li, H. Lu, Novel analysis–forecast system based on multi-objective optimization for air quality index. J. Clean. Prod. **208**, 1365–1383 (2019)
21. P. Du, J. Wang, W. Yang, T. Niu, Container throughput forecasting using a novel hybrid learning method with error correction strategy. Knowl.-Based Syst. **182**, 104853 (2019)
22. A. Mansour-Saatloo, A. Moradzadeh, B. Mohammadi-Ivatloo, A. Ahmadian, A. Elkamel, Machine learning based PEVs load extraction and analysis. Electronics (Switzerland) **9**(7), 1–15 (2020). https://doi.org/10.3390/electronics9071150
23. H. Yang, Z. Jiang, H. Lu, A hybrid wind speed forecasting system based on a 'decomposition and ensemble'strategy and fuzzy time series. Energies **10**(9), 1422 (2017)
24. X. Kong, X. Liu, R. Shi, K.Y. Lee, Wind speed prediction using reduced support vector machines with feature selection. Neurocomputing **169**, 449–456 (2015)
25. A. Moradzadeh, A. Mansour-Saatloo, B. Mohammadi-Ivatloo, A. Anvari-Moghaddam, Performance evaluation of two machine learning techniques in heating and cooling loads forecasting of residential buildings. Appl. Sci. (Switzerland) **10**(11), 3829 (2020). https://doi.org/10.3390/app10113829
26. D.H. Lim, S.H. Lee, M.G. Na, Smart soft-sensing for the feedwater flowrate at PWRs using a GMDH algorithm. IEEE Trans. Nucl. Sci. **57**(1), 340–347 (2010)
27. G.-R. Ji, P. Han, and Y.-J. Zhai, Wind speed forecasting based on support vector machine with forecasting error estimation. In *2007 International Conference On Machine Learning and Cybernetics*, (2007), vol. 5, pp. 2735–2739
28. N. Shabbir, R. AhmadiAhangar, L. Kütt, M. N. Iqbal, and A. Rosin, Forecasting short term wind energy generation using machine learning. In *2019 IEEE 60th International Scientific Conference on Power and Electrical Engineering of Riga Technical University (RTUCON)*, 2019, pp. 1–4
29. C. Wan, Z. Xu, P. Pinson, Z.Y. Dong, K.P. Wong, Probabilistic forecasting of wind power generation using extreme learning machine. IEEE Trans. Power Syst. **29**(3), 1033–1044 (2013)
30. X. Luo et al., Short-term wind speed forecasting via stacked extreme learning machine with generalized correntropy. IEEE Trans. Indust. Inform. **14**(11), 4963–4971 (2018)
31. M. Khodayar, J. Wang, Spatio-temporal graph deep neural network for short-term wind speed forecasting. IEEE Trans. Sustain. Energy **10**(2), 670–681 (2018)
32. A. Ahmadi, M. Nabipour, B. Mohammadi-Ivatloo, A.M. Amani, S. Rho, M.J. Piran, Long-term wind power forecasting using tree-based learning algorithms. IEEE Access **8**, 151511–151522 (2020). https://doi.org/10.1109/ACCESS.2020.3017442
33. D. Zhang, X. Peng, K. Pan, Y. Liu, A novel wind speed forecasting based on hybrid decomposition and online sequential outlier robust extreme learning machine. Energy Convers. Manag. **180**, 338–357 (2019)
34. L. Zhang, F. Tian, Performance study of multilayer perceptrons in a low-cost electronic nose. IEEE Trans. Instrum. Meas. **63**(7), 1670–1679 (2014)
35. A. Moradzadeh, K. Khaffafi, Comparison and evaluation of the performance of various types of neural networks for planning issues related to optimal management of charging and discharging electric cars in intelligent power grids. Emerging Sci. J. **1**(4), 201–207 (2017). https://doi.org/10.28991/ijse-01123

36. A. Moradzadeh, K. Pourhossein, B. Mohammadi-Ivatloo, F. Mohammadi, Locating inter-turn faults in transformer windings using isometric feature mapping of frequency response traces. IEEE Trans. Indust. Inform, 1–1 (2020). https://doi.org/10.1109/tii.2020.3016966
37. I. H. Witten, E. Frank, and Mark A. Hall, *Data Mining: Practical Machine Learning* (2011)
38. A. Moradzadeh and K. Pourhossein, Early detection of turn-to-turn faults in power transformer winding: an experimental study. In *Proceedings 2019 International Aegean Conference on Electrical Machines and Power Electronics, ACEMP 2019 and 2019 International Conference on Optimization of Electrical and Electronic Equipment, OPTIM 2019* (2019), pp. 199–204, doi: https://doi.org/10.1109/ACEMP-OPTIM44294.2019.9007169
39. S. Souahlia, K. Bacha, A. Chaari, MLP neural network-based decision for power transformers fault diagnosis using an improved combination of Rogers and Doernenburg ratios DGA. Int. J. Electr. Power Energy Syst. **43**(1), 1346–1353 (2012). https://doi.org/10.1016/j.ijepes.2012.05.067
40. A. Moradzadeh and K. Pourhossein, Early detection of turn-to-turn faults in power transformer winding: an experimental study. In *Proceedings 2019 International Aegean Conference on Electrical Machines and Power Electronics, ACEMP 2019 and 2019 International Conference on Optimization of Electrical and Electronic Equipment, OPTIM 2019*, (2019), pp. 199–204, doi: https://doi.org/10.1109/ACEMP-OPTIM44294.2019.9007169
41. A.G. Ivakhnenko, New methods of control-system investigation. Control **3**(30), 96–99 (1960)
42. A. G. Ivakhnenko, *Polynomial Theory of Complex Systems: IEEE Transactions on Systems, Man and Cybernetics*. (ISYMAW, 1971)
43. A. Moradzadeh and K. Pourhossein, Application of support vector machines to locate minor short circuits in transformer windings. In *2019 54th International Universities Power Engineering Conference (UPEC)*, (2019), pp. 1–6

Chapter 13
Effective Load Pattern Classification by Processing the Smart Meter Data Based on Event-Driven Processing and Machine Learning

Saeed Mian Qaisar and Futoon Alsharif

13.1 Introduction

At the present, the demand for intelligent energy management is increasing. It is due to the growing world population, declining oil reserves, and rising energy consummation, happening because of increased usage of modern gadgets. In this framework, the smart grid idea has arisen from attempts to make power grids more efficient, cleaner, environmentally sustainable, and functional by creating technical possibilities spanning all processes from electricity generation to delivery and use [1]. Smart grid analyzes and monitors the transfer of electricity from the generation unit to the customer. Thanks to current power networks and advanced metering network, this scenario is more realistic for future technology deployment [2]. The smart grid includes the advanced metering infrastructure (AMI). Smart meters are the most important aspect of modern grid infrastructures.

All have become digital with the major advances in the world of the Internet and technology. The Internet has been a huge part of our lives. In this picture, a novel technology, namely, the Internet of Things (IoT), is developed. This is a network that contains several electronic devices and sensors that are wired together to share such

S. Mian Qaisar (✉)
Electrical and Computer Engineering Department, College of Engineering, Effat University, Jeddah, Saudi Arabia

Communication & Signal Processing Lab, Energy & Technology Center, Effat University, Jeddah, Saudi Arabia
e-mail: sqaisar@effatuniversity.edu.sa

F. Alsharif
Electrical and Computer Engineering Department, College of Engineering, Effat University, Jeddah, Saudi Arabia

M. Nazari-Heris et al. (eds.), *Application of Machine Learning and Deep Learning Methods to Power System Problems*, Power Systems,
https://doi.org/10.1007/978-3-030-77696-1_13

Fig. 13.1 Components of the smart meter data intelligence chain [5]

information across the Internet. IoT-based apps tend to converse and exchange data with each other. As sophisticated infrastructures, they link a variety of electrical devices that can communicate with each other. The IoT is effectively incorporated to develop the AMIs. Communication among devices is realized via a variety of techniques. Home area network is used to connect the smart meter to the smart devices. It monitors and manages and the micro generation network. It offers unified control, resources, and facility maintenance. The linkage between the smart meter and appliances can be realized via a wired or a wireless protocol such as Wi-Fi.

AMI distinguishes automatic smart reading (AMR) schemes. AMRs work with less evolved technology such as manually switching off equipment, gathering data offline, and so on. However, AMIs include real-time processes, customer/pricing/usage options, remote-controlled and automated maintenance systems, etc. With recent developments, the incorporation of AMI is crucial in smart grids. Neighborhood area network is used for data sharing between smart meters in the neighborhood. This supports communications for diagnosis, system updates, and real-time notifications. ZigBee protocol is commonly used in NAN thanks to fast data transfer rates and low cost. Many smart meters are attached to a central server over the wide area network. The networking systems GSM, GPRS, 3G, and WiMAX will be used to link these meters to the server.

Home smart meters measure fine granular energy usage in real time and are seen as the basis for a future smart power grid. Technological developments have increased the use of smart meters as a replacement of the traditional ones [3, 4]. These meters are key components of smart grids that provide significant civic, environmental, and economic benefits to various stakeholders [5]. The massive smart meter installations require a huge amount of data collection with the desired granularity [6]. Automated data acquisition, storage, processing, and interpretation are the main factors behind the performance of smart meters. The process is demonstrated in the block diagram in Fig. 13.1.

A fine-grained metering data is required to attain realistic benefits in terms of performance and resilience to multiple smart grid stakeholders [5, 6]. Every stakeholder has a different objective from the other: suppliers, which are companies buying electricity from the wholesale market and selling it back to consumers in the retail market, want to lower the operational overheads associated with manual meter reading and enhance customer loyalty potentially. The transmission system and distribution network operators are expecting to get benefited from a more flexible demand side to allow more low-carbon technology penetration. Governments are expecting the enhancements of energy efficiency in the end-use side, which are provided by smart meters, and will help to reach the goals of binding carbon reduction. End users are hoping to get benefited from the lower bills of electricity as they become more aware of energy. With such expectations, it is no surprise that smart meters are experiencing a period of rapid growth.

During peak hours, the local consumption of electricity is about 40% of the total residential sector consumption and more than a third of the total power demand [7]. Consumer awareness will help in predicting the demand and adapt future investments. Giving regular feedback on electricity usage is helpful for the end users to have more details about their consumption of electricity which supports the measures for effective use of energy [5, 7, 8].

In this framework, several automated appliance classification techniques have been reported [9–11]. In [9], authors have used zero-normalization consumption time series with delta and delta-delta coefficients as attributes of intended appliances. The classification is performed by using the Gaussian mixture model (GMM) classifier. In [10], authors have used entropy of consumption time series with delta and delta-delta coefficients as attributes of intended appliances. The classification is performed with hidden Markov models (HMM). In [11], authors have used zero-normalization of real power, reactive power, and root mean square (RMS) current time series with delta and delta-delta coefficients as attributes of intended appliances. The classification is performed with hidden Markov models (HMM).

Conventionally, the smart meters data is collected in a time-invariant fixed-rate manner [9–11], resulting in the collection, storage, processing, and classification of a significant amount of unnecessary data [12]. In the suggested metering framework:

- Event-driven sensing is incorporated to attain significant real-time data compression.
- A novel adaptive rate concept is presented for data collection, processing, and classification.
- It improves the designed solution's compression and processing effectiveness compared to classical fix-rate counterparts.

13.2 Background and Literature Review

13.2.1 *Smart Metering*

Nowadays, the traditional grids of power are being replaced by smart grids around the world [13]. Solar and wind are renewable energy sources which the smart grid includes [13]. The smart grid allows the two-way communication and exchange of data between the electricity providers and end users [13]. The smart grid allows the two-way communication and exchange of data between the electricity providers and end users. They are devices that monitor energy usage to the consumer for consumption reduction purposes [14]. In the residential area, these meters determine the consumption of power at fine granularities in real time and are being extensively deployed worldwide because of their significant advantages to the industry of electricity supply and its customers [5, 7, 8]. There are many smart metering

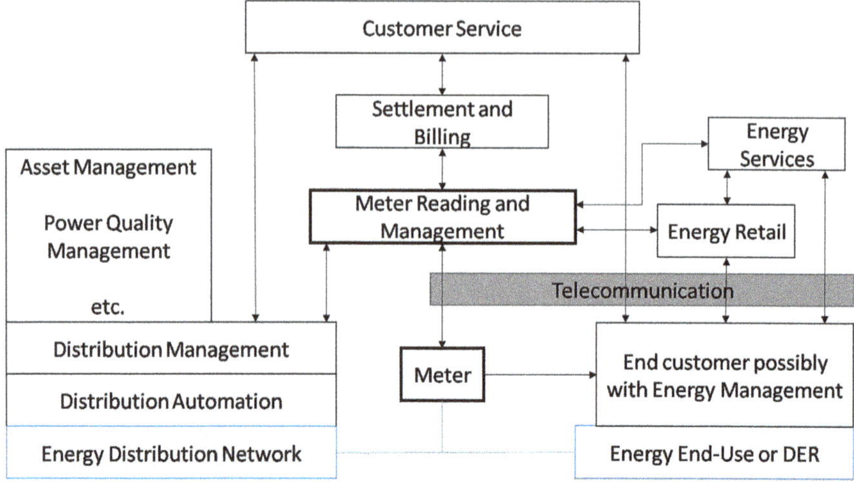

Fig. 13.2 Applications and advantages of smart meters

applications such as quality of power and reliability monitoring, analysis, modeling and forecasting of loads, consumer energy management, and much more [6]. The following block diagram provided by the authors of [15] shows some examples of the applications and advantages of smart meters:

The following points represent the characteristics and advantages of smart metering mentioned in the above Fig. 13.2 [15]:

- Automatic collection, processing, transmitting, monitoring, and usage of metering data.
- Automatic meter control.
- Two-way communication.
- Gives accurate and up-to-the-minute consumption data to the related people.
- Provides help to the services which enhance the efficiency of energy for the consumption and billing.

The smart meter database represents the consumption data of typical home appliances provided by a smart meter. Many studies have been done using smart meter data, such as [16, 17]. The authors of [16] addressed aggregated demand decomposition methodology which was based on sub metering that is allowing smart meter measurements and artificial neural network. The approach in general results in an estimate of the shares of various load classes and a combination of controllable and uncontrollable load within the overall expected load, with the implementation of forecast in different demand response systems [16]. Figure 13.3 shows its load disaggregation flow chart. It represents the demand decomposition technique in the system of the smart meter with just a few consumers who can monitor each appliance they have, as a relatively practical situation for the potential smart distribution grid. The authors of [16] had made two assumptions in this regard.

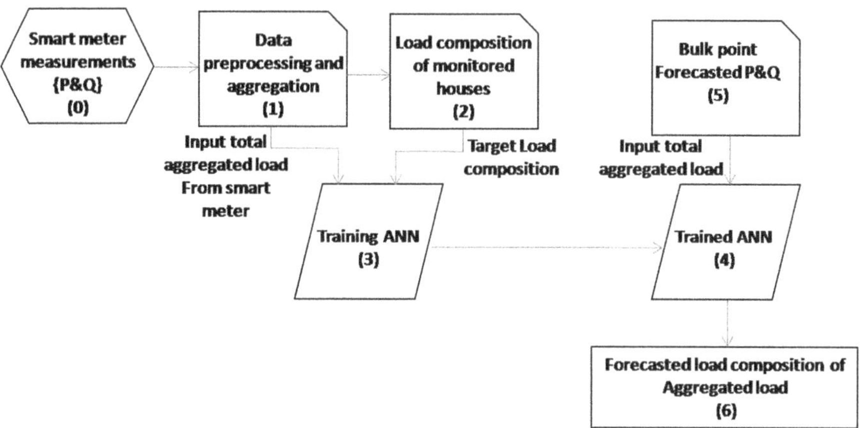

Fig. 13.3 Flow chart for load disaggregation

The real power of each appliance can be recorded in the smart meters, whereas reactive load is extracted based on probabilities [16].Total consumption (actual and reactive power) is already estimated in the substation which is in block 5 [16]. In block 1, the smart meter data is being preprocessed and collected at a concentrator stage [16]. In block 2, the section of the consumption that is submetered can be broken down into groups or manageable and unmanageable load by directly adding up the consumption of the devices which belong to the same group [16]. In the following stage, this sub-metric data trains the artificial neural network (ANN) classifier, so that it gets the ability to identify the load classification using the active and reactive load graph of the controlled consumers [16]. When the ANN is trained, in block 4, the ANN utilizes the prediction of the total active and reactive loads at the concentrator stage in block 5 as input and then provides the associated load classification, that is the weighting factors of each load group as output in block 6 [16].

13.2.2 Smart Meter Data Acquisition

In [17], the researchers presented a three-phase model that was beneficial for the suppliers of electricity because of its demand management flexibility and energy control effectiveness. The solution is an ensemble of unsupervised clustering and supervised classification [17]. It can classify the consumers of electricity, recognizing their characteristics of consumption, and identifying recent customers [17]. In the presented model, to perform load pattern clustering and characteristic recognition, all load patterns obtained from the extraction of load patterns are used in customer categorization [17].

To extract its relevant features, the metering data is processed and analyzed. Authors have reported the methods of extraction of the features in [6, 16]. The principle is to construct a collection of features that can reflect the information that is essential for interpretation and classification, in the most practical way. There are several tactics for feature extraction of the load profiles of appliances. Certain examples are the short-time Fourier transform (STFT), wavelet transformation, and K-means algorithm [18–20]. The derived features are peak values, average values, the consumption's root mean square (RMS) values, and their harmonics.

13.2.3 Feature Extraction

Feature selection techniques can be classified into two categories: classical techniques and biologically oriented techniques as the authors of [21] had divided them. The classical techniques are represented in statistics or syntactic nature, and the biologically oriented techniques can be represented in a neural or genetic-based algorithm [21]. A feature is a pattern with reduced-dimension representation [21]. To lower the pattern representation dimensions, feature selection, and extraction in pattern recognition depends on obtaining mathematical techniques [21]. To achieve the reduction of dimensions, either feature selection is performed or feature extraction [21]. Some factors can be affected by the choice of features, attributes, or measurements [21]. One is the class satisfaction accuracy, the second is the classification required to tie them, the third is the required number of examples for learning, and the last is classification performing price [21]. For a feature to be good, it needs to meet two conditions, and one should not change if any differences happen within a class [21]. The second is that it should show the significant differences when distinguishing between patterns of different classes [21]. There are many feature extraction methods such as non-transformed signal characteristics, transformed signal characteristics, structural descriptors, etc., and the following Fig. 13.4 shows the main feature extraction methods that were mentioned in [21].

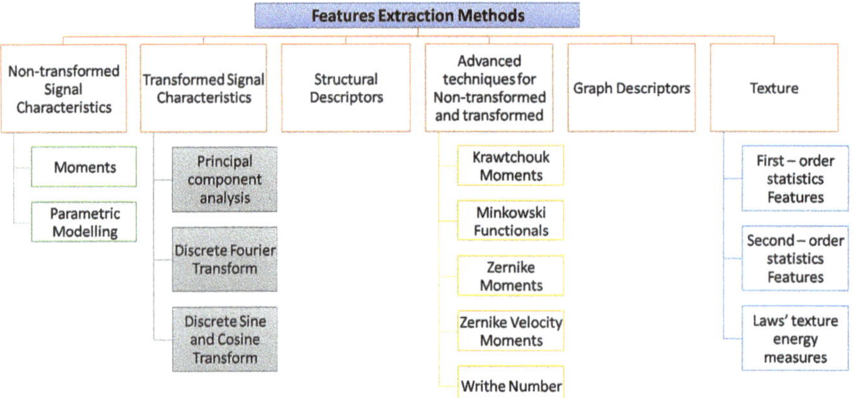

Fig. 13.4 Different methods of feature extraction

13.2.4 Pattern Recognition

High-resolution smart meter data offer a variety of knowledge on customer habits and preferences of energy use. Besides this, deregulation of the power market, especially on the supply side, in many countries around the world continues to push forward. Exploiting the massive smart meter data can promote and boost efficiency and profitability on the demand side worldwide. In this sense, in recent years, the electricity industry has experienced significant breakthroughs in machine learning [6].

Recognition of concerned appliance patterns is attained by using the mined features from their consumption time series. In [22], the classification of patterns has been described by authors as a study of how machines can detect, learn to discern, and make logical decisions about patterns. In [18], the authors used load disaggregation pattern recognition. The idea is to collect the consumption information of each concerned device separately without interference. The prepared database is composed of the attributes, which can be explored and processed to classify the concerned devices [18]. In [19], authors employed the automated pattern detection, to exploit data which is collected from smart meters, to disaggregate overall energy usage. The precession of pattern identification is dependent on the performance of data collection, extraction of features, and classification. A tool used to apply a class description to some given instances in the function space is called a classifier [22].

In [23], the authors proposed a fast approach for electrical appliance recognition based on analyzing the signatures of the smart home load. These signatures are defined by active and reactive power and fast Fourier transform of the signals of the current [23]. The recognition algorithm they used to extract a decision tree was C4.5 [23]. One of the advantages of their approach was the automated decision tree generation which enables the rapid deployment of the detection algorithm irrespective of the smart home [23]. They tested 13 appliances, which were toaster, electric kettle, two different stove burners, microwave, vacuum, oven, coffee maker, blender, two different dryers with different modes, range hood fan, and mixer. The classification rate achieved was 93.4397% of true positive [23].

The authors of [19] had proposed an approach of automatically identifying the devices according to the distributed power measurement and actuation units. They had two contributions to their paper. One is introducing their trace-base repository of actual power consumption traces, and the other is evaluating the electrical appliances' automatic recognition based on the traces collected [19]. They extracted 517 features from the traces collected to prepare them for classifier training [19]. They used different classification algorithms such as bagging, Bayesian network, Naïve Bayes, random committee, etc. The highest classification accuracy they got was 95.5% using the random committee classification method [19].

Bad power quality detection addresses may be lost data or irregular trends triggered by unplanned occurrences or data collection, communication, or entry failure. Detection of bad power quality can be broken down into predictive, machine learning, and probabilistic approaches [6]. These methods and tools can be used directly on the collected smart meter data. Such studies are summed up as time series-based methods, low-rank matrix technique, and time-window-based approach as a function of the modeling techniques. Data on smart meters are simply time series. An optimally weighted average approach for data denoising and classification has been suggested. It can be incorporated for offline or online power issues detection [6]. An ensemble of autoregressive with exogenous inputs (ARX) and artificial neural network (ANN) is also used for the identification of power supply issues [6]. The use of energy is correlated with both spatially and temporally. Exploring the spatiotemporal connection may help to recognize and recuperate outliers. A low-rank matrix fitting-based approach for performing data cleaning and imputation was proposed in [6]. Because smart meter data is obtained in real-time or near real-time fashion, an online mechanism of identification of anomalies using the Lambda architecture has been developed in [6]. The suggested method of online identification can be performed in parallel, providing high performance when dealing with big datasets.

Under other techniques, energy theft will change the smart meter information. Detection of energy theft can be achieved using smart meter data and status information of the power grid, such as voltages of nodes. Supervised classifiers are efficient energy theft detection techniques. They usually consist of two stages, namely, extraction of features and classification. In [6], an approach is reported, where the nontechnical failure was first measured to train a theft identification classifier. The load profiles were clustered using K-means clustering. Various potential destructive samples are used to train the classifier. Following a series of suspicious detections, an energy theft warning is issued. The approach suggested could also classify the category of energy theft. In addition to clustering-based extraction of features, an encoding technique is first conducted on the load data in [24], which served as classifier inputs such as SVM and a rule engine-based algorithm for detecting energy theft. In [25], the authors have introduced an independent variable that is implemented as a top-down structure. It is founded on the principles of the decision tree and the SVM algorithms. The decision tree estimated the expected energy consumption based on the number of appliances, the individuals, and the ambient temperature. Then, the decision tree production was fed into the SVM to decide whether the user is ordinary or abusive.

Load profiling corresponds to the characterization of the appliances' consumption manners or consumers as a function of the energy utilization manner. Specific cluster-based approaches, such as K-means, hierarchical clustering algorithms, and self-organizing maps (SOM), are utilized in this framework. They are applied directly to the time series of consumption behavior, collected from the smart meters [6]. In [26], authors investigated how the resolution of energy consumption time series, collected via using smart meters, influences clustering results. The well-known clustering techniques were conducted on the smart meter dataset. In this case, the dataset is composed of a time series, collected at various sampling rates, ranging from 1 min to 2 h. In [27], deep learning-based stacked sparse auto-encoders were implemented to compress the load profile and extract the feature. A locality sensitive hashing approach is further suggested to identify the load profiles and extract the descriptive load profiles based on the decreased and encoded load profile. A profound analysis of the time series of consumption patterns, collected via smart metes, is required to extract the pertinent local and global classifiable features. It enhances the precision of the classification of load profiles. Three novel categories of local and global features are suggested in this framework. These features are produced via the effective application of conditional filters on the time series of energy consumption. Moreover, the calibration and normalization, profile errors, and shape signatures are also employed to realize an effective load profiling [28].

Using existing statistical information, the load forecasting methodology is often used by electricity utilities for grid planning, generating capacity, demand control, financial modeling, and potential energy use. Precise forecasting of electricity charges is important for electricity providers to reduce financial risk, design power grids, and maximize operating performance. The process of load forecasting could be categorized into short-term, medium-term, and long-term predictions. Here the short term refers to a period between 1 day and 1 week. The medium term refers to a period from 1-week to 1-month, and the long term refers to the duration from months to several years [6]. In this context, time series models with neural networks have been used for short-term forecasting. Artificial neural networks have also been employed for short-term forecasting [6]. Clustering approaches with support vector machines have been used for medium-term forecasting [6].

13.3 Materials and Methods

Figure 13.5 shows the principle of the proposed system. The different system modules are being described in the followed subsections.

13.3.1 Smart Meter Database

In this chapter, the evaluation results of the suggested solution are prepared while studying its performance on the dataset ACS-F2 [9]. This dataset comprises of time series of energy consumption parameters of 15 major home appliances [9]. In total 225 different appliances from 15 classes are considered. Two 1-h data collection and recording sessions are conducted for each intended appliance [9]. Six different time series are recorded for each recording session and from each considered appliance. These are, respectively, real power, reactive power, RMS current, RMS voltage, frequency, and power factor. Each time series is recorded in a disaggregated manner

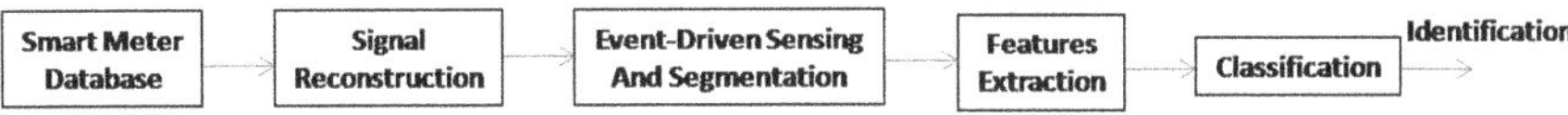

Fig. 13.5 The system block diagram

at an acquisition rate of 0.1 Hz. The recordings are made in a disaggregated fashion [29]. In this study, only six categories of appliances are considered. It includes kettles, fridges, freezers, microwave ovens, fans, monitors, and televisions. For each intended device, real and reactive power consumptions are taken into account.

13.3.2 Signal Reconstruction

The reconstruction process is the process of producing a continuous-time signal that goes along with the points of the discrete-time signal (i.e., it simply takes a group of samples and converts them back to a continuous function) [30, 31]. The first step to understand the reconstruction process is to generate a train of continuous-time impulse from a sampled signal, which can be represented mathematically as follows:

$$x_{imp}(t) = \sum_{n=-\infty}^{\infty} x_s(n)\delta(t - nT_s). \quad (13.1)$$

where $x_s(n)$ is the sampled signal and T_s is the sampling period. A low-pass filter with specific conditions is then applied to generate an output signal $\widetilde{x}(t)$. The result of the construction process is as follows:

$$\widetilde{x}(t) = \sum_{n=-\infty}^{\infty} x_s(n)g(t - nT_s). \quad (13.2)$$

The considered consumption parameters waveforms are up-sampled with a factor of 10,000. It is performed to evaluate the event-driven sensing module. A cascaded arrangement of four stages of cubic-spline interpolators and anti-imaging filters is utilized to conduct the up-sampling. In this way, the quasi-analog version of the incoming signal $y(t_n)$ is obtained. It is named $\widetilde{x}(t)$. The process of obtaining $\widetilde{x}(t)$ up-sampling of $y(t_n)$ is given by Eq. (13.3), where U is the up-sampling factor.

$$\widetilde{x}(t) = y\left(t_{\frac{n}{U}}\right)\vec{\mathrm{x}}(\mathrm{t}) = \mathrm{y}\left(\mathrm{t}_{\frac{\mathrm{n}}{\mathrm{U}}}\right). \quad (13.3)$$

13.3.3 Event-Driven Sensing (EDS)

In the proposed case, the event-driven ADCs (EDADCs) are utilized for the acquisition of the intended appliances' energy consumption patterns [20, 32]. These data converters are realized by using the principle of event-driven sensing (EDS). They can change the rate of acquisition as a function of the incoming signal time-variations [33, 34]. In this case, a time series sample is recorded only when the intended analog signal traverses one of the predefined threshold levels, placed across the amplitude of the signal. Therefore, samples are divided in time in a nonuniform manner. The size of the samples taken depends on the variations of the $\widetilde{x}(t)$ $\vec{x}(t)$ [33, 34]. The procedure can be represented mathematically with Eq. (13.4), where t_n represents the current sampling instance, t_{n-1} is the previous sampling instance, and dt_n represents the time distance between these two instances:

$$t_n = t_{n-1} + dt_n. \tag{13.4}$$

The EDADC acquires only the relevant or active portions of the incoming analog signal. On other hand, the remaining portion of the signal-like baseline is disregarded. Therefore, relative to traditional counterparts, the obtained number of samples is dramatically diminished. It renders a significant decrease in the amount of collected information and could produce a notable real-time compression. It contributes to the reduction of post-processing operations and improves the performance of the system in terms of the time of execution and the usage of overhead power and energy [33–36].

13.3.4 Event-Driven Segmentation

The segmentation operation separates the signal into several fixed-length portions [37]. It permits the relevant parameters of the incoming signal to be identified and mined effectively. The EDS output is placed in a nonuniform time-amplitude plane. Therefore, the traditional segmentation mechanisms cannot be utilized in this case [35–37]. In this context, an original event-driven segmentation mechanism is employed. It is named the activity selection algorithm (ASA) [35, 36]. This partitions the output of EDADC into portions of variable lengths. The ASA is conducted by exploiting the nonuniformity of the event-driven acquired signal. It preserves the valuable details, such as repartitioning among the consecutive sampling instants, count of samples, etc. [34]. The EDADC provides the nonuniform data to the ASA which selects the appropriate parts. Within the whole length of the signal, a series of selected windows would be defined [36]. The activity selection is extremely important to minimize the processing operation and hence the power consumption. It permits the realization of post-adaptive rate feature extraction in an adaptive rate manner. In this way, the useful features are mined solely in the time domain and are utilized afterward by the classifiers for an automated load profiling.

13.3.5 Feature Extraction

Pertinent features are mined from each portion. Thanks to the event-driven acquisition we can extract interesting and pertinent information regarding the frequency content of the signal in the time domain [38]. In this way, the used technique of feature extraction permits the extraction of important information from the power consumption time series without requiring the complicated frequency domain transformation. On another hand, the conventional counterparts are based on the fixed-rate acquisition and processing methods, and therefore they are obliged to employ the time-frequency analysis while performing the feature extraction. It confirms a notable processing load effectiveness of the suggested solution in contrast to the counterparts [37]. Therefore, the necessary classifiable features are mined by using the nonuniformly recorded power consumption time series in only time domain.

Instantaneous real power consumption and instantaneous RMS current are considered for each designated appliance. Each considered time series is reconstructed by using the suggested technique. Onward, the quasi-analog waveforms are recorded by using the EDADC and segmented by using the ASA. Afterward, four different parameters for each segment are extracted. Let i is indexing the ith selected segment, W^i. Then, $C^i$$C^i$, ΔA^i,aÜA^i, $A^i_{max}$$A^i_{max}$, and dt^i_{mean} are respectively the extracted number of threshold crossings, the peak-to-peak amplitude, the maximum amplitude, and the average sampling step for $W^i$$W^i$. The EDS delivered time series of real and reactive power are also considered. For each instance the collected version of the incoming instances features are used to strengthen and improve the system precision.

13.3.6 Classification Techniques

The derived characteristics are used to distinguish targeted appliances. In this context, we utilized the known algorithms of classification such as k-nearest neighbor (KNN), artificial neural network (ANN), and Naïve Bayes.

13.3.6.1 K-Nearest Neighbor (KNN)

The KNN is regarded as an easy but powerful classifier and its ability to deliver high-quality results even for applications that are known for their complexity [39]. It depends on the nearest feature space training examples [22]. It classifies the concerned object according to its majority vote by its neighbors [22]. In a dataset, the features' distance is used by KNN to decide which data belongs to what class. When the distance in the data is near, a group is formed, and when the distance in the data is far, other groups are formed. A category membership might be the output of the KNN classifier. The categorization of an object is done through the majority vote by its neighbors, the object being assigned to the most common classification among its k nearest neighbors (k could generally be a small positive whole number). The object is assigned solely to the nearest neighbor's single classification if the k equals one [39].

13.3.6.2 Artificial Neural Network (ANN)

Processing data full of features is what the ANN classification method is known for [39]. ANN is a model of neural network which is part of artificial intelligence. Instead of programming the procedure system to do a certain number of tasks, this classification method teaches the system to perform tasks [18]. Artificial intelligence system (AI) is developed to execute these tasks. It is a useful model that can understand the patterns hidden in the data that duplicate useful information fast and accurately. Neural networks are one case of these AI models. AI systems should be constantly discovered from knowledge [39]. The most available approaches are artificial intelligence techniques in the fields of evaluation in relationships with dissimilar information. A man-made neural network consists of several artificial neurons that, according to requirements, are correlative together. The neural network's goal is to turn the inputs into critical outputs [6]. The teaching mode may be monitored or not controlled. In the presence of noise, neural networks learn.

13.3.6.3 Naïve Bayes

The Naïve Bayes classification method is used. Naïve Bayes can be described as a short version of the Bayes theorem [5]. It functions in a way that the probability of one attribute is not affected by the probability of another attribute. It develops independent assumptions of 2Q for a series of Q attributes. Authors of [6] examined some of the Naïve Bayes classification method key performance criteria. They concluded that the classifier accuracy is a function of training data noise, bias, and variability. Only by choosing good training data, the noise can be reduced. Bias is the error that is caused by very large groupings in the training data. Variance is the error caused by the fact that these groupings are too small.

13.4 Results

The categories of appliances include kettles, fridges and freezers, microwave ovens, fans, monitors, and televisions. Fifteen appliances of different brands from each category are considered. Therefore a total of 90 appliances are taken into consideration. For each appliance two 2-h retrieval sessions are held. It results in 180 multi-dimensional time series of the power consumption-related characteristics of the considered appliances. The classification techniques used in this case are KNN, SVM, ANN, and Naïve Bayes. Examples of the intended instances of real power

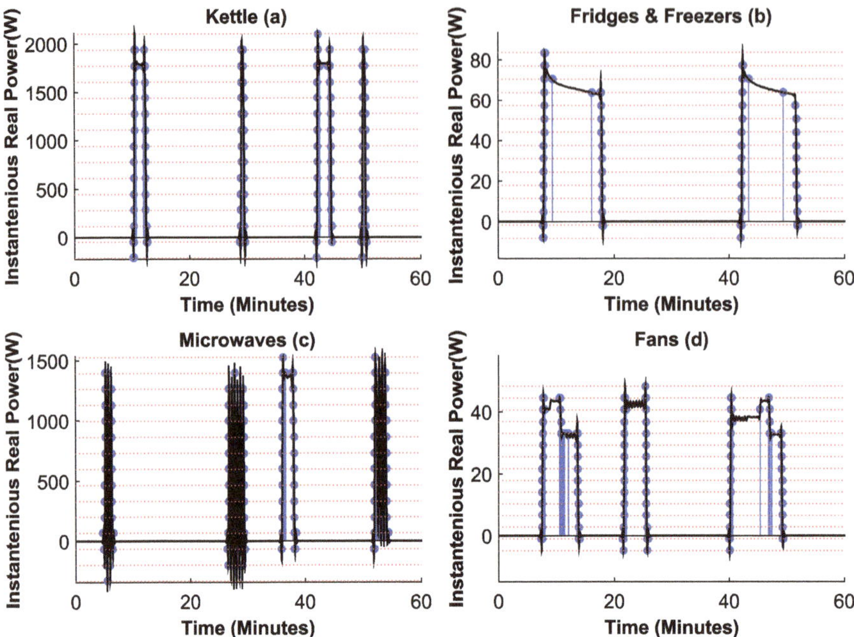

Fig. 13.6 The real power instances digitized with a 4-bit resolution EDADC for kettles, fridges and freezers, microwaves, and fans

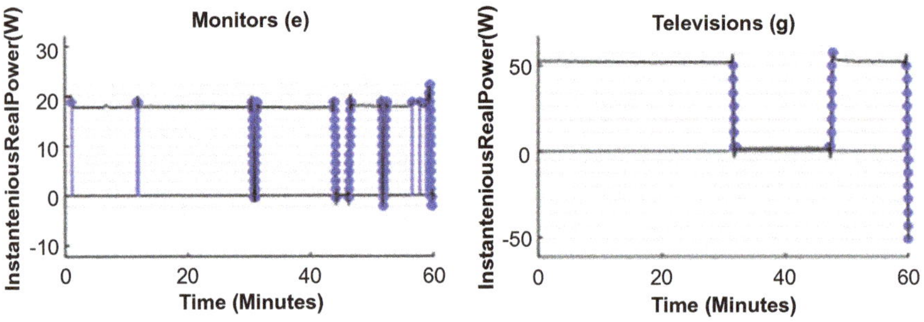

Fig. 13.7 The real power instances digitized with a 4-bit resolution EDADC for monitors and televisions

consumption and RMS current, obtained with the EDS mechanism, are shown respectively in Figs. 13.6, 13.7, 13.8, and 13.9.

The current cannot be measured explicitly in actual application and requires the current to voltage conversion. This is because most real-life data converters are built on the principle of acquiring voltage waveforms. It could be realized by using charge resistance or transimpedance mountings. However, for this simulation-based study, the realization and design of this module are not considered.

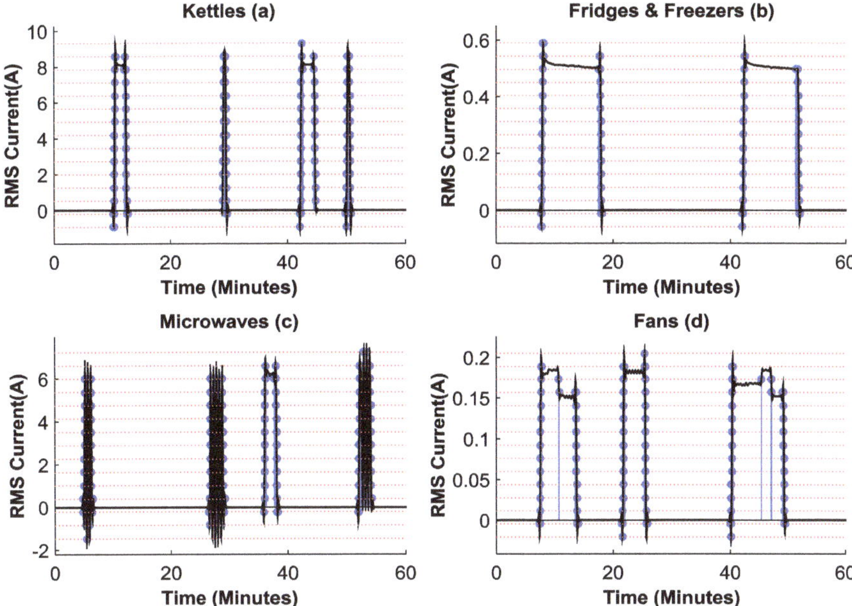

Fig. 13.8 The RMS current instances digitized with a 4-bit resolution EDADC for kettles, fridges and freezers, microwaves, and fans

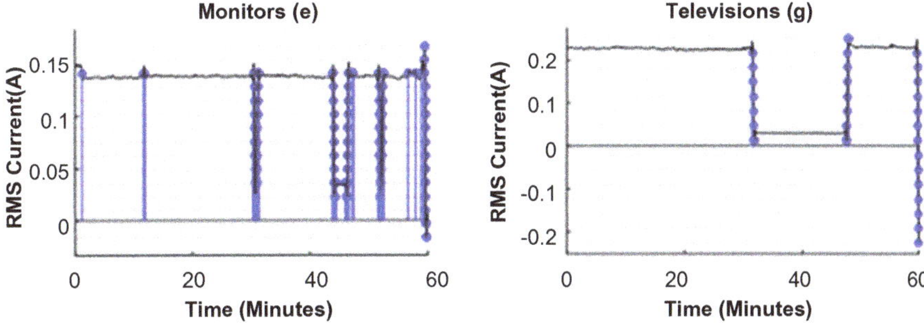

Fig. 13.9 The RMS current instances digitized with a 4-bit resolution EDADC for monitors and televisions

The obtained compression gains are summed up in Table 13.1. These are computed as ratios between the collected number of samples, obtained in the conventional and sugegsted approaches. It demonstrates that the suggested method achieves a complete 3.1 times, 6.6 times, 1.6 times, 3.4 times, 2.2 times, and 6.0 times compression gains, respectively, for the case of kettles, fridges and freezers, microwaves, fans, monitors, and televisions. The compression gain shows how much we are going to gain in terms of the reduction in the amount of the collected information and in the amount of information that is going to be processed and transmitted. It aptitudes a noticeable reduction in the arithmetic complexity and the

Table 13.1 Summary of the compression gains

Appliances	Compression gain	Average compression gain
Kettles	3.1	3.8
Fridges and freezers	6.6	
Microwaves	1.6	
Fans	3.4	
Monitors	2.2	
Televisions	6.0	

Table 13.2 Accuracy for the six-class appliances consumption pattern recognition (KNN)

Appliances	Classification accuracy (%age)	Average classification accuracy (%age)
Kettles	91.5	90.9
Fridges and freezers	91.4	
Microwaves	90.6	
Fans	91.8	
Monitors	89.9	
Televisions	90.3	

Table 13.3 Accuracy for the six-class appliances consumption pattern recognition (ANN)

Appliances	Classification accuracy (%age)	Average classification accuracy (%age)
Kettles	94.6	94.4
Fridges and freezers	95.5	
Microwaves	93.3	
Fans	95.2	
Monitors	94.2	
Televisions	93.8	

power consumption of the proposed solution compared to the conventional approach. Also, embedding the EDS in smart meters significantly decreases the activity of data storage and collection compared to standard methods. It additionally enhances the effectiveness of the designed strategy over the concurrent ones.

This case yields an overall 3.8-fold compression ratio. Compared to the traditional method, the proposed tactic has greatly reduced arithmetic complexity and power consumption. Additionally, embedding the EDS into smart meters greatly decreases data collection and transfer operation relative to conventional approaches. This further increases the efficacy of the designed approach over the conventional ones.

180 instances are considered for the 6 categories of appliances. For equal representation, 30 instances are taken from each category. The limited dataset could lead to biased classification evaluation. To prevent any bias in performance, the tenfold cross-validation is utilized [22]. The accuracies in terms of the percentage of correct recognition are listed respectively in Table 13.2, Table 13.3, and Table 13.4 for the KNN, ANN, and Naïve Bayes classifiers.

Table 13.4 Accuracy for the six-class appliances consumption pattern recognition (Naïve Bayes)

Appliances	Classification accuracy (%age)	Average classification accuracy (%age)
Kettles	87.6	86.9
Fridges and freezers	88.2	
Microwaves	85.9	
Fans	88.4	
Monitors	84.8	
Televisions	86.3	

Table 13.5 Summary for the average classification techniques accuracy

Appliances	KNN classifier	ANN classifier	Naïve Bayes classifier
Kettles	90.9	94.4	86.9
Fridges and freezers			
Microwaves			
Fans			
Monitors			
Televisions			

Table 13.6 Summary for the average F-measure values

Appliances	KNN classifier	ANN classifier	Naïve Bayes classifier
Kettles	0.907	0.937	0.865
Fridges and freezers			
Microwaves			
Fans			
Monitors			
Televisions			

To overall summaries of results for different classifiers are presented by Table 13.5 and Table 13.6, respectively, for the accuracy scores and F-measure values.

From the above table, it is clearly shown that the ANN classifier achieves the best performance. The best consumption pattern recognition of appliances is achieved by employing the combination of EDADC, ASA, event-driven feature extraction, and the ANN classification method.

Consequently, the suggested solution reduces considerably the complexity and power consumption, lowers data transmission operation compared to the conventional approach, and increases system efficiency.

13.5 Discussion

Home smart meters are used in real time for calculating power usage at the finest granularities and are seen as the bases of the future smart grid technological developments have evolved the utilization of smart meters, instead of traditional ones.

Such meters are the essential elements of smart grids and provide important advantages for various stakeholders. These multitudes of advantages can be categorized as social, environmental, and economic. The massive installations of smart meters are producing a large volume of data collection with the desired granularity.

Automated information collection, storage, processing, and classification are the main factors behind the performance of smart meters.

The collection of fine-grained metering data is important to give practical advantages to several stakeholders of the smart grid in terms of performance and sustainability. There are different goals for different stakeholders: vendors for example would like to reduce the operating expenses involved with traditional meter reading and significantly increase the loyalty of the consumer. The operators of the transmission systems and distribution networks want to take advantage of a more robust demand side that enables lower carbon technology penetration. Governments aim to achieve the objectives of reducing carbon emissions by increasing the energy efficiency in the consumer side, which is given by smart meters. Consumers are having a better awareness of energy. Therefore, they are expecting to benefit from the reduced electricity bills. With these goals, it is not surprising that smart meters are having a faster growth period.

From a generalized viewpoint, identifying the type of device is a difficult task because of several reasons. Firstly, there are possible overlaps among a wide variety of types, for example, laptops and tablets. Secondly, there is a wide range of devices that fall under the same group because of their unlike operating mechanisms and technical changes that occur among appliances. Generally, the appliance identification must guarantee the precise ability to simplify without exceeding a certain number of appliances.

In many uses, the data collected through smart meters are analyzed. A smart meter's performance is dependent on algorithms that can run and perform real-time intelligent operations. A smart meter chipset module can perform the appropriate function by programming within hardware limitations. The smart meter processor typically performs multiple tasks such as measuring electricity, showing electrical parameters, reading smart cards, handling data and power, detecting malfunctions, and interacting with other devices.

To consumers, one of the key benefits of smart metering is to help them conserve more money. Comprehensive knowledge of usage would allow consumers to make better decisions on their schedules for energy use. Despite the recent increases in electricity prices, consumers should by moving their high-load household appliances to off-peak hours to reduce their energy costs and billing. One feature that would help the consumers directly is the applications that can offer details on intermittent electricity and overall energy consumption. In this context, the data can be collected on an hourly basis, daily basis, or weekly basis. Another useful use of energy disaggregation is to break down the gross power consumption of a household into individual appliances. In this way, the consumer will be aware of how much energy every appliance uses to help them control their consumption better.

Smart metering services would allow the distribution system operators (DSOs) to properly control and sustain its network and more effectively deliver the electricity, while reducing its running costs. Automatic billing systems will decrease billing-related concerns and visits to the site. It helps the dealer to remotely read the consumer meter and to submit correct, timely bills without daily on-site meter readings. This also offers a two-way link that allows the detection and management of the remote issue possible. Energy stealing is one of the DSOs' most critical issues, which causes a major profit deficit. A variety of methods for the prevention of energy theft have been suggested by incorporating the AMI into electrical power systems [40].

The installation of smart meters often poses many obstacles, while adding important benefits to the community. One of the challenges in deploying modern metering technology is from an economic standpoint. The AMI's construction, implementation, and servicing entail several problems and include budgets of many billion dollars for deployment and servicing. Therefore, a cost-benefit study will be a fair starting point for potential smart metering infrastructures to develop. The advantages can be distinguished between those who are primary and secondary. Primary benefits are that which will specifically influence the bills of consumers, while indirect benefits are in terms of efficiency and changes in environmental standards and would have potential economic impacts. In [41], the current value of potential profits concerning costs is analyzed while presuming a project life of 13 years, including 3 years of execution and 10 years of service. It is reported that, by considering both primary and secondary economic advantages, smart grid provides a favorable economic benefit if they obtained profit to cost ratio is between 1.5 and 2.6. There are also protection and privacy issues. Customers collaborate with the utilities through the installation of smart meters and two-way communication capability to control electricity consumption. The details they exchange show consumer preferences and behaviors, how they use electricity, the number of people in their homes, and the devices in use that subject them to privacy breaches. In a smart grid context, a stable and scalable distributed computing network is required. In this viewpoint, one of the big challenges is to make real-time data accessible from smart meters to all the stakeholders that use these details to meet those requirements.

The classical way of acquisition of data is time-invariant. The data is collected at a rate of Nyquist, regardless of their rate of information, which results in a substantially large amount of unwanted data being collected, transmitted, processed, and analyzed. The classical analog-to-digital converters sample and process the data based on the principle of Nyquist. The parameters of design of these classical analog-to-digital converters are therefore chosen for the most unfavorable scenario. This implies that these analog-to-digital converters are not efficient for arbitrary signals like appliance usage parameters.

Table 13.7 Comparison with state-of-the-art methods

Study	Feature extraction	Classification	Accuracy (%)
[9]	Time series of power consumption and its derivated coefficients	GMM	89.8
[10]	Entropy of time series of power consumption and its derivated coefficients	HMM	93.6
[11]	Time series of multiple consumption parameters and their derivated coefficients	HMM	93.9
This study	EDS delivered time series of real and reactive powers + C^i, ΔA^i, A^i_{max}, and dt^i_{mean}	ANN	94.4

The event-driven sensing mechanism is employed in this framework at the stage of data acquisition to attain real-time data compression. In the following step, the essential adaptive data processing, segmentation, and extraction techniques are suggested. Compared with classical counterparts, it confirms a significant compression and computational efficiency of the method proposed. In this study, EDADCs were used. They work based on event-driven sampling, and they can adjust their frequency of sampling according to the variations of the incoming signal. Based on the results, previously described in the section of results, it is obvious that the different compression gains for the various studied cases guarantee a significant diminishing in the computational complexity of and the power consumption of the suggested solution in contrast with the traditional fixed-rate methods. Additionally, embedding the EDS into smart meters greatly decreases data collection and transfer operation relative to conventional approaches. It additionally enhances the effectiveness of the designed strategy over the concurrent ones. To conclude, the proposed solution results in a significant reduction in the complexity and consumption of power, lowering the data transmission operation compared with the classical method, and boosting the system performance while securing high automatic appliance identification precision.

Implementing the event-driven signal acquisition and processing technique in smart meters is still quite a new idea. It is not evident to make a comparison between the suggested technique and the state-of-the-art techniques as they are based on classical sampling and processing. Nevertheless, a comparison has been made among the main previous researches using the same database of appliances [9–11]. The following table represents the comparison among the classification accuracies for all the researches considered.

Table 13.7 shows that the proposed technique ensures greater accuracy compared to the previous ones [6, 9, 11]. The key benefit of the proposed system over the counter equivalents is the significant compression gain. Moreover, it aptitudes a significant efficiency in the processing and power consumption of the proposed solution over the equivalents. Also, a significant reduction in data storage and transmission activities is ensured. On the other hand, the disadvantage of this method is that it is still a new method; it is not tested on a variety of appliances or extended databases of appliances.

13.6 Conclusion

A new approach for automatically profiling the power consumption time series of major household appliances is presented. It is based on event-driven processing, extraction of features, and classification. In comparison to conventional methods, the proposed solution does not need the complex computation of frequency domain-based feature extraction. The final results showed that the implementation of EDADC and ASA considerably decreased the count of samples to be processed. A case of six categories of appliances has been studied. Different classification techniques have been performed for the six appliances. It has been found that a 3.8-fold reduction in the count of collected samples is achieved over the classical approach. In comparison to classical techniques, the suggested method affirms a significant decrease in the complexity of the system. The suggested methodology has been approved to acquire an average of 94.4% precision of appliance consumption pattern identification using the ANN classifier. It assures the desire in using the method proposed in current dynamic load management and detailed electricity bills.

A potential expansion of this work is to explore the current solution when considering broader appliance categories. The performance of the devised tactic depends on the selected system parameters such as resolution, reference segment length, quantization scheme, feature extraction, and classification algorithms. The development of an automatic mechanism to choose the optimal system parameters for a targeted application is another future task. Other research angles are the usage of higher resolution converters, adaptive quantization, and other robust classifiers such as rotation forest and random forest.

Acknowledgments This project is funded by the Effat University of Jeddah, under the grant number UC#9/29 April.2020/7.1-22(2)2.

References

1. R. Ullah, Y. Faheem, B.-S. Kim, Energy and congestion-aware routing metric for smart grid AMI networks in smart city. IEEE Access **5**, 13799–13810 (2017)
2. H. Mohammed, S. Tonyali, K. Rabieh, M. Mahmoud, and K. Akkaya, Efficient privacy-preserving data collection scheme for smart grid AMI networks, (2016), pp. 1–6
3. S. Darby, Smart metering: What potential for householder engagement? Build. Res. Inf. **38**(5), 442–457 (2010)
4. Q. Sun et al., A comprehensive review of smart energy meters in intelligent energy networks. IEEE Internet Things J. **3**(4), 464–479 (2015)

5. D. Alahakoon, X. Yu, Smart electricity meter data intelligence for future energy systems: A survey. IEEE Trans. Ind. Inform. **12**(1), 425–436 (2015)
6. Y. Wang, Q. Chen, T. Hong, C. Kang, Review of smart meter data analytics: Applications, methodologies, and challenges. IEEE Trans. Smart Grid **10**(3), 3125–3148 (2018)
7. G. Barnicoat, M. Danson, The ageing population and smart metering: A field study of householders' attitudes and behaviours towards energy use in Scotland. Energy Res. Soc. Sci. **9**, 107–115 (2015)
8. N. Uribe-Pérez, L. Hernández, D. De la Vega, I. Angulo, State of the art and trends review of smart metering in electricity grids. Appl. Sci. **6**(3), 68 (2016)
9. A. Ridi, C. Gisler, and J. Hennebert, ACS-F2—A new database of appliance consumption signatures. In *2014 6th International Conference of Soft Computing and Pattern Recognition (SoCPaR)*, 2014, pp. 145–150
10. A. Ridi, C. Gisler, and J. Hennebert, Appliance and state recognition using Hidden Markov Models. In *2014 International Conference on Data Science and Advanced Analytics (DSAA)*, 2014, pp. 270–276, doi: https://doi.org/10.1109/DSAA.2014.7058084
11. "Processing smart plug signals using machine learning - IEEE Conference Publication." https://ieeexplore.ieee.org/document/7122532. Accessed 2 June 2020
12. M. Verhelst, A. Bahai, Where analog meets digital: Analog? To? Information conversion and beyond. IEEE Solid-State Circuits Mag. **7**(3), 67–80 (2015)
13. X. Liu, L. Golab, W. Golab, I.F. Ilyas, S. Jin, Smart meter data analytics: Systems, algorithms, and benchmarking. ACM Trans. Database Syst. TODS **42**(1), 1–39 (2016)
14. N. Mogles et al., How smart do smart meters need to be? Build. Environ. **125**, 439–450 (2017). https://doi.org/10.1016/j.buildenv.2017.09.008
15. P. Koponen et al., *Definition of Smart Metering and Applications and Identification of Benefits*. (2008)
16. J. Ponoćko, J.V. Milanović, Forecasting demand flexibility of aggregated residential load using smart meter data. IEEE Trans. Power Syst. **33**(5), 5446–5455 (2018). https://doi.org/10.1109/TPWRS.2018.2799903
17. Z. Jiang, R. Lin, F. Yang, A hybrid machine learning model for electricity consumer categorization using smart meter data. Energ. Basel **11**(9), 2235 (2018). http://dx.doi.org.effatuniversity.idm.oclc.org/10.3390/en11092235
18. S. Biansoongnern, B. Plungklang, Non-intrusive appliances load monitoring (nilm) for energy conservation in household with low sampling rate. Procedia Comput. Sci. **86**, 172–175 (2016)
19. A. Reinhardt et al., On the accuracy of appliance identification based on distributed load metering data. In *2012 Sustainable Internet and ICT for Sustainability (SustainIT)*, (2012), pp. 1–9
20. M. Weiss, A. Helfenstein, F. Mattern, and T. Staake, Leveraging smart meter data to recognize home appliances. In *2012 IEEE International Conference on Pervasive Computing and Communications*, (2012), pp. 190–197
21. *Pattern Recognition and Signal Analysis in Medical Imaging*. (Elsevier, 2014)
22. K.-S. Fu, *Applications of Pattern Recognition* (CRC, New York, 2019)
23. C. Belley, S. Gaboury, B. Bouchard, A. Bouzouane, An efficient and inexpensive method for activity recognition within a smart home based on load signatures of appliances. Pervasive Mob. Comput. **12**, 58–78 (2014)
24. S.S.S.R. Depuru, L. Wang, V. Devabhaktuni, R.C. Green, High performance computing for detection of electricity theft. Int. J. Electr. Power Energy Syst. **47**, 21–30 (2013)
25. A. Jindal, A. Dua, K. Kaur, M. Singh, N. Kumar, S. Mishra, Decision tree and SVM-based data analytics for theft detection in smart grid. IEEE Trans. Ind. Inform. **12**(3), 1005–1016 (2016)
26. R. Granell, C.J. Axon, D.C. Wallom, Impacts of raw data temporal resolution using selected clustering methods on residential electricity load profiles. IEEE Trans. Power Syst. **30**(6), 3217–3224 (2014)

27. E.D. Varga, S.F. Beretka, C. Noce, G. Sapienza, Robust real-time load profile encoding and classification framework for efficient power systems operation. IEEE Trans. Power Syst. **30**(4), 1897–1904 (2014)
28. R. Al-Otaibi, N. Jin, T. Wilcox, P. Flach, Feature construction and calibration for clustering daily load curves from smart-meter data. IEEE Trans. Ind. Inform. **12**(2), 645–654 (2016)
29. S. Kiran, H. A. Khattak, H. I. Butt, and A. Ahmed, Towards Efficient Energy Monitoring Using IoT. In *2018 IEEE 21st International Multi-Topic Conference (INMIC)*, (2018), pp. 1–4
30. Signal Reconstruction. http://pilot.cnxproject.org/content/collection/col10064/latest/module/m10788/latest. Accessed 05 Dec 2019
31. Physically Based Rendering - 3rd Edition. https://www.elsevier.com/books/physically-based-rendering/pharr/978-0-12-800645-0. Accessed 05 Dec 2019
32. Y. Hou et al., A 61-nW level-crossing ADC with adaptive sampling for biomedical applications. IEEE Trans. Circuits Syst. II Express Briefs **66**(1), 56–60 (2018)
33. S. M. Qaisar, D. Dallet, S. Benjamin, P. Desprez, and R. Yahiaoui, Power efficient analog to digital conversion for the Li-ion battery voltage monitoring and measurement. In *2013 IEEE International Instrumentation and Measurement Technology Conference (I2MTC)*, (2013), pp. 1522–1525
34. S. M. Qaisar, R. Yahiaoui, and D. Dominique, A smart power management system monitoring and measurement approach based on a signal driven data acquisition. In *2015 Saudi Arabia Smart Grid (SASG)*, (2015), pp. 1–4
35. S. M. Qaisar, A Computationally Efficient EEG Signals Segmentation and De-noising Based on an Adaptive Rate Acquisition and Processing. In *2018 IEEE 3rd International Conference on Signal and Image Processing (ICSIP)*, (2018), pp. 182–186
36. S.M. Qaisar, L. Fesquet, M. Renaudin, Adaptive rate filtering a computationally efficient signal processing approach. Signal Process. **94**, 620–630 (2014)
37. V.K. Ingle, J.G. Proakis, *Digital Signal Processing Using Matlab: A Problem Solving Companion* (Cengage Learning, Boston, 2016)
38. B.A. Moser, Similarity recovery from threshold-based sampling under general conditions. IEEE Trans. Signal Process. **65**(17), 4645–4654 (2017)
39. M. Paluszek, S. Thomas, *MATLAB Machine Learning* (Apress, New York, 2016)
40. S.K. Singh, R. Bose, A. Joshi, Energy theft detection for AMI using principal component analysis based reconstructed data. IET Cyber-Phys. Syst. Theory Appl. **4**(2), 179–185 (2019)
41. S. Pawar and B. Momin, Smart electricity meter data analytics: A brief review. (2017), pp. 1–5

Chapter 14
Prediction of Out-of-Step Condition for Synchronous Generators Using Decision Tree Based on the Dynamic Data by WAMS/PMU

Morteza Abedi, Mohammad Reza Aghamohammadi, Sasan Azad, Morteza Nazari-Heris, and Somayeh Asadi

Nomenclature

ANN	Artificial neural network
AR	Autoregressive
CCT	Critical clearance time
CDDT	Clearance detector decision tree
DT	Decision tree
FDDT	Fault detector decision tree
FDT	Fault duration time
IPDT	Instability predictor decision tree
HV	High voltage
L-G	Line to ground
LLG	Line to line to ground
LLL	3 line
MLPANN	Multilayer perceptron artificial neural network
NW	Number of moving time windows
O.S	Out of step
PMU	Phasor measurement unit
R5-LLL	3 line with resistive fault 5 ohm

M. Abedi (✉) · M. R. Aghamohammadi · S. Azad
Department of Electrical Engineering, Shahid Beheshti University, Tehran, Iran
e-mail: sa_azad@sbu.ac.ir

M. Nazari-Heris · S. Asadi
Department of Architectural Engineering, Pennsylvania State University, University Park, PA, USA
e-mail: mun369@psu.edu; m.nazariheris.2015@ieee.org; sxa51@psu.edu; asadi@engr.psu.edu

M. Nazari-Heris et al. (eds.), *Application of Machine Learning and Deep Learning Methods to Power System Problems*, Power Systems,
https://doi.org/10.1007/978-3-030-77696-1_14

R15-LLL	3 line with resistive fault 15 ohm
SNR	Signal to noise ratio
SVD	Sampling vector data
SVM	Support vector machine
TSTM	Transient stability time margin
WAMS	Wide-area management system

14.1 Introduction

Power systems are designed to be compatible with various disturbances such as faults, significant and sudden changes in loads, and losing the generators. They should keep system's crucial parameters, like the voltage and frequency, within the permissible range. However, a system is not always secure against all the consequences, and some unexpected events may occur that lead to instability in the rotor angle, frequency, and voltage, which may cause the instability of the whole system [1]. By looking at power systems from a historical point of view, it is evident that researchers have been working on the issue of the rotor angle, and they have tried to tackle this problem. Also, in recent years with increasing the number of consumers, the flowing power in the lines has increased; however, due to economic and environmental situations, the number of transmission lines remained constant. Therefore, it is evident that an increase in the flowing power in low-voltage lines will threaten the power system security. The instability of the whole system may occur due to a little disturbance [2, 3].

Rotor angle stability is an essential issue in power system stability problems that plays a considerable role in the power system's planning and operation. The transient stability of the rotor angle is the system's capability to maintain synchronism in the event of a severe disturbance, such as a short circuit appearance in the transmission line. The system's response to such disturbances causes substantial changes in the rotor angle of the generators, flowing power, and other variables. If the angular distance between the machines of the system remains within certain boundaries, the system will maintain its synchronism. Nevertheless, if synchronism is lost due to transient instability, the infinite angular deviation between the system's machines will appear 2–3 s after the initial disturbance occurs. Transient stability is based on the system's initial state and disturbance severity. The most crucial form of disturbance usually studied in the stability research topics contains faults that bring severe kinetic energy on the system at once, consequently affecting the system [4]. A power system is a vast interconnected nonlinear system in which failure or instability occurs due to the toleration of severe stresses to the system. Therefore, early detection of instability, preventive actions, and on-time control are the power system operator's goals. Unless a proper response is made to the occurrence of a significant disturbance in the system, the generators' unintentional outage may happen, followed by the subsequent outage of the system elements, and eventually to a complete blackout. Procedures such as the controlled outage of generators, load shedding, and isolating a section of the network are handled to maintain the system's synchronization and stability. Therefore, a fast and accurate method that can identify the transient steady state from instability is required [5]. The instability of the rotor angle of synchronous generators in the event of severe disturbance is considered a significant threat to system security and reliability. In the event of a rotor angle transient instability, the generator will get out of synchronism, which results in thermal damage and mechanical on the generator. In order to prevent any damage to the generator, the generator must be disconnected from the grid immediately after the detection of the O.S conditions. Therefore, rapid detection and identifying of O.S conditions of an unstable generator is essential to maintain the safety of the generator. In order to achieve this goal, it is very necessary to use equipment such as WAMS and PMU in monitoring the whole system and receiving operating variables in order to predict the transient instability [6]. Using the PMU and receiving data in telecommunications and processing it in a control center are one of the tasks that can improve data transferring and processing speed, hence the speed of the decision-making process. Transient stability monitoring is shown in Fig.14.1.

In order to maintain the safety of the generator, the O.S relay is installed on all synchronous generators of power system, which is responsible for detecting the status of O.S after its occurrence and rapid interruption of the generator from the power system. Therefore, early generator instability prediction is necessary to maintain the generator's safety and prevent possible damage to the generator. Hence, using intelligent machine learning such as SVM, ANN, and DT compared to the conventional analytical methods can improve the predictive ability.

The protection issue of the generator against unstable conditions can be investigated in two parts: identifying and predicting the O.S conditions of a generator. In recent years, in the field of research and industry, the evaluation, identification, and prediction of the rotor angle's transient stability have been considered an essential issue in the power system [8]. In [9], the energy function-based method is used to determine the stable operating area for the synchronous generator using the data after a fault in the power system. In [10], using the measurement of rotor angle and the Prony analysis-Hilbert transform method, a scheme for online monitoring and prediction O.S status of the power system generator is presented. In [11], the fuzzy logic theory is utilized to create a protection framework to detect O.S conditions of generators. In [12], the use of a nonlinear classifier based on SVM learning is proposed to detect the transient instability in a high-dimensional power system. In [13], a new strategy for improving the interpretability of traditional SVM results, avoiding the problem of false alarms and missed alarms, and solving the shortcomings of SVM is proposed. In this strategy, two improved and consecutive SVMs are used, which ensure the stability or instability of the power system in most cases. In [14], a method for online detection of generator O.S is presented based on measuring the voltage and current in a line and the energy function. In [15], the transient stability is monitored and predicted by a relative angle and predefined thresholds of the relative angle obtained with online data from PMUs and off-line simulations, respectively; in [16], for fast transient stability evaluation, fuzzy logic classifiers

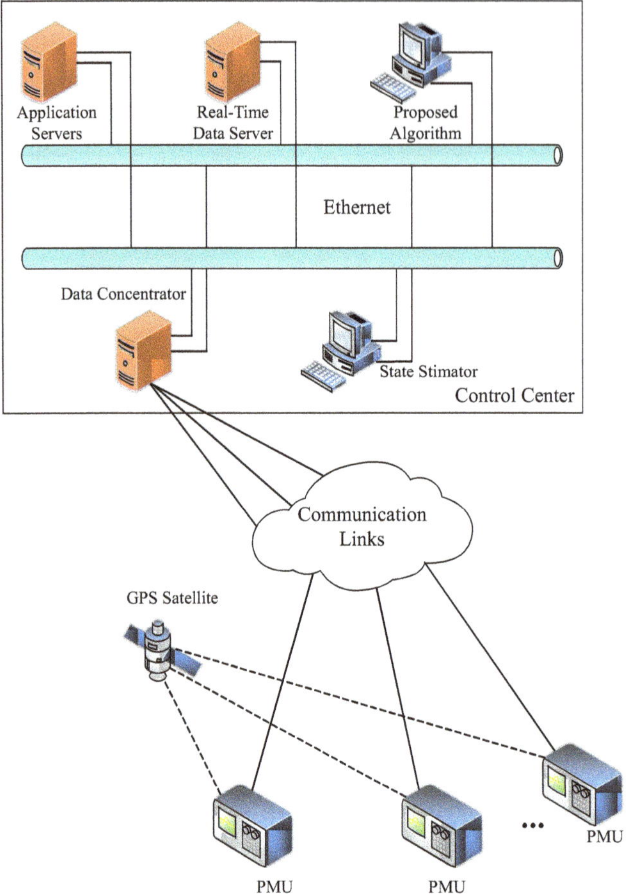

Fig. 14.1 Transient stability monitoring by WAMS/PMU [7]

obtained strategic bus information based on WAMS information in real-time conditions, and they brought it to the DT. DT can decide about the O.S status with an interval of 1–2 s from the time of clearance. In [17], a DT-based approach is proposed to detect inter-area instability by WAMS/PMU data. The inter-area instability is detected using WAMS data, including voltage and voltage phase angle measurements obtained by PMU. In [18], a rapid real-time detection method is proposed to detect transient instability using WAMS, where trajectory characteristics and transient energy are combined for improving the fast and accuracy of transient instability detection. In [19], CCT and TSTM have been utilized as transient stability identification criteria. Two multilayer perceptron artificial neural networks are used to estimate CCT and TSTM, and generator stability status assessment has been suggested. In [20], a model based on radial basis function and MLPANN is presented in order to detect the transient instability, in which transient stability is evaluated by the prediction of rotor angles and angular velocities of generators. In [21], to estimate the generator transient stability state, a normalized transient stability margin is presented by a multilayer perceptron neural network. The presented neural network uses pre-fault operation information for predicting the generator stability state. In [22], a real-time dynamic security assessment scheme is presented using PMUs and DTs. DTs perform online security assessment operations based on real-time measurements. In [7], a method for determining power fluctuations resulting in power system instability is proposed for real-time monitoring of rotor angle stability, utilizing WAMS/PMU data and based on Lyapunov exponentials. In [23], a real-time cluster prediction and identification algorithm have been developed without depending on the network model. The proposed method classifies generators as unstable generators by dividing them into coordinated groups behaviorally. The authors in [24] proposed an SVM classifier to predict the generator stability based on the generator's dynamic trajectories such as voltage, rotor angle, electromagnetic power, speed, and other electrical parameters. In [25], O.S prediction logic is presented based on the AR model. The proposed method is based on the time domain, using the mth order AR model and the voltage phase difference between the buses as input variables of the autoregressive model. The power fluctuations after clearance of a severe fault are modeled to predict out of step in the proposed method. In [5], a hybrid classifier is presented to predict the transient instability of the power system following a disturbance. An index is made as input data for classification by the relative rotor angles of the generators in the center of inertia scheme. In [2], the authors proposed a method to predict transient stability after the fault occurrence utilizing a post-fault voltage measurement at the generator bus of a multi-machine power network and the SVM classifier. The proposed method always initiates the fault event that causes the transient stability state's evaluation process after the voltage drop. In [3], an angular instability predictor is presented based on the bus voltage's measured trajectories after the occurrence of a severe disturbance. In [26], an intelligent wide-area synchrophasor system with ANN is introduced for real-time prediction of transient instability. In [27], a generator rotor angle instability predictor is proposed by a DT-based method to predict the O.S of synchronous generators. The proposed method can distinguish between stability state and O.S based on input characteristics measured before the moment of out of step. In [28], using an adaptive ANN, an instability rotor angle predictor is proposed. In [29], a new algorithm is proposed to predict the generator's O.S status using the measured information of rotor acceleration and speed by PMU, based on the change rate of speed and acceleration. In all the proposed methods, assuming that the moment of occurrence of the fault and clearance of the fault has already been detected, they rely mainly on the information after clearance of the fault. Some methods predict instability only for a group of generators instead of a single generator. The performance of the proposed methods due to noise, bad data, or miss data, changes in the grid configuration, and asymmetric faults have not been investigated and remain unanswered.

In this chapter, to quickly predict the O.S state of an unstable generator, an intelligent O.S prediction approach is proposed that includes a FDDT, a CDDT, and an IPDT. In this three-step scheme, using local online operating data obtained by

PMU, which is installed on the generator terminal, in the first step, FDDT detects the fault occurrence; in the second step, CDDT detects the fault clearance; and finally, in the third step, IPDT predicts O.S. The input vector data for all DTs is a moving time window containing several sampled vectors data measured in consecutive time steps. Every SVD may contain operating variables such as voltage magnitude and phase angle, active and reactive power, current magnitude and phase angle, generator rotor speed, angle, and acceleration. In this method, first, the time of occurrence and clearance is detected based on the variables of the generator's electrical operation, and then using IPDT and electromechanical operating variables after clearance, the stability of the generator is predicted.

14.2 The Framework of the Proposed Method

In this section, the initial framework of the presented algorithm for predicting transient stability is presented. The proposed scheme's main structure consists of three learning machines, which are used to classify the data. The machines apply the measured data before, during, and after the disturbance as FDDT, CDDT, and IPDT inputs, respectively, to precisely predict the system's transient stability. The proposed scheme's primary attitude is based on the fact that the instability of a generator could be distinguished from its transient behavior after-fault clearance condition. It is worth noting based on the generator's dynamic attributes; its stability status is mostly visible in the performance after the fault clearance. Therefore, the behavior of the generator after the fault clearance is categorized between stability and instability status. In other words, the dynamic behavior of generator after-fault clearance contains precious information about its stability nature that can be used to predict the instability state.

Indeed, it should be noted that the use of a learning machine reduces the time in determining the status of the system and the system operator has more chances to achieve control duties. The studies performed to determine the transient stability of the power system are based on the basic concept of fault CCT such that for the duration of the fault in the system more than CCT, the generated scenario leads to instability. Also, the generated scenarios are not concentrated on exact faults, and they contain faults at different places and different loading levels. Significant and evident signs in determining the state of power system instability for a synchronous generator are the variables of rotor angle and speed.

The behavior of the generator rotor angle for different conditions is shown in Fig. 14.2. According to Fig. 14.2, the typical trajectory of generator rotor angle is shown for different fault clearance times according to the fault CCT = *317* ms. In Fig. 14.2, the rotor angle variations are investigated for four different scenarios containing two unstable scenarios ($tclear3 = 320\,ms > CCT$, $tclear4 = 325\,ms > CCT$) and two stable scenarios ($tclear1 = 310\,ms < CCT$, $tclear2 = 315\,ms < CCT$). As it is observable, the change in generator rotor angle is relatively similar and slightly different for stable and unstable scenarios until the fault is cleared. However, after

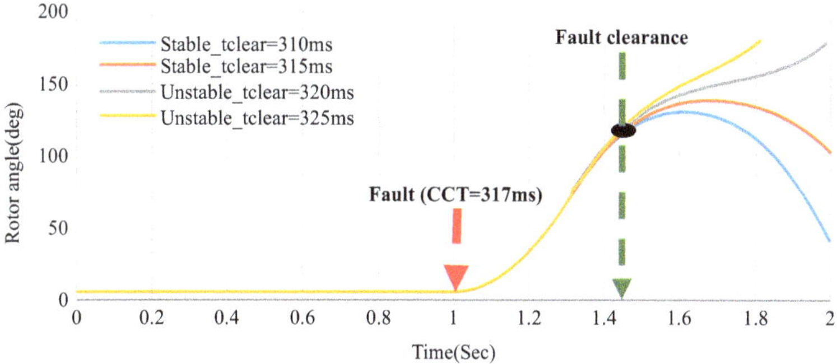

Fig. 14.2 Generator rotor angle fluctuation for four different scenarios

the clearance of the fault, the generator's rotor angle starts to increase the distance from the stable scenario and makes more distinction. The distinction made between stable and unstable scenarios after clearance can be appropriate in discriminating stable from unstable scenarios. It is worth noting that earlier studies and research assume that the occurrence of a fault and its clearance is known by generator and predictor of instability using generator information after the fault clearance. However, in the actual operation status, the fault occurrence and clearance affecting the generator's stability cannot be detected by the generator's local protection.

Fault occurrence and clearance outside the generator protection zone cannot activate the intelligent O.S relay, which operates according to post-fault information. Therefore, for an intelligent O.S prediction relay that operates based on data, which sampled after-fault clearance, detecting the fault occurrence and clearance using local generator information are necessary. To this end, the motivation of proposing an intelligent three-step scheme consisting of three DTs is to address all issues related to fault occurrence and clearance detection and instability prediction in the form of an O.S protection strategy.

The conceptual structure of the introduced intelligent O.S relay is shown in Fig. 14.2. Three DTs operate based on actual local information obtained from the PMU installed in the generator terminal in the proposed structure.

In the first step, the FDDT is used for detecting faults occurrence in the generator. Suppose the FDDT detects any faults or abnormal conditions similar to the fault that could be disadvantageous to the generator's stability. In that case, it sends output signal 1 to the CDDT to activate it to verify that the fault has been cleared or still exists. When the protection relays clear the fault, the CDDT detects this circumstance and sending output signal 1 to IPDT to activate it.

As shown in Fig. 14.3, during actual operation, a time window consisting of n data is provided to the FDDT at any given time by the fluctuating behavior of the generator tasted by the PMU. If the FDDT detects any faults or abnormalities, it sends output signal 1 to activate the CDDT. The next time window of the sampled data goes into the CDDT; otherwise, the time window containing the sampled data will still be available to the FDDT until it detects a fault or abnormal conditions.

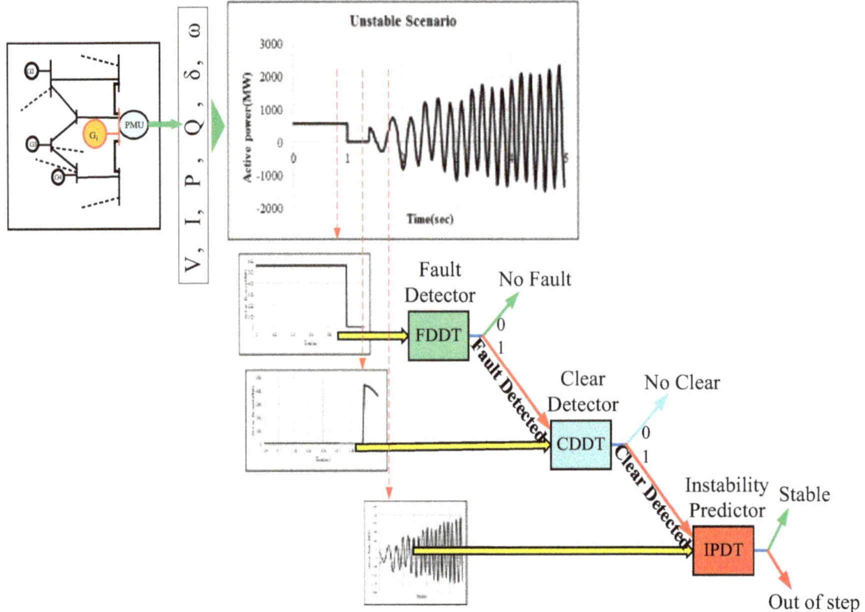

Fig. 14.3 The proposed 3step scheme for predicting O.S

Instantly after detecting any faults, the sampling window information is transferred to the CDDT. If the fault clearance detects by CDDT, the next time window of the sampling data will be transferred to the IPDT; otherwise, the time window information will be provided to the CDDT until the CDDT detects a fault clearance. All three DTs, which are introduced above, work in a design called an intelligent O.S prediction relay scheme that works for a generator.

14.3 The Proposed Method

The basis of the presented algorithm is that the instability of a generator is reflected in the electromechanical behavior after clearance of its fault, including the generator rotor angle, speed, and acceleration. Following the occurrence of a fault, the rotor angle and speed of a generator are modeled during the fault and after the fault is cleared by Eq. (14.1), which is called oscillation equations:

$$\begin{aligned} \frac{d\omega}{dt} &= \frac{\omega_0}{2H}(P_{\mathrm{m}} - P_{\mathrm{e}}) = \frac{\omega_0}{2H}P_{\mathrm{acc}} \\ \frac{d\delta}{dt} &= \omega - \omega_0 \end{aligned} \tag{14.1}$$

where ω, δ, and H are the speed and angle of the rotor and the constant inertia of the generator before the fault occurs and is the accelerator power. By merging the swing equations during the fault and after the fault clearance, the generator's rotor angle and speed after the fault clearance can be estimated by Eq. (14.2):

$$\begin{aligned} \delta_{\text{pos}} &= \delta_0 + \frac{\omega_0}{4H} P_{\text{acc}}^{\text{f}} \cdot t_{\text{f}}^2 + \frac{\omega_0}{4H} P_{\text{acc}}^{\text{pos}} \cdot t_{\text{p}}^2 \\ \omega_{\text{pos}} &= \omega_0 + \frac{\omega_0}{2H} P_{\text{acc}}^{\text{f}} \cdot t_{\text{f}} + \frac{\omega_0}{2H} P_{\text{acc}}^{\text{pos}} \cdot t_{\text{p}} \end{aligned} \quad (14.2)$$

which $\delta_0$and ω_0 are the rotor angle and speed before the fault, respectively. δ_{pos} and ω_{pos} are the rotor angle and speed after the fault clearance, respectively. $P_{\text{acc}}^{\text{f}}$ and $P_{\text{acc}}^{\text{pos}}$ are the average accelerator power during fault and after the fault clearance, respectively. t_{f} and $t_{\text{p}}$are the duration of the fault and the moment after the fault clearance, respectively, which indicates the severity of the disturbance. In Eq. (14.2), the fault duration t_{f} and the average accelerator power $P_{\text{acc}}^{\text{f}}$ indicate the severity of the disturbance, which affects the stability of the generator. Equation (14.2) shows that the severity of the fault is directly affected by changes in the rotor angle and speed of the generator after-fault clearance. In fact, in power system stability studies, the trajectory trend of changing the rotor angle after-fault clearance is used as a criterion for evaluating the stability of the generator. Therefore, the generator's behavioral information after clearance of the fault is prosperous and useful to distinguish the generator's stable state from the unstable state. In other words, the information after-fault clearance in the generator behavior has properties that can be used to predict the instability of the generator. Thus, intelligent methods play a prominent role in extracting the characteristics of such precious information. The DT is an intelligent, powerful tool for extracting features from complex data that can be used to predict the synchronous generator's instability in power systems.

14.3.1 Decision Tree

In this study, the DT is used to evaluate the system's transient stability. Using the DT learning machine is simple, fast, and appropriate for classifying complex issues. These intelligent machines are suitable for problems with a low output. One of the most prominent features of the DT compared to other learning machines is its decision-making process algorithm, which is well-defined and transparent. On the other hand, the decision-making process in other learning methods is not transparent and is like a black box.

The advantages of the DT over other learning machines, such as the neural network, are:

- The DT explains its prediction in the form of a set of rules, while in neural networks, the only prediction is expressed and how it is hidden in the network.

- Also, in the DT, unlike neural networks, the data do not need to be numerical. It can support numeric variables and can work with categorical variables such as type and gender.
- Also, DT can determine outstanding operating variables concerning output. In other words, by placing a prominent variable in the central node, the significance of this variable in classification can be estimated.

In this chapter, for predicting the generator's rotor angle instability based on fault clearance data, the DT learning machine is used as an O.S predictor.

The introduced intelligent predictor scheme includes three consecutive DTs that use local online data from the generator oscillation produced by the PMU. The presented predictor scheme is a combination program of the DT learning machine as the intelligent predictor and the PMU as the local dynamic data generator. In this method, the task of fault occurrence and clearance detection and instability prediction is performed by three consecutive DTs. IPDT has been used for performing the main task of predicting O.S, for which the entrance data is provided by the PMU and after-fault clearance. Therefore, there needs to be a difference between the faulty data and after the fault clearance. For this purpose, in the previous two steps, the two DTs are used as fault occurrence detector and the fault clearance detector to distinguish the data after the fault clearance and during the fault.

14.3.2 Fault Detection by FDDT

The FDDT is the first step in predicting the presented O.S method. The main task of FDDT is detecting faults or any abnormal conditions from the generator's point of view. FDDT uses electrical operating variables provided online by the PMU. Input data of FDDT is a moving time window including six samples of a generator's electrical operating variables such as current and voltage magnitude, and their phase angle, active and reactive power. The FDDT ability to detect a fault is mainly due to the sudden change that occurs between the electrical operating variable values before the fault occurrence and during the fault (e.g., I, V, θ_I, θ_V, P, Q). SVD at time t includes some sampled operating variables (e.g., I, V, θ_I, θ_V, P, Q). In Eq. (14.3) the sampled operating electrical variables at time t are shown:

$$\mathrm{SVD}(t) = [I(t), V(t), \theta_\mathrm{I}(t), \theta_\mathrm{V}(t), P(t), Q(t)] \tag{14.3}$$

Figure 14.4a represents a time window structure, including six consecutive SVDs created with a sampling interval of ΔT. A time window is defined as a precious time window for detecting the occurrence of fault when it contains a combination of pre-fault and during-fault sampled operating data. Whenever all the sampled operating data of a time window is completely about before the fault or entirely during the fault, that time window is defined as a weak time window with no information about the fault. Thus, the FDDT cannot detect the fault occurrence.

Time Window	Sample 1	Sample 2	Sample 3	Sample 4	Sample 5	Sample 6
TWf 1	SVD (Tf-5Δt)	SVD (Tf-4Δt)	SVD (Tf-3Δt)	SVD (Tf-2Δt)	SVD (Tf-Δt)	SVD (Tf) (fault)
TWf 2	SVD (Tf-4Δt)	SVD (Tf-3Δt)	SVD (Tf-2Δt)	SVD (Tf-Δt)	SVD (Tf) (fault)	SVD (Tf+Δt)
TWf 3	SVD (Tf-3Δt)	SVD (Tf-2Δt)	SVD (Tf-Δt)	SVD (Tf) (fault)	SVD (Tf+Δt)	SVD (Tf+2Δt)
TWf 4	SVD (Tf-2Δt)	SVD (Tf-Δt)	SVD (Tf) (fault)	SVD (Tf+Δt)	SVD (Tf+2Δt)	SVD (Tf+3Δt)
TWf 5	SVD (Tf-Δt)	SVD (Tf) (fault)	SVD (Tf+Δt)	SVD (Tf+2Δt)	SVD (Tf+3Δt)	SVD (Tf+4Δt)
TWf 6	SVD (Tf) (fault)	SVD (Tf+Δt)	SVD (Tf+2Δt)	SVD (Tf+3Δt)	SVD (Tf+4Δt)	SVD (Tf+5Δt)

a. six precious time windows including pre-fault and during fault occurrence data for FDDT

Time Window	Sample 1	Sample 2	Sample 3	Sample 4	Sample 5	Sample 6
TWC1	SVD (Tc-5Δt)	SVD (Tc-4Δt)	SVD (Tc-3Δt)	SVD (Tc-2Δt)	SVD (Tc-Δt)	SVD (Tc) (clear)
TWC2	SVD (Tc-4Δt)	SVD (Tc-3Δt)	SVD (Tc-2Δt)	SVD (Tc-Δt)	SVD (Tc) (clear)	SVD (Tc+Δt)
TWC3	SVD (Tc-3Δt)	SVD (Tc-2Δt)	SVD (Tc-Δt)	SVD (Tc) (clear)	SVD (Tc+Δt)	SVD (Tc+2Δt)
TWC4	SVD (Tc-2Δt)	SVD (Tc-Δt)	SVD (Tc) (clear)	SVD (Tc+Δt)	SVD (Tc+2Δt)	SVD (Tc+3Δt)
TWC5	SVD (Tc-Δt)	SVD (Tc) (clear)	SVD (Tc+Δt)	SVD (Tc+2Δt)	SVD (Tc+3Δt)	SVD (Tc+4Δt)
TWC6	SVD (Tc) (clear)	SVD (Tc+Δt)	SVD (Tc+2Δt)	SVD (Tc+3Δt)	SVD (Tc+4Δt)	SVD (Tc+5Δt)

b. 6 precious time windows including during fault and post fault clearance data for CDDT

Time Window	Sample 1	Sample 2	Sample 3	Sample 4	Sample 5	Sample 6
SVD (Tc) (clear)	SVD (Tc+Δt)	SVD (Tc+2Δt)	SVD (Tc+3Δt)	SVD (Tc+4Δt)	SVD (Tc+5Δt)	SVD (Tc+6Δt)

C. Time window including post fault clearance data for IPDT

Fig. 14.4 The moving time windows structure for DTs. (**a**) Six precious time windows including pre-fault and during-fault occurrence data for FDDT. (**b**) Six precious time windows including during-fault and post-fault clearance data for CDDT. (**c**) Time window including post-fault clearance data for IPDT

During the generator's operation, the moving time windows are continuously updated with new sampled operating variables. Thus, due to the occurrence of a fault, six consecutive time windows have precious information about the moment of the fault occurrence and the moment before the fault occurrence.

Figure 14.4a indicates the structure of six consecutive precious time windows, each including six sampled operating data, in which the yellow samples belong to the during-fault data, while the blue samples belong to pre-fault data. According to Fig. 14.4a, SVD(*Tf*) (red) shows the operating data sampled at the moment of fault occurrence, SVD($Tf\text{-}k\Delta T$) belongs to k sample data before fault occurrence, and SVD($Tf + k\Delta T$) refers to the k sample data after occurrence of the fault.

As can be seen, after the fault occurrence time, the number of SVDs containing the fault data in the time window increases. Adopting six SVDs per time window enables FDDT to view precious information in six consecutive time windows that increase FDDT detection reliability. Equation (14.4) displays the conceptual performance of FDDT for the precious and weak time window:

$$\text{FDDT}: \begin{cases} 0 = f\left(\text{SVD}_{\text{pre}}\right) & : \text{weak data} \\ 1 = f\left(\text{SVD}_{\text{pre}}, \text{SVD}_{\text{f}}\right) & : \text{precious data} \\ 0 = f(\text{SVD}_{\text{f}}) & : \text{weak data} \end{cases} \tag{14.4}$$

After detection of fault occurrence by FDDT, it sends output signal 1 to the CDDT to activate it.

14.3.3 Fault Clearance Detection by CDDT

After the FDDT detects the fault, the CDDT is activated for detecting the fault clearance. The CDDT uses electrical operating variables provided online by the PMU. Input information for CDDT is a moving time window including six samples of generator electrical operating variables such as current and voltage magnitude, and their phase angle, active and reactive power. The CDDT ability to detect a fault's clearance is mainly due to the sudden change that occurs between the electrical operating variables values during fault and after the fault clearance (e.g., I, V, θ_{I}, θ_{V}, P, Q).

Figure 14.4b demonstrates the structure of a time window that includes six consecutive SVDs produced with the same time step ΔT. A time window is a precious time window for detecting the clearance of fault when it contains a combination of during-fault and after-fault clearance sampled operational data.

The window is a weak time window when all sampled operating data are entirely gathered during the fault or after the fault clearance. This time window does not have any clearance information, and CDDT cannot detect clearance. During the generator operating, the moving time windows are continuously updated with newly sampled operating data. Therefore, due to the fault clearance, six consecutive time windows have precious information.

Figure 14.4b demonstrates the structure of six consecutive precious time windows, each including six sampled operating data. In these time windows, the yellow samples belong to the during-fault data, while the blue samples belong to the data which is sampled after-fault clearance.

According to Fig. 14.4b, SVD(TC) (green) shows the sampled data at the time of fault clearance, SVD(TC-$k\Delta T$) belongs to k sample data before fault clearance, and SVD($TC + k\Delta T$) refers to the k sample data after clearance the fault. According to the time windows, it is observed that after the fault clearance time, the number of SVDs with data that are sampled after-fault clearance increases.

Adopting six SVDs per time window makes CDDT capable of viewing precious data through six consecutive time windows. This design enhances the reliability of the CDDT detection performance. Equation (14.5) displays the conceptual performance of CDDT concerning the precious and weak time window:

$$\mathrm{CDDT}:\begin{cases} 0 = f(\mathrm{SVD_f}) & : \text{weak data} \\ 1 = f(\mathrm{SVD_f}, \mathrm{SVD_{pos}}) & : \text{precious data} \\ 0 = f(\mathrm{SVD_{pos}}) & : \text{weak data} \end{cases} \tag{14.5}$$

After detecting fault clearance by CDDT, it sends output signal 1 to the IPDT to activate it.

14.3.4 Prediction of Instability by IPDT

The final step of the proposed intelligent relay is to predict the O.S status of the generator by IPDT that is the presented scheme's major objective. The IPDT works based on the generator's online data. IPDT input vector data is a moving time window containing six sampled operating data, including generator oscillation and electromechanical data such as rotor angle, speed, and acceleration.

Figure 14.4c demonstrates the structure of a time window, including six consecutive SVDs produced with the same time step ΔT. The time window is continuously created and submitted to the IPDT. However, IPDT is only activated by CDDT when the fault is cleared.

If the IPDT predicts the generator's potential to O.S, it will send a "1" signal on the output as a warning of an unstable status of the generator. It is necessary to mention; if the IPDT fails for predicting the generator stability by the first time window, the next time window will continue to enter the IPDT until the prediction of the generator stability status is made even with a delay.

The proposed intelligent predictor relay is considered as a backup for the main O. S relay. Equation (14.6) displays the conceptual performance of IPDT concerning data that are sampled after-fault clearance:

$$\text{IPDT}:\begin{cases} 0 = f(\text{SVD}_{\text{pos}}) & : \text{Stable} \\ 1 = f(\text{SVD}_{\text{pos}}) & : \text{Unstable} \end{cases} \tag{14.6}$$

The input data of DTs which consists of six consecutive time windows provides strong performance for DTs against any disturbances may occur in samplings such as miss data or bad data.

14.3.5 Nature of Operating Variables Used by DT

Usually, in order to train DTs, various operating variables of generators can be used, such as active and reactive output power, voltage magnitude and phase angle, rotor angle, and speed. In the moment of fault occurrence and fault clearance, variations in operating variables such as active power, voltage, and current are more precise compared to variations in operating mechanical variables such as speed and rotor angles.

Therefore, electrical operating variables that change suddenly at the moment of occurrence and clearance of the fault are the most appropriate and precious variables for FDDT and CDDT input data. The electromechanical variables such as rotor angle, speed, and acceleration, which indicate the generator rotor angle's stability, are the most appropriate and precious variables for using as input data of IPDT.

Since the magnitude of changes in the electrical operating variables in the time step ΔT is better than the magnitude of the variables, changes in the specified time step's electrical variables are used as the input vectors to the DTs.

14.4 Simulation

To evaluate the presented model's effectiveness to predict the O.S of synchronous generators following a fault occurrence, the proposed protection scheme has been implemented on the IEEE 39-bus power system. The presented model for generator #7 which is connected to bus #36 will be investigated.

In order to select the appropriate operating variables of the power system for training FDDT, CDDT, and IPDT, all electrical and electromechanical operational variables such as active and reactive power, voltage phasor, current phasor, generator rotor angle, speed, and acceleration are considered and have thoroughly been reviewed. Finally, an appropriate number of operating variables are selected based on DTs performance and detection/prediction. In this research, a set of specific operating variables is introduced for each DT.

14.4.1 Producing Training Scenarios of DT

In this subsection, training scenarios from generator G7, located in bus #36 in the IEEE New England 39-bus power system, have been produced. Since the proposed O.S scheme is built to predict a particular generator's stability status, the diversity of events in which the generator experience unstable condition is not very wide. That is, the generator may become unstable in many scenarios. However, the oscillator behavior pattern of the generator is relatively limited for all of the instability scenarios. This means that by adequately selecting different fault scenarios, the main feature of the generator oscillation behaviors that lead to instability can be obtained. Therefore, to prepare appropriate training scenarios, six different load levels (4000 MW, 5000 MW, 6000 MW, 7500 MW, 9000 MW, 10500 MW) have been selected.

Concerning transient stability, the criterion should be used to determine stability in order to identify the status of the power system in terms of transient stability and instability. The proposed method criterion is based on CCT to determine a stable and unstable scenario. It is assumed that a fault occurs in the network and is cleared after a certain period (relay operation time). This fault has its own CCT attribute that indicates how long the network has time to clear the fault. CCT determines the significance of the fault. The smaller the CCT related to a fault, the fault has more significance, and the less time available to the network to clear it. As a result of this fault, network instability is more likely to occur. Therefore, for the network to remain stable after-fault clearance, the clearance time (relay operation) must be less than the CCT.

In order to produce various stable/unstable scenarios for training three DTs in each operating load level, the symmetrical LLL-faults with different fault clearance times, according to generator G7 CCT (i.e., 0.1*CCT, 0.5*CCT, 0.8*CCT, 0.95*CCT, 1.2*CCT, 1.4*CCT), is applied to the power system under study. Scenarios with less fault clearance time than CCT are defined as stable scenarios, while scenarios with fault clearance duration higher than CCT are defined as unstable scenarios. These faults are applied to buses No. 21, No. 22, No. 23, and 24 and lines 16–24, 21–22, 22–23, 21–16, 16–24, and 23–24 around generator G7. This method creates a total of 324 disturbance scenarios.

The transient stability simulation is performed using DIgSILENT Power Factory® software, in which the generators are modeled in full detail. The stability status of generator G7 is obtained for all scenarios. Among the designed scenarios, 220 scenarios, 70 stable scenarios, and 150 unstable scenarios are acceptable.

The generator operating variables transfer rate, which are provided by PMU for is 10 ms. Therefore, the sampling data's sampling step is updated in the moving time windows which is considered equal to 10 ms ($\Delta t = 10$ ms). For each kind of stable/unstable scenario, according to the FDT, the NW is determined by the speed of movement of a sample based on Eq. (14.7):

Table 14.1 Number of time windows generated for training/testing of DTs

Time window	FDDT and CDDT		IPDT	
	Training	Test	Training	Test
Stable state	2510	1124	503	209
Unstable state	5211	2108	1034	421
Total	7721	3232	1537	630

$$\mathrm{NW} = \underbrace{\frac{\mathrm{FDT}}{0.01}}_{I} + \underbrace{10}_{II} \tag{14.7}$$

The first part of Eq. (14.7) defined by (*I*) is the number of time windows that belongs to the fault duration time used for FDDT and CDDT training/testing process. In the second part of Eq. (14.7) defined by (*II*), "10" is the number of time windows after the fault is cleared for IPDT training/testing process.

According to Table 14.1, by applying Eq. (14.7) to 220 scenarios, a total of 13,390 time windows for stable/unstable training and testing are generated, of which 70% and 30% of the time windows for training and testing DTs are used, respectively.

14.4.2 Time Windows Data

In preparing training scenarios, the number of samples that should be gathered after clearance of the fault as the IPDT input time window to predict O.S is significant. The duration time between clearance of the fault and O.S depends on the generator, controllers, and location in the power system. In most severe instability conditions, this duration time is short. There is a minimum duration for each generator, which is related to the most severe instability conditions for the generator. The IPDT sampling time window should not exceed this minimum duration time because the time window may contain after O.S data, which is practically worthless data. Therefore, for each generator, this minimum duration time must be specified first. Then, the sampling time window is determined based on it.

14.4.3 FDDT Training

Table 14.2 presents the FDDT performance of training concerning different input operating variables that provide the maximum accuracy for the detection of the fault occurrence.

As seen in Table 14.2, the set with the highest accuracy is selected as the FDDT input variables, which include all three variables of $\Delta\theta_{\mathrm{v}}$, ΔP, and ΔQ. The FDDT

Table 14.2 FDDT performance results for different input variables

Variables	Train (7721)			Test (3232)		
	Acc. %	Error detection		Acc. %	Error detection	
		%	No. of misclassification		%	No. of misclassification
ΔP, $\Delta\theta_v$, ΔQ	99.71	0.29	19	99.75	0.25	8

input vector consisting of three selected variables along with samples taken at specified time steps is shown in Eq. (14.8):

$$\begin{aligned}
&\text{Sampling vector data}: \text{SVD}(t) = [\Delta\theta_V(t), \Delta P(t), \Delta Q(t)]\\
&\Delta\theta_V(t) = [\Delta\theta_V(t - k\Delta t)], \quad k = 1, \ldots, 5\\
&\Delta P(t) = [\Delta P(t - k\Delta t)], \quad k = 1, \ldots, 5\\
&\Delta Q(t) = [\Delta Q(t - k\Delta t)], \quad k = 1, \ldots, 5
\end{aligned} \tag{14.8}$$

The structure of trained FDDT is shown in Fig. 14.5a. The numbers in parentheses refer to the sample number of Fig. 14.4a.

14.4.4 CDDT Training

Table 14.3 presents the CDDT performance of training concerning different input operating variables, which provide maximum accuracy for detecting the fault clearance.

Different combinations of variables are examined. As observed in Table 14.3, the set has been selected as the CDDT input variables with the highest accuracy, which includes all three variables of $\Delta\theta_I$, ΔP, ΔQ. The CDDT input vector consisting of three selected variables, along with samples taken at specified time intervals, is presented in Eq. (14.9):

$$\begin{aligned}
&\text{Sampling vector data}: \text{SVD}(t) = [\Delta\theta_I(t), \Delta P(t), \Delta V(t)]\\
&\Delta\theta_I(t) = [\Delta\theta_I(t - k\Delta t)], \quad k = 1, \ldots, 5\\
&\Delta P(t) = [\Delta P(t - k\Delta t)], \quad k = 1, \ldots, 5\\
&\Delta V(t) = [\Delta V(t - k\Delta t)], \quad k = 1, \ldots, 5
\end{aligned} \tag{14.9}$$

The structure of trained CDDT is shown in Fig. 14.5b. The numbers in parentheses refer to the sample number of Fig. 14.4b.

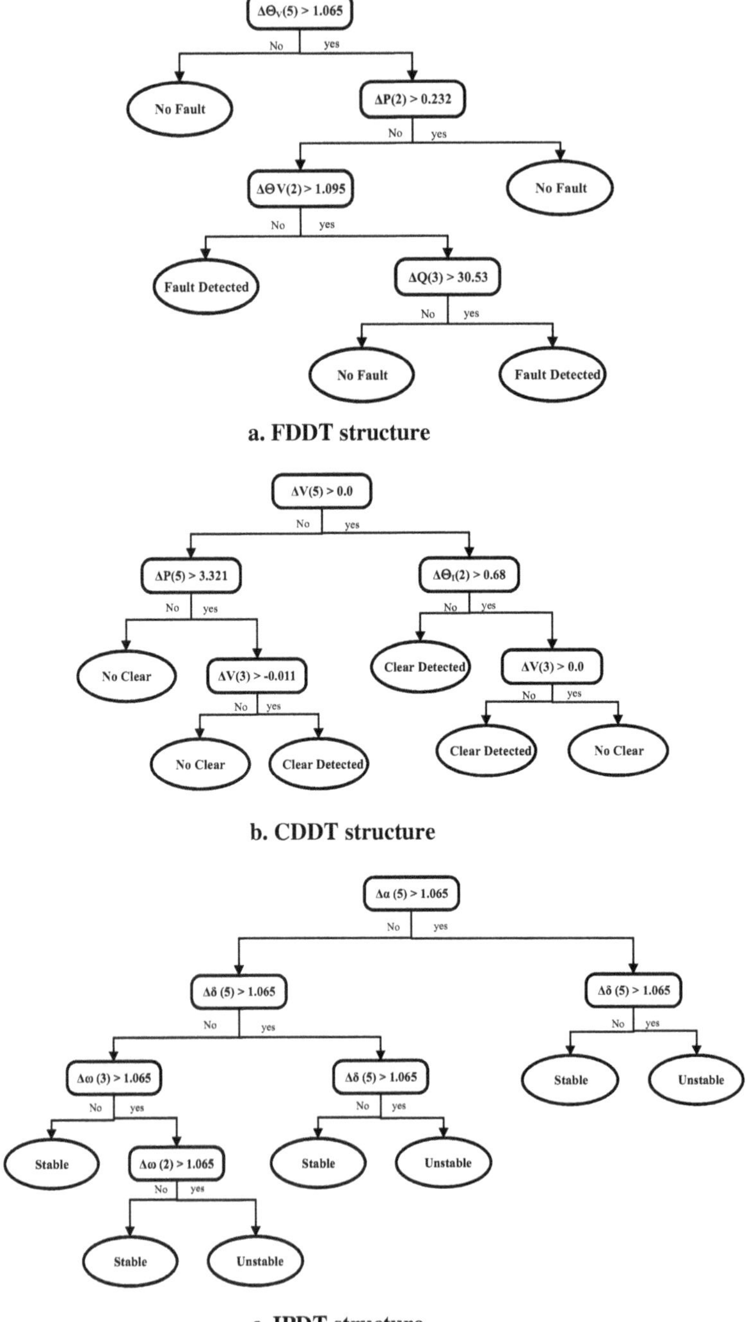

Fig. 14.5 DTs structure (**a**) FDDT structure, (**b**) CDDT structure, (**c**) IPDT structure

Table 14.3 CDDT performance results for different input variables

Variables	Train (7828)			Test (3995)		
	Acc. %	Error detection		Acc. %	Error detection	
		%	No. of misclassification		%	No. of misclassification
ΔP, $\Delta\theta_I$, ΔQ	99.71	0.29	22	98.95	1.05	34

Table 14.4 IPDT performance results for different input variables

Variables	Training (1537)			Testing (630)		
	Acc. %	Error detection		Acc. %	Error detection	
		%	No. of misclassification		%	No. of misclassification
$\Delta\alpha$, Δw, $\Delta\delta$	98.73	1.27	20	89.96	10.03	63

14.4.5 IPDT Training

This is the last step of the proposed method, which is responsible for predicting the instability. The stability status of the generator is predicted based on data that are sampled after-fault clearance. For this purpose, changes in rotor angle, speed, and acceleration magnitude are used as input variables. The phase angle of the generator's terminal voltage is estimated as the generator rotor angle [30]. Therefore, the rotor speed and rotor acceleration are calculated by the generator rotor angle using Eqs. (14.10) and (14.11):

$$\omega(t+1) = \frac{\delta(t+1) - \delta(t)}{\Delta t} \tag{14.10}$$

$$\alpha(t+1) = \frac{\omega(t+2) - \omega(t+1)}{\Delta t} \tag{14.11}$$

Due to the angle, speed, and acceleration of the rotor changes as input variables, a different combination of these variables is used as the IPDT sampled data vector. The performance of IPDT training concerning different combinations of variables $\Delta\alpha$, Δw, and $\Delta\delta$ are examined.

As shown in Table 14.4, the set has been selected as the IPDT input variables with the highest accuracy, including all three variables of $\Delta\alpha$, Δw, and $\Delta\delta$. The IPDT input vector consisting of three selected variables along with samples taken at specified time steps is shown in Eq. (14.12).

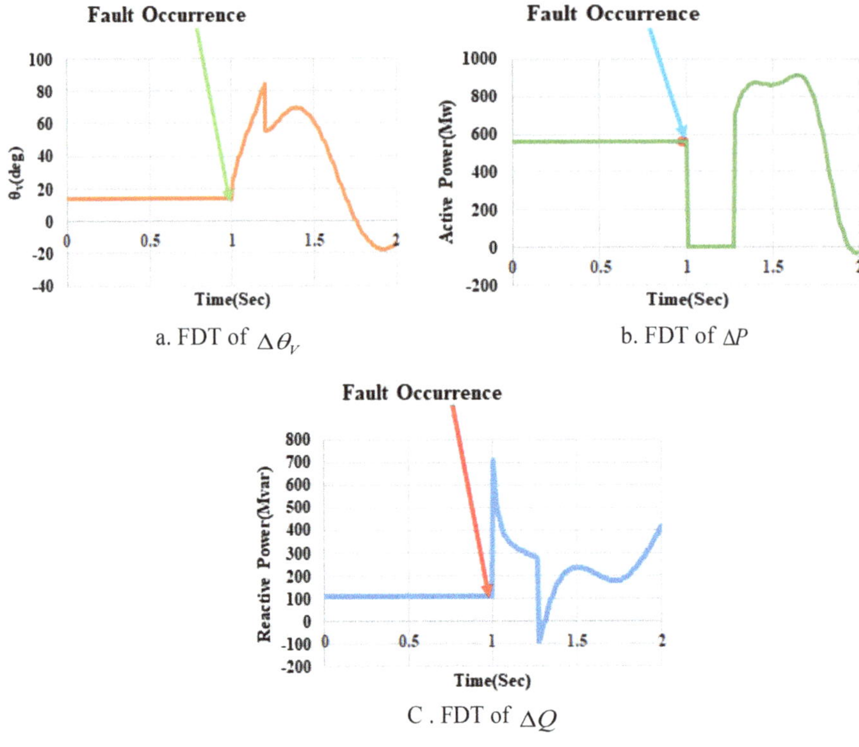

Fig. 14.6 Operating variables of FDDT. (**a**) FDT of $\Delta\theta_V$. (**b**) FDT of ΔP, (**c**) FDT of ΔQ

$$\begin{aligned}
&\text{Sampling vector data}: \text{SVD}(t) = [\Delta\delta(t), \Delta\omega(t), \Delta V(t)] \\
&\Delta\delta(t) = [\Delta\delta(t - k\Delta t)], \quad k = 1, \ldots, 5 \\
&\Delta\omega(t) = [\Delta\omega(t - k\Delta t)], \quad k = 1, \ldots, 5 \\
&\Delta\alpha(t) = [\Delta\alpha(t - k\Delta t)], \quad k = 1, \ldots, 5
\end{aligned} \tag{14.12}$$

The structure of trained IPDT is shown in Fig. 14.5c. The numbers in parentheses refer to the sample number of Fig. 14.4c. According to Fig. 14.5, the threshold values of each DT are determined during the training process.

In Fig. 14.6a–c, a three-phase short circuit fault at $t = 1$ s is applied to the generator's HV bus under study. As can be seen, the operating variables used in FDDT have significant changes at the time of the fault.

In Fig. 14.7a–c, a three-phase short circuit fault at $t = 1$ s is applied to the HV bus of the generator under study and is cleared at $t = 1.27$ s. As can be seen, the operating variables used in CDDT have significant changes at the time of clearance.

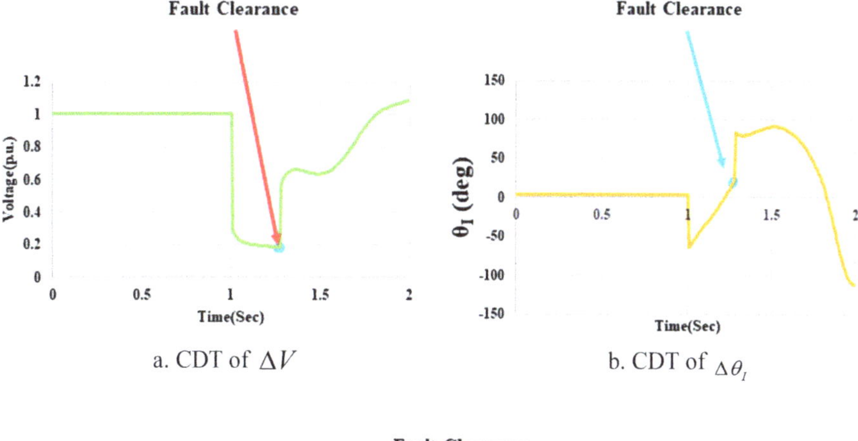

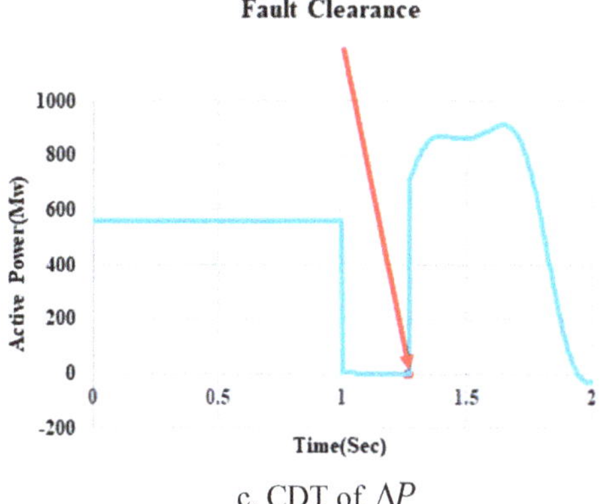

Fig. 14.7 Operating variables of CDDT. (**a**) CDT of ΔV, (**b**) CDT of $\Delta\theta_I$, (**c**) CDT of ΔP

14.4.6 Over-/Under-Fitting Evaluation of IPDT

Since DTs' performance is threatened by over-/under-fitting in the training procedure, the number of training scenarios is gradually decreased/increased. The training/test performance of IPDT is also evaluated. According to Table 14.5, the number of training/testing scenarios uses in this study is optimal and provides the best training/testing accomplishment for IPDT. According to the optimal number of scenarios, any increase/decrease in the number of scenarios causes over-/under-fitting in the training process, which is a factor to reduce the accuracy of training.

It should be noted that since the intelligent O.S predictor relay is designed individually for each generator with respect to the generator behavior, so the training process includes the number of training scenario and training time independent of the size of the power system.

Table 14.5 IPDT performance concerning over-/under-fitting

Training cases	No. of training scenario	Training result		Testing result	
		Acc. %	No. of Misclassification	Acc. %	No. of Misclassification
1	4985	96.3	184	87.9	110
2	7846	95.8	329	88.7	93
3	10320	97.35	273	88.81	96
4	13390	98.73	170	89.96	63
5	14890	97.21	564	81.02	162
6	16281	94.05	968	85.73	138

Table 14.6 Comparison of DTs performance with ANN and SVM

	DT performance		SVM performance		MLPANN performance	
Acc.%	Train	Test	Train	Test	Train	Test
Fault occurrence detection	99.71	99.75	99.34	98.96	99.08	99.54
Fault clearance detection	99.71	99.75	99.34	98.83	99.19	99.1
Instability occurrence prediction	98.73	89.96	94.93	85.72	96.65	85.9

14.4.7 Comparison of DT Performance with SVM and ANN

The proposed intelligent relay performance is compared with other learning machines. Two learning machines consist of MLPANN and SVM trained with the same data and then compared with the result of the DT-based algorithm. In this comparison, ANN with ten neurons in the hidden layer and a Levenberg-Marquardt algorithm is used. The other intelligent method is SVM with a standard deviation of 0.1 and a regulating parameter of 10. The accuracy of training and testing for the presented scheme compared to other intelligent schemes is shown in Table 14.6. According to Table 14.6, the ANN and SVM performances are almost the same, while the proposed intelligent predictor relay based on DT performs better.

14.5 Validation of the Proposed Scheme Performance

To validate the proposed intelligent relay scheme, its performance is examined for seven stable/unstable scenarios that were not used in training. The characteristics of the seven scenarios are shown in Table 14.7. In addition to the 8 developed scenarios, 57 other stable and unstable scenarios have been developed to evaluate IPDT

Table 14.7 Characteristics of scenarios for validation of DTs performance

Scenario	Load level (MW)	Stability status	Fault type	Location of fault		CCT (ms)	FDT (ms)
				Bus/line	Line %		
1	6832	Unstable	LLL	Lines 23–24	50	352	400
2	5040	Stable	LLL	Bus 24	–	495	300
3	6350	Unstable	L-G	Lines 22–23	40	428	550
5	6350	Unstable	LLG	Lines 22–23	60	318	450
5	6350	Unstable	R_5-LLL	Bus 24	–	363	450
6	6350	Stable	R_{15}-LLL	Bus 24	–	624	550
7	6350	Unstable (change configuration)	LLL	Lines 23–24	40	276	350

Table 14.8 DTs validation results

Scenario (status)	Actual time			FDDT	CDDT	IPDT
	Fault occurrence (s)	Fault clearance (s)	O.S (s)	Fault detection (s)	Clear detection (s)	O.S (s)
1 (U)	1.00	1.4	1.49	1.0	1.4	1.45
2 (S)	1.00	1.3	–	1.0	1.3	–
3 (U)	1.00	1.55	1.81	1.0	1.55	1.6
4 (U)	1.00	1.45	1.71	1.0	1.45	1.5
5 (U)	1.00	1.45	1.68	1.0	1.45	1.5
6 (S)	1.00	1.55	–	1.0	1.55	–
7 (U)	1.00	1.35	1.49	1.0	1.35	1.4

performance. According to each of the scenarios, the relevant CCT is calculated using a time domain simulation to show the scenario's stability and compare it with the fault clearance time. The validation results of FDDT, CDDT, and IPDT for seven scenarios compared to the real-time fault occurrence/clearance and O.S obtained by time domain simulation studies are shown in Table 14.8. According to Table 14.8, for all instability scenarios, IPDT predicts instability before the actual O.S time. Figure 14.8 shows the analytical results of IPDT performance for 44 unstable scenarios among 57 scenarios. Figure 14.8 shows the bar chart which the horizontal axis is the length of time that the IPDT predicted O.S earlier than real-time. For example, in 13 unstable scenarios, an O.S is predicted with a duration of 150–200 ms earlier than the real time.

Scenario 1: In this scenario, the generator is unstable, and IPDT predicts an unstable state. Figure 14.9a shows the changes in the active power of the generator from the time before the fault occurrence to the time of after-fault clearance. Before the fault occurrence time to the time of after-fault clearance, consecutive time windows of continuously sampled data enter FDDT, CDDT, and IPDT.

According to Fig. 14.9b, in FDDT, the sudden change at the moment of the fault occurrence in the sampled data is known as a fault occurrence and generates a "1" activation signal to activate CDDT at its output. As long as this sudden change of

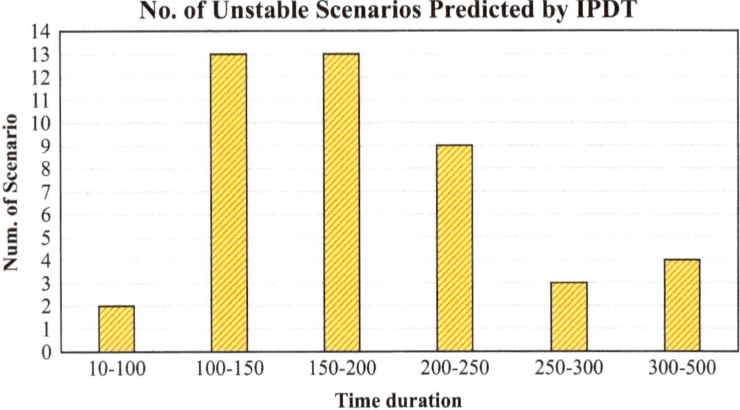

Fig. 14.8 IPDT statistical performance concerning different scenarios

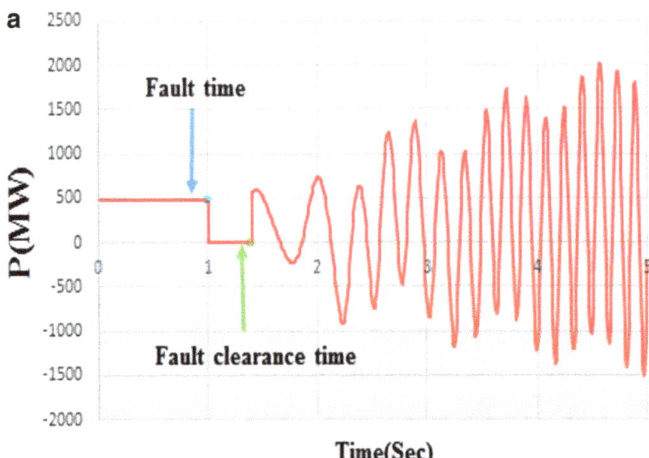

Fig. 14.9 scenario 1—Time windows and output of DTs. (**a**) The fluctuation of generator electrical active power following a fault occurrence and clearance in an unstable scenario, (**b**) time windows for FDDT and CDDT and their output for the unstable scenario, (**c**) the fluctuation of rotor angle and output of IPDT in an unstable scenario

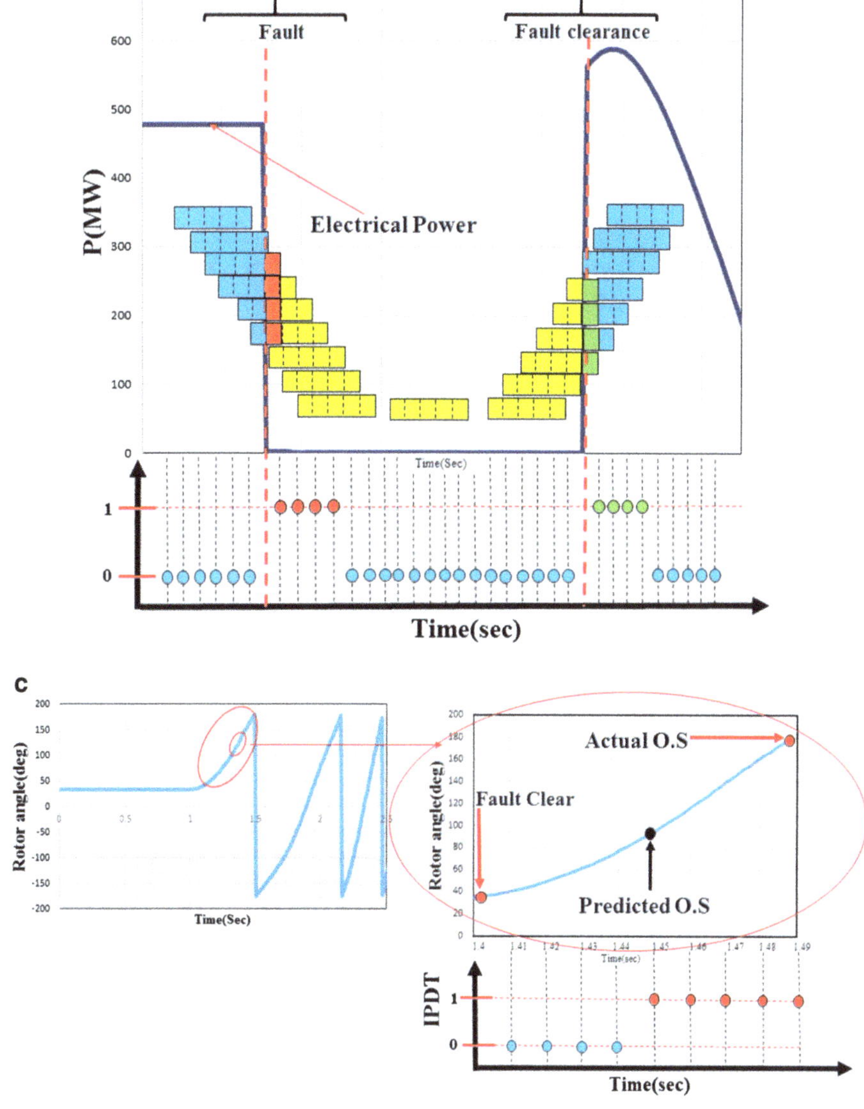

Fig. 14.9 (continued)

data is seen in the sampled time window, the FDDT output remains "1" for six consecutive time windows. At the moment of fault clearance, a similar sudden change of data (blue) occurs in the sampling window. The CDDT knows this sudden change in data as a fault clearance and generates a "1" signal at its output to activate IPDT. So long as the sudden change of data is in the sampling time window, the CDDT output remains "1" for six consecutive windows.

The result of IPDT performance of this scenario is shown in Fig. 14.9c. The IPDT performance starts when the CDDT performance is finished. In other words, the IPDT performance starts when all time window samples contain post-clearance data, depending on the length of the time window, the duration it takes 50 ms.

Table 14.7 presents the time performance of DTs compared to actual time. In this scenario, IPDT predicts the generator's instability, 50 ms earlier than the actual time of the generator O.S.

Scenario 2: The studied scenario is a stable case in which, according to Table 14.7, IPDT predicts a stable state of the generator. In this scenario, the IPDT continuously reflects the "0" signal as no instability at the output. FDDT and CDDT clearly show the fault occurrence and fault clearance at t = 1.0 and t = 1.3 s, respectively, which are the same as the time's actual values. Figure 14.10a shows the changes in the generator's active power from the time before the fault occurs after the time of the fault clearance. In this period, consecutive time windows of continuously sampled data enter FDDT, CDDT, and IPDT.

As shown in Fig. 14.10b, in the moment of the fault occurrence owing to a sudden change in electrical power, a sample of faulty data (yellow) enters the sampling

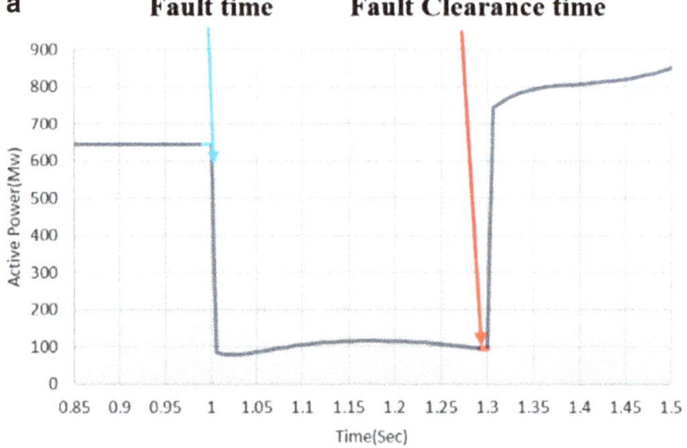

Fig. 14.10 scenario 2—Time windows and output of DTs. (**a**) The fluctuation of generator electrical active power following a fault occurrence and clearance in stable scenario. (**b**) Time windows for FDDT and CDDT and their output for the stable scenario. (**c**) The fluctuation of rotor angle and output of IPDT

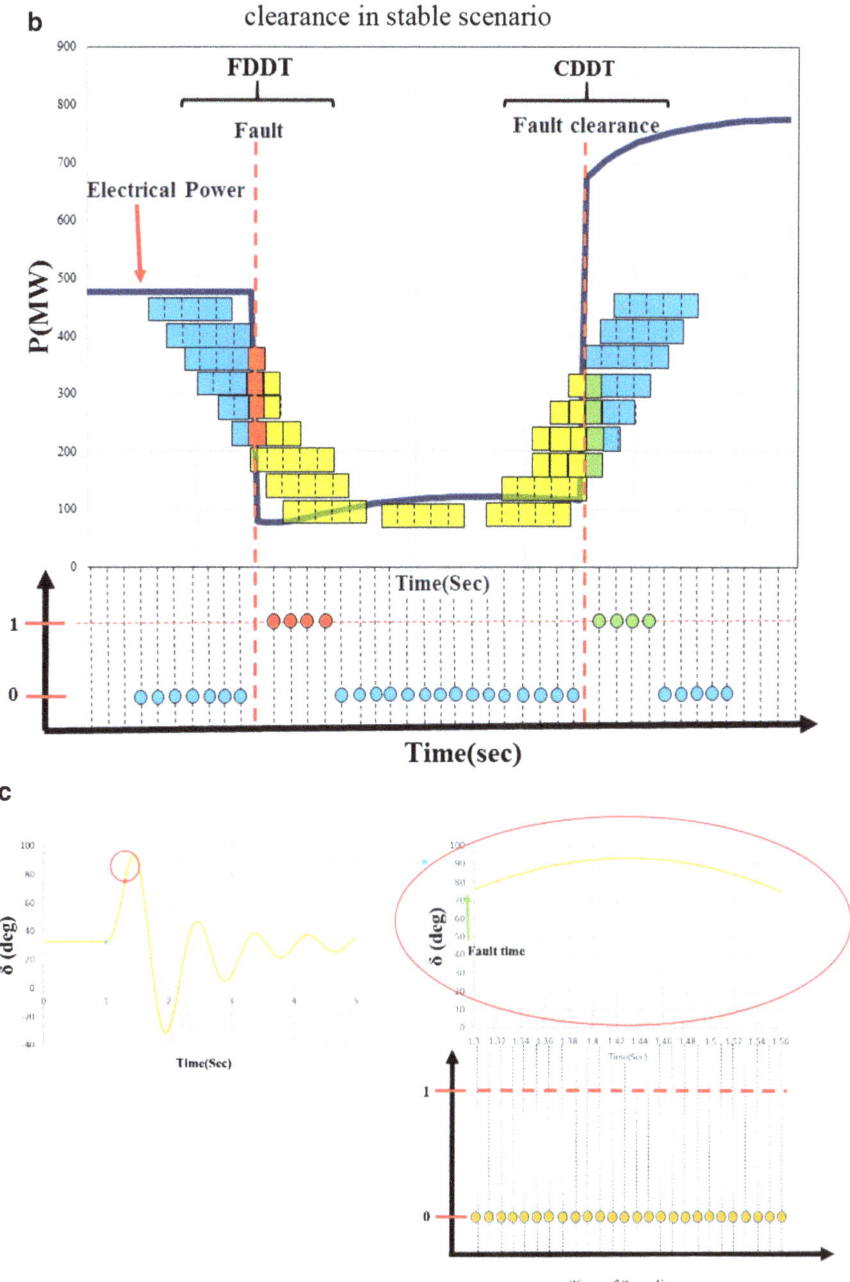

Fig. 14.10 (continued)

window. In FDDT, this change in the sampled data is known as a fault occurrence and generates a "1" activation signal to start CDDT activity at its output. As long as this sudden change of data is present in the sampled time window, the FDDT output remains "1" for six consecutive time windows. At the moment of fault clearance, a similar sudden change of data (blue) occurs in the sampling time window. The CDDT knows this sudden change in data as a fault clearance that generates a "1" signal at its output to activate IPDT. So long as the sudden change of data is in the sampling time window, the CDDT output remains "1" for six consecutive windows. In Fig. 14.10c, the IPDT performance result to predict the generator's stable state is demonstrated.

Scenario 3: This scenario shows an unstable case, and the performance of the DTs is evaluated for L-G in lines 22–23 with the fault characteristics listed in Table 14.7. As shown in Table 14.8, all DTs detect and predict the fault occurrence/fault clearance and instability of the generator accurately.

Scenario 4: This scenario shows an unstable case, and the performance of the DTs is evaluated for a LLG in lines 22–23 with the fault characteristics listed in Table 14.7. All DTs accurately detect and predict the fault occurrence/fault clearance and O.S of the generator. Based on results of Scenarios 3 and 4 based on Table 14.8, the DTs can perform their tasks properly against asymmetric LG and LLG faults, though DTs training with three-phase short circuit fault scenarios.

Scenario 5/6: These scenarios show unstable/stable cases, and the performance of DTs is evaluated for the occurrence of a fault with resistance (RF $= 5\ \Omega$/RF $= 15\ \Omega$). FDDT and CDDT detected fault occurrence and fault clearance for both stable and unstable states, respectively. IPDT predicts the stable case for scenario 5(RF $= 15\ \Omega$) and then shows a "0" signal at the output and also predicts the unstable case for scenario 6(RF $= 5\ \Omega$) at $t = 1.5$ s correctly.

Scenario 7: Regarding the change in the network configuration, the performance of the DTs for an unstable fault occurred in the lines 23–24 is evaluated. All DTs detect occurrence/clearance and correctly predict O.S, respectively.

14.5.1 Validation with Noisy Data

In this section, FDDT, CDDT, and IPDT training/testing is evaluated for noise-free data in training and testing. Therefore, four noise levels (SNR = 20, 30, 40, 50 dB) are examined in the following conditions:

1. Train with noise-free data and test combined with noise
2. Test and train data combined with noise

According to Table 14.9, the performance of DTs accuracy in the case where DTs are tested with noise data results in a relatively high prediction error in IPDT. In Table 14.10, DTs are both tested and trained data combined with noise. The IPDT prediction error is lower in this case, indicating that the presence of noise in the test and train data can improve and make IPDT performance more robust. Also, by

Table 14.9 Performance of FDDT, CDDT, IPDT with noisy data in test

SNR(dB)	No-noise = 0	20	30	40	50
FDDT acc. (%)	99.75	99.45	99.47	99.47	99.52
CDDT acc. (%)	99.75	99.45	99.47	99.47	99.52
IPDT acc. (%)	89.96	73.78	79.46	85.78	88.9

Table 14.10 Performance of FDDT, CDDT, IPDT with noisy data in train and test

SNR(dB)	No-noise = 0	20	30	40	50
FDDT acc. (%)	99.75	99.79	99.8	99.79	99.81
CDDT acc. (%)	99.75	77.79	99.8	99.79	99.81
IPDT acc. (%)	70	76.52	76.52	81.37	89.85

examining the performance of FDDT and CDDT, it is determined that the error rates are negligible because the noise cannot prevent the data from changing abruptly in the time windows of the fault occurrence or fault clearance.

14.6 Conclusion

In this study, to predict a synchronous generator's transient instability, an intelligent relay scheme based on DT is presented. In the proposed scheme, fault occurrence, fault clearance, and generator O.S are detected and predicted by FDDT, CDDT, and IPDT. According to the results, IPDT relies on post-clearance data generated by FDDT and CDDT in the previous steps to predict instability and O.S status. It was discovered that electrical operating variables are the most appropriate variables for fault occurrence/clearance detection. In contrast, electromechanical operational variables are the most relevant variables for predicting O.S. Adopting six data samples in each time window, including data of fault occurrence and fault clearance as input vector data, leads to increased reliability and provides the opportunity for DTs to detect fault conditions and clearance in six consecutive time windows. In addition, this type of time window ensures the proper performance of DTs against missed data and bad data that may be present in the sampled time windows. According to results it can be concluded that the FDDT, CDDT, and IPDT performance accuracy for prediction and detection is 99.71, 99.71, and 98.73, respectively. The performance of the presented framework has been assessed by various scenarios not observed in the DT learning process. It is necessary to mention that the presented intelligent relay can be used as a supportive relay for the main O.S relay. In other words, the presented intelligent O.S relay can operate individually or in association with prevalent relays. In cooperation status, its function can be utilized as a support for a prevalent relay, so that in the final decision about the generator stability status, if the presented relay predicts instability, it will act accordingly or if the output impedance relay synchronous also detects instability and acts following its detection.

References

1. F.R. Gomez Lezama, *Prediction and Control of Transient Instability Using Wide Area Phasor Measurements*, Thesis, Doctor of Philosophy, Department of Electrical and Computer Engineering University of Manitoba (2011)
2. A.D. Rajapakse, F. Gomez, Rotor angle instability prediction using post-disturbance voltage trajectories. IEEE Trans. Power Syst. **25**(2), 947–956 (2009)
3. F.R. Gomez, A.D. Rajapakse, U.D. Annakkage, I.T. Fernando, Support vector machine-based algorithm for post-fault transient stability status prediction using synchronized measurements. IEEE Trans. Power Syst. **26**(3), 1474–1483 (2010)
4. N. Amjady, S.A. Banihashemi, Transient stability prediction of power systems by a new synchronism status index and hybrid classifier. IET Gener. Transm. Distrib. **4**(4), 509–518 (2010)
5. A.G. Bahbah, *Power System Transient Stability Assessment for Real-Time Applications*, Thesis, Doctor of Philosophy, Department of Electrical and Computer Engineering University of Clemson (2000)
6. W.Xiaochen, et al., Review on transient stability prediction methods based on real time wide-area phasor measurements, in *2011 4th International Conference on Electric Utility Deregulation and Restructuring and Power Technologies (DRPT)* (2011)
7. Y. Jie et al., PMU-based monitoring of rotor angle dynamics. IEEE Trans. Power Syst. **26**(4), 2125–2133 (2011)
8. C.-X. Dou, J. Yang, X. Li, T. Gui, Y. Bi, Decentralized coordinated control for large power system based on transient stability assessment. Int. J. Electr. Power Energy Syst. **46**(March), 153–162 (2013)
9. D.P. Wadduwage, C.Q. Wu, U.D. Annakkage, Power system transient stability analysis via the concept of Lyapunov exponents. Electr. Power Syst. Res. **104**(November), 183–192 (2013)
10. F. Shi, H. Zhang, G. Xue, Instability prediction of the inter-connected power grids based on rotor angle measurement. Int. J. Electr. Power Energy Syst. **88**, 21–32 (2017)
11. H. Talaat, Predictive O.S relaying using fuzzy rule-based classification. Electr. Power Syst. Res. **48**, 143–149 (1999)
12. L.S. Moulin, A.P.A. da Silva, M.A. El-Sharkawi, R.J. Marks II, Support vector machines for transient stability analysis of large-scale power systems. IEEE Trans. Power Syst. **19**(2), 818–825 (2004)
13. W. Hu, Z. Lu, S. Wu, et al., Real-time transient stability assessment in power system based on improved SVM. J. Mod. Power Syst. Clean Energy **7**, 26–37 (2019)
14. K.R. Padiyar, S. Krishna, Online detection of loss of synchronism using energy function criterion. IEEE Trans. Power Deliv. **21**(1), 46–55 (2006)
15. A.F. Diaz-Alzate, J.E. Candelo-Becerra, J.F. Villa Sierra, Transient stability prediction for real-time operation by monitoring the relative angle with predefined thresholds. Energies MDPI Open Access J. **12**(5), 1–17 (2019)
16. I. Kamwa, S.R. Samantaray, G. Joos, Development of rule-based classifiers for rapid stability assessment of wide-area post-disturbance records. IEEE Trans. Power Syst. **24**(1), 258–270 (2009)
17. Q. Gao, S. Rovnyak, Decision trees using synchronized phasor measurements for wide-area response-based control. IEEE Trans. Power Syst. **26**(2), 855–861 (2011)
18. J.Z. Hui Deng, X. Wu, K. Men, Real time transient instability detection based on trajectory characteristics and transient energy, in *Power and Energy Society General Meeting*, (2012), pp. 1–7
19. A. Karimi, S.Z. Esmaili, Transient stability assessment of power systems described with detailed models using neural networks. Int. J. Electr. Power Energy Syst. **45**(1), 279–292 (2013)
20. Bahbah, A.A. Girgis, New method for generators' angles and angular velocities prediction for transient stability assessment of multimachine power systems using recurrent artificial neural network. IEEE Trans. Power Syst. **19**(2), 1015–1022 (2004)

21. A. Karimi, Power system transient stability margin estimation using neural networks. Int. J. Electr. Power Energy Syst. **33**(4), 983–991 (2011)
22. S. Kai, S. Likhate, V. Vittal, V.S. Kolluri, S. Mandal, An online dynamic security assessment scheme using phasor measurements and decision trees. IEEE Trans. Power Syst. **22**(4), 1935–1943 (2007)
23. Y. Wang, J. Yu, Real time transient stability prediction of multi-machine system based on wide area measurement, in *Power and Energy Society General Meeting*, (2009), pp. 1–4
24. D. You, K. Wang, L. Ye, J. Wu, R. Huang, Transient stability assessment of power system using support vector machine with generator combinatorial trajectories inputs. Int. J. Electr. Power Energy Syst. **44**(1), 318–325 (2013)
25. K. Yamashita, H. Kameda, O.S prediction logic for wide area protection based on an autoregressive model, in *Proceedings of IEEE PES Power Systems Conference Expo*. New York, 10–13 Oct 2004
26. F. Hashiesh, H.E. Mostafa, A.-R. Khatib, M. Ibrahim Helal, M.M. Mansour, An intelligent wide area Synchrophasor based system for predicting and mitigating transient instabilities. IEEE Trans. Smart Grid **3**(2), 645–652 (2012)
27. T. Amraee, S. Ranjbar, Transient instability prediction using decision tree technique. IEEE Trans. Power Syst. **28**(3), 3028–3037 (2013)
28. A.N. Al-Masri et al., A novel implementation for generator rotor angle stability prediction using an adaptive artificial neural network application for dynamic security assessment. IEEE Trans. Power Syst. **23**(3), 2516–2525 (March 2013)
29. A.R. Sobbouhi, M.R. Aghamohammadi, A new algorithm for predicting O.S condition in large-scale power systems using rotor speed–acceleration. Int. Trans. Electr. Energy Syst. **26**, 486–508 (2016)
30. A.R. Sobbouhi, M.R. Aghamohammadi, A new algorithm for predicting O.S using rotor speed-acceleration based on phasor measurement units (PMU) data. Electr. Power Components Syst. **43**(13), 1478–1486 (2015)

Chapter 15
The Adaptive Neuro-Fuzzy Inference System Model for Short-Term Load, Price, and Topology Forecasting of Distribution System

Mehrdad Setayesh Nazar and Ashkan Eslami Fard

Nomenclature

Abbreviations

ANFIS	Adaptive neuro-fuzzy inference system
ANN	Artificial neural network
ARIMA	Autoregressive integrated moving average
ARMA	Autoregressive moving average
CANFIS	Co-active neuro-fuzzy inference system
CSI	Contingency severity index
DG	Distributed generation
GA	Genetic algorithm
GO	Gravitational search optimization
LMP	Locational marginal prices
LPF	Load and price forecasting
MAPE	Mean absolute percentage error
MLP	Multilayer perceptron
SVM	Support vector machine
SVR	Support vector regression
WT	Wavelet transform

Indices

l	Decomposition index of wavelet decomposition
k	Scaling index of wavelet decomposition

M. Setayesh Nazar (✉) · A. Eslami Fard
Faculty of Electrical Engineering, Shahid Beheshti University, Tehran, Iran
e-mail: m_setayesh@sbu.ac.ir; a_eslamifard@sbu.ac.ir

M. Nazari-Heris et al. (eds.), *Application of Machine Learning and Deep Learning Methods to Power System Problems*, Power Systems,
https://doi.org/10.1007/978-3-030-77696-1_15

Parameters

NB Number of buses
NCC Number of critical contingencies
NL Number of lines
$P_{\max}$ Maximum active power flow of line
$V_{\min}$ Minimum voltage of bus
α, β, χ Coarse and fine-scale coefficients of wavelet decomposition machine
W Weighting factor

Variables

P Active power flow of line
V Voltage of bus

15.1 Introduction

The load and price-forecasting (LPF) procedure is one of the most important tools for the distribution system operational processes. The utilization of fossil-fueled distributed generation (DG) units can highly change the load and price of system patterns with respect to the cases that the DGs are not committed. The distributed energy generation can change the nodal energy generation/consumption values. Further, the DGs electricity transactions with the upward system can change the hourly load and price of the distribution system buses.

In recent years, many kinds of research have carried out to optimize the LPF process that can be categorized into different groups. The first group of LPF process employs statistical algorithms such as autoregressive integrated moving average (ARIMA) method [1]. The second group utilizes feature selection algorithms and clustering machines such as Kalman filter [2], Box-Jenkins models [3], support vector regression (SVR) [4], support vector machine (SVM) [5], fuzzy models [6], artificial neural network (ANN) [7], and expert systems [8]. The third group combines the forecasting machines of the first and second groups to increase the speed and accuracy of the forecasting process. The wavelet transform (WT), co-active neuro-fuzzy inference system (CANFIS), and adaptive neuro-fuzzy inference system (ANFIS) are examples of the categories methods, which are used for price and load forecasting [9, 10]. Ref. [11] presented a hybrid LPF process that utilized a feature selection method and gravitational search optimization (GSO) algorithm. The ANN-based LPF method maximum mean absolute percentage error (MAPE) of PJM market was about 1.1849% for the worst case. Ref. [12] proposed a multistage ANFIS-based LPF process that used feature selection, WT machine, and Kalman filter. The first stage of LPF used the WT and Kalman machines to decompose the input data into three frequency components and predict the decomposed signals. Then, a Kohonen-based network found similar days, and the ANFIS and multilayer perceptron (MLP) neural networks carried out the second stage of LPF process.

Finally, in the third stage, the ANFIS and MLP neural networks were employed to forecast the price and load curves. The average MAPE value of the proposed method for the Spanish electricity market was about 3.88%. Ref. [13] introduced a LPF model that forecasting procedures were carried out using SVM and WT machines, and the optimization of learning parameter was performed by the GSO method. The simulation of the algorithm showed that the proposed method increased the accuracy of prediction by about 15.21% for the PJM market. Ref. [14] introduced a time series decomposition and sin-cosine optimization process to predict the day-ahead price of the market. The method used a deterministic and probabilistic forecasting algorithm to increase the accuracy of the procedure. The MAPE of the introduced method took on a value of 21.471% for the Australian electricity market price. Ref. [15] evaluated a four-stage price-forecasting machine that consisted of preprocessing, optimization, forecasting, and evaluation machines. The first stage used the signal-processing machine to detect and remove white noises. The second stage optimization was carried out by the grey wolf meta-heuristic approach to optimize the forecasting parameters. The third stage utilized Elman neural network model for price forecasting, and the fourth-stage machine evaluated the effectiveness of forecasting machines. The MAPE of the forecasting method for the data of New South Wales market, Australia, was about 3.331%. Ref. [16] proposed a two-stage price-forecasting machine that correlated the wind power generation with the market price. The direct method and rerouting method assessed the real-time prices and wind electricity generation correlation. The rerouted method utilized SVM, ANN, and regression machines, and the mean square error value for the ERCOT market was about 22.25%. Ref. [17] introduced a load-forecasting algorithm that used the relevance vector machine to predict the load of the electricity market. The feature selection and WT were carried out as preprocessing procedures, and the method was assessed for New York and New England electricity markets. The values of the MAPE for the hour-ahead and day-ahead price forecasting were about 0.86% and 1.58%, respectively. Ref. [18] proposed a stochastic autoregressive moving average (ARMA) model for price forecasting that was assessed for the Australian electricity market. The results showed that the developed model predicted the spark values of prices and considered the nonlinearity of regional markets. Ref. [19] presented an SVM-based load-forecasting algorithm that used the grey wolf optimization process to optimize the parameters of the forecasting machine. The proposed algorithm MAPE for the New South Wales electricity load forecast was about 0.5204%. Ref. [20] assessed a gradient boosting machine for load forecasting of Chinese electric power systems. The inputs of forecasting machines were the daily, average, maximum, and minimum temperature parameters, and the day of the week and the calculated value of MAPE were about 0.032%. Ref. [21] introduced a dynamic decomposition process for load forecasting that decomposed the error of forecasting. The error snapshot matrices were calculated and the series of errors were generated. Then, the load sequences were formulated based on the estimated values of errors. The MAPE of the proposed method was about 4.452% for the Chinese power system. Ref. [22] proposed a feature selection-based price-forecasting algorithm that clustered the price data using a noncooperative game model. The game theory was used to determine the winning neurons, and the forecasting was carried out using Bayesian recurrent ANN. The MAPE of forecasting for the New York power system was about 4.14%. Ref. [23] introduced a load-forecasting algorithm that employed Boltzmann and mutual information machines. The forecasting parameter optimization was carried out using genetic wind driven method. The case study was performed for the PJM electricity market, and the MAPE value for the proposed method was about 0.492%. Ref. [24] proposed a three-part algorithm that used WT, mutual information machine, ARIMA machine, and SVM process. The SVM machine parameter selection was determined using a meta-heuristic optimization procedure. The MAPE values for load and price forecasting of New South Wales electricity market were about 2.11% and 7.86%, respectively.

The described references do not consider the optimal topology forecasting of the distribution system for the day-ahead horizon. This book chapter is about the forecasting algorithm that considers the load, locational marginal prices (LMPs), and topology forecasting of the distribution system for day-ahead scheduling horizon.

15.2 Problem Modeling and Formulation

The distribution system operator (DSO) forecasts the day-ahead values of distribution system aggregated nodal load and the LMPs of buses. As shown in Fig. 15.1, the DSO transacts energy with distributed generation (DG) units that are the nonutility energy generation facilities.

The LMPs have nonconstant average and variance and depend on the marginal costs of electricity generation/load reduction, congestion, and loss of lines. However, the load of buses may have similar patterns. The LMPs are dependent on the loads and vice versa, based on the fact that the higher values of loads may increase the congestion, loss, and energy procurement costs of the system. Further, the higher values of LMPs may reduce the volume of responsive loads based on their price elasticity [25].

An iterative multistage LPF process is introduced, which its block diagram is presented in Fig. 15.2.

At first, a primary LPF is carried out using a load, LMP, and topology historical database. Then, the similar days clustering for load forecasting is performed using Kohonen machine. The second stage load forecasting is processed using ANN machine and daily temperature database. The feature selection procedure processes the price data to find their relevancy, and the second stage price forecasting is carried out using ANFIS process.

The topology-forecasting process uses the contingency analysis database and determines the hourly available decision control variables. Then, the optimal topology of the system is determined using a genetic algorithm (GA) optimization procedure.

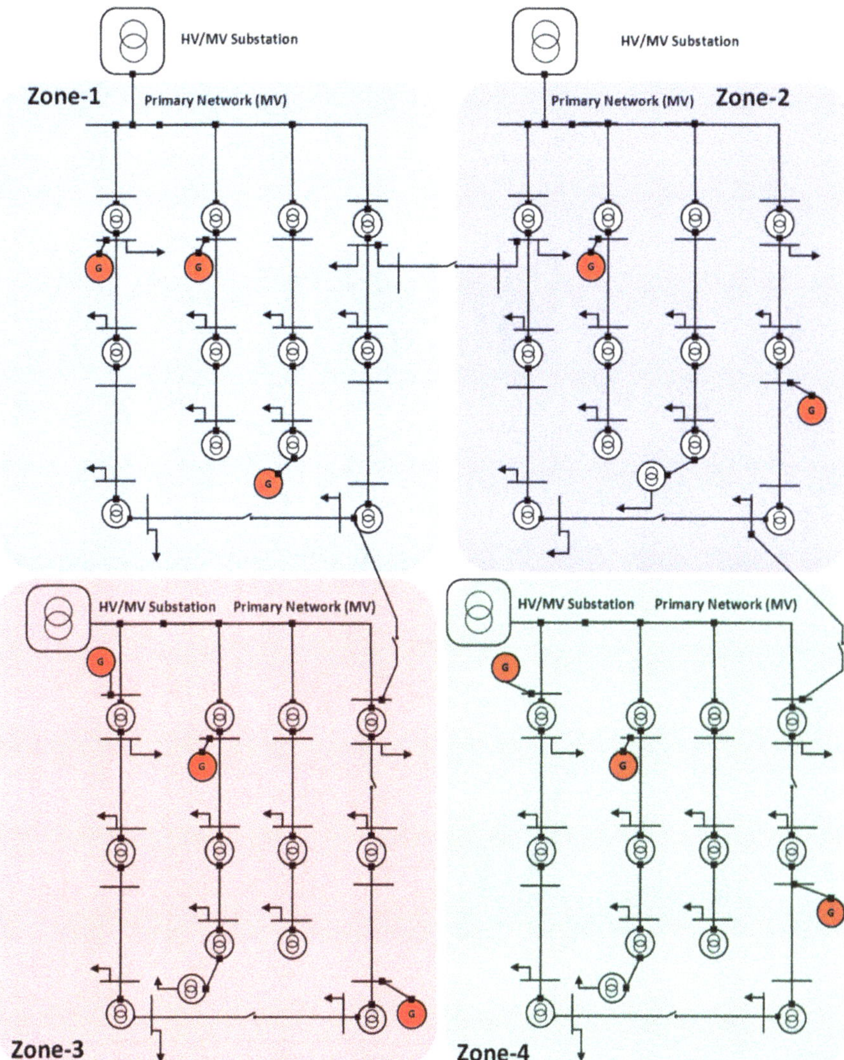

Fig. 15.1 The schematic diagram of the distribution system

The first stage carries out WT process to decompose the data into different frequencies. The forecasting processes of decomposed frequencies are easier than the undecomposed data. The WT decomposition process is carried out in two levels, as shown in Fig. 15.3. The high-frequency components (D_2 and D_1) and low-frequency component (A_1) are decomposed from the original data.

The WT formulation can be presented as Eq. (15.1):

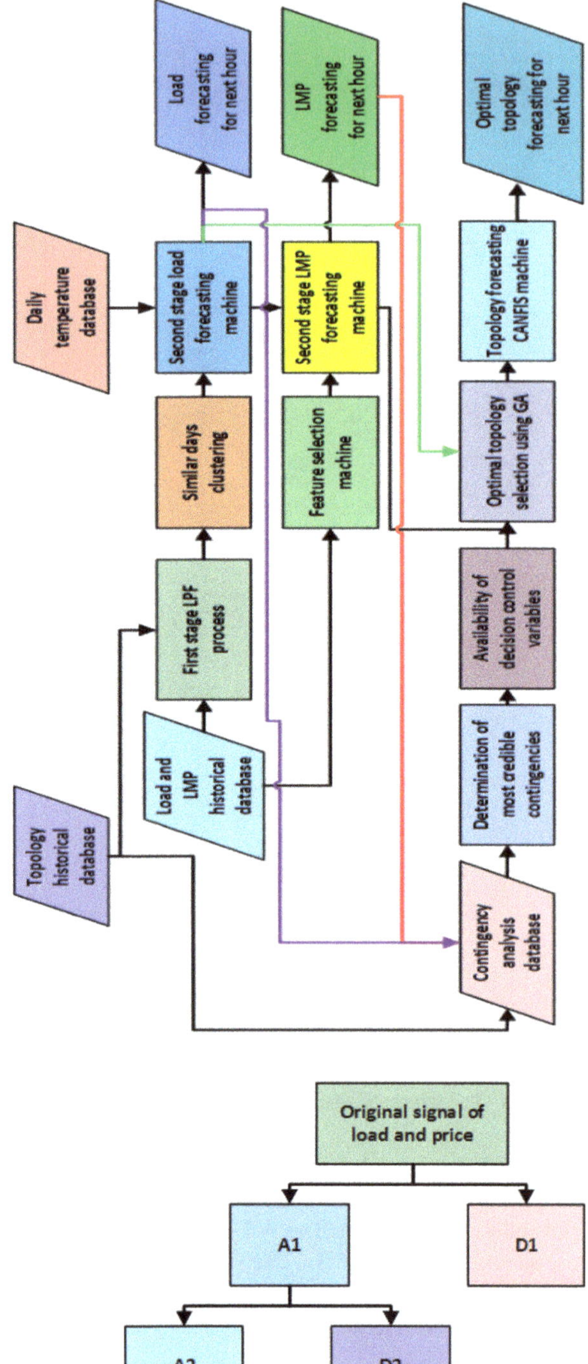

Fig. 15.2 The proposed iterative multistage LPF process

Fig. 15.3 The decomposition of the original signal by wavelet transform

$$f(t) = \sum_{k} c_{l_0,k} \alpha_{j0,k}(t) + \sum_{l<l_0} \sum_{k} \beta_{l,k} 2^{\frac{l}{2}} \chi(2^l t - k) \tag{15.1}$$

where, l, k are decomposition and scaling index, respectively. Further, α, β, and χ are coarse and fine-scale coefficients, respectively.

Thus, six components of the original load and price data are available. Then, a Kalman forecasting machine is utilized for the first-stage LPF that can be formulated as Eq. (15.2):

$$\text{load}(\nu) = \sum_{i=1}^{t} \varphi_i \cdot \text{load}(\nu - \nu_i) + \sum_{i=1}^{j} \psi_i \cdot \text{price}(\nu - \theta_i) \tag{15.2}$$

where, price($\nu - \theta_i$) and load($\nu - \nu_i$) are the past price and load parameters, respectively. Further, φ and ψ are calculated from transition matrices, and θ_i and ν_i are the indices of previous prices and loads, respectively.

The similar day clustering process is carried out by the Kohonen machine and the processed data are delivered to the ANN-based load-forecasting procedure. The decomposed frequency components of the original load signal are forecasted by the ANN machines in the second stage of the forecasting process. Further, the mutual information machine finds the dependency of LMPs. The second stage LMP forecasting utilizes the ANFIS machine. The detailed formulations of mutual information and ANFIS machines are presented in [12]. The forecasted values of A_1, A_2, and D_2 signals are reconstructed as shown in Fig. 15.4.

The optimal topology-forecasting process utilizes the off-line calculation engine and consists of the following stages:

1. Off-line contingency analysis and ranking for different load patterns are carried out.

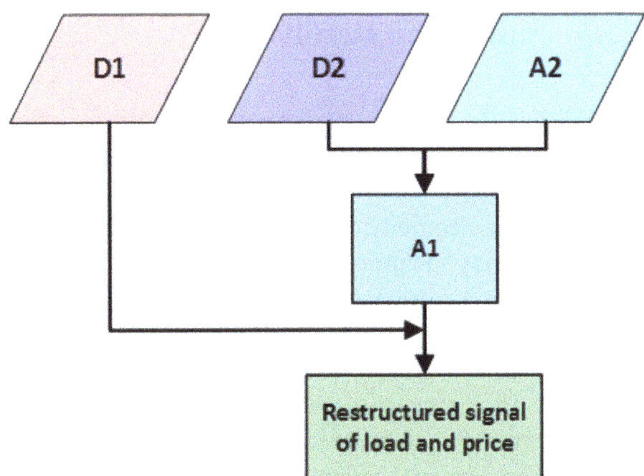

Fig. 15.4 The restructuring of decomposed signals

2. Optimal switching of the distribution system using the GA is performed. The objective function of GA optimization process is defined as Eq. (15.3):

$$\min Z = W_1 \cdot \sum_{\mathrm{NB}} \left(\frac{V - V_{\min}}{V_{\min}} \right)^2 + W_2 \cdot \sum_{\mathrm{NL}} \left(\frac{P - P_{\max}}{P_{\max}} \right)^2 \tag{15.3}$$

   where, W is the weighting factor. Equation (15.3) is subjected to the AC load flow and distribution system device loading constraints. The details of GA are available in [26].
3. The homogeneous topologies are determined for different contingency and load conditions.
4. The learning of CANFIS machine is performed. The detailed formulation of CANFIS machine is presented in [27].

The online optimal topology-forecasting procedure consists of the following steps:

1. Contingency Severity Index (CSI) is calculated using Eq. (15.4)

$$\mathrm{CSI} = \sum_{\mathrm{NCC}} \left[\sum_{\mathrm{NB}} \left(\frac{V - V_{\mathrm{normal}}}{V_{\mathrm{normal}}} \right)^2 + \sum_{\mathrm{NL}} \left(\frac{P - P_{\max}}{P_{\max}} \right)^2 \right] \tag{15.4}$$

2. The CANFIS machine forecasts the optimal topology of the distribution system based on the CSI, LMPs, and load of the system.

The CANFIS machine is the extension of ANFIS machine that consists of multi-outputs that are shown in Fig. 15.5.

The number of CANFIS machine outputs is equal to the number of distribution topologies that are determined in the off-line studies.

15.3 Simulation Results

The 33-bus IEEE test system was considered for simulating the proposed method. Figure 15.6 presents the topology of the 33-bus IEEE test system, and its data is available in [28].

Numerous electricity consumption scenarios were generated, and the LMPs were calculated using the proposed method of [29]. The Nord Pool per unit load patterns were considered for generating of the 33-bus IEEE system hourly load pattern for different days and seasons. Then, the system load and LMP data were delivered to the proposed LPF machines.

Figure 15.7 depicts the WT decomposition levels. The A_2 signal has the highest contribution value. However, the D_2 and D_1 signals have values lower than 0.045 that are very low with respect to the A_2 signal value.

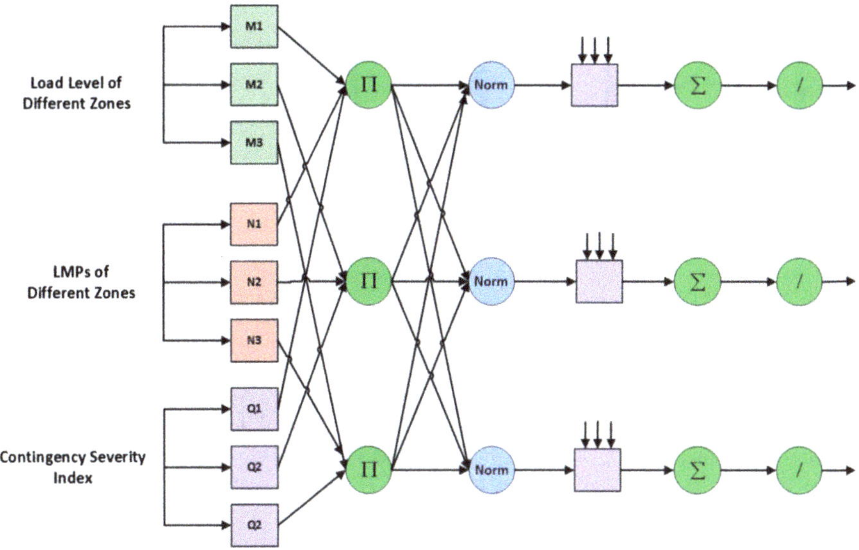

Fig. 15.5 The schematic diagram of CANFIS machine

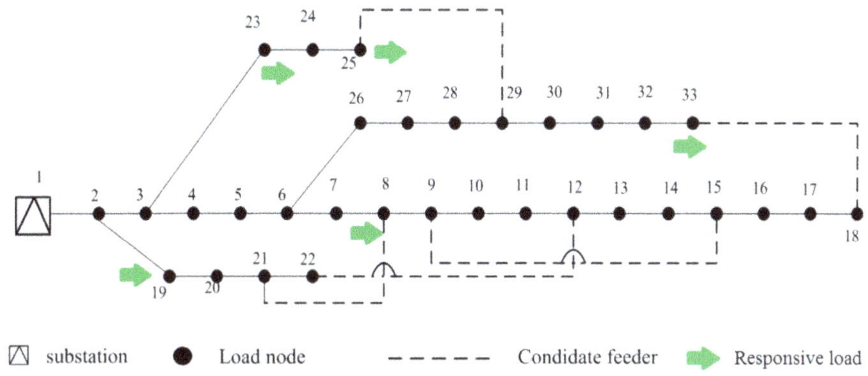

Fig. 15.6 The single line diagram of the 33-bus IEEE test system

Figure 15.8a, b present the correlation between load and the historical load database. There is a linear correlation between the load of forecasting day and the load of the previous day. Further, a linear correlation between the load of forecasting day and the load of the 7 days before can be concluded.

The first-stage load-forecasting input data were selected as the following set:

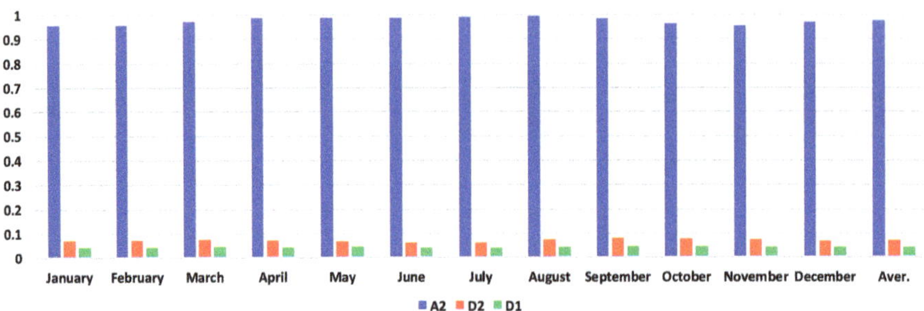

Fig. 15.7 The wavelet transform decomposition levels of the original signal

$$\{L(d,t-1), L(d,t-2), L(d,t-3), L(d,t-4), L(d,t-5), L(d-1,t-1), L(d-1,t), L(d-1,t+1), L(d-7,t-1), L(d-7,t), L(d-7,t+1)\} \quad (15.5)$$

Further, the first-stage price-forecasting input data were selected as the following set:

$$\{P(d,t), P(d,t-1), P(d,t-2), P(d,t-3), P(d,t-4), P(d,t-5), P(d-1,t-1), P(d-1,t), P(d-1,t+1), P(d-7,t-1), P(d-7,t), P(d-7,t+1)\} \quad (15.6)$$

The described data were considered as the input of Kalman filters, and the A_2, D_2 and D_1 signals were forecasted by the Kalman filters as shown in Fig. 15.9.

An average LMP (ALMP) was defined, which was the hourly average of LMP for the specified day-ahead horizon.

Figure 15.10 presents the ALMP A_2, D_2, and D_1 signals and the final value of ALMP for the day-ahead horizon.

Then, the Kalman forecasted LMPs, and loads were fed to the Kohonen machine to cluster the input data. Figure 15.11 presents the clustered load data for the A_2 signal, which was determined by the Kohonen machine. More details of parameter selection are presented in [12].

The Kohonen process found the following patterns for the A_2, D_2, and D_1 signals:

Similar days D_2 component for load data:
{109،122،125،230،235،236،237،244،411،594،599}.
Similar days A_2 component for load data:
{116،117،125،228،229،234،238،475،481،287،501،591،592،593،594،598}
Similar days D_1 component for load data:
{131،159،178،250،279،496،549،556}.

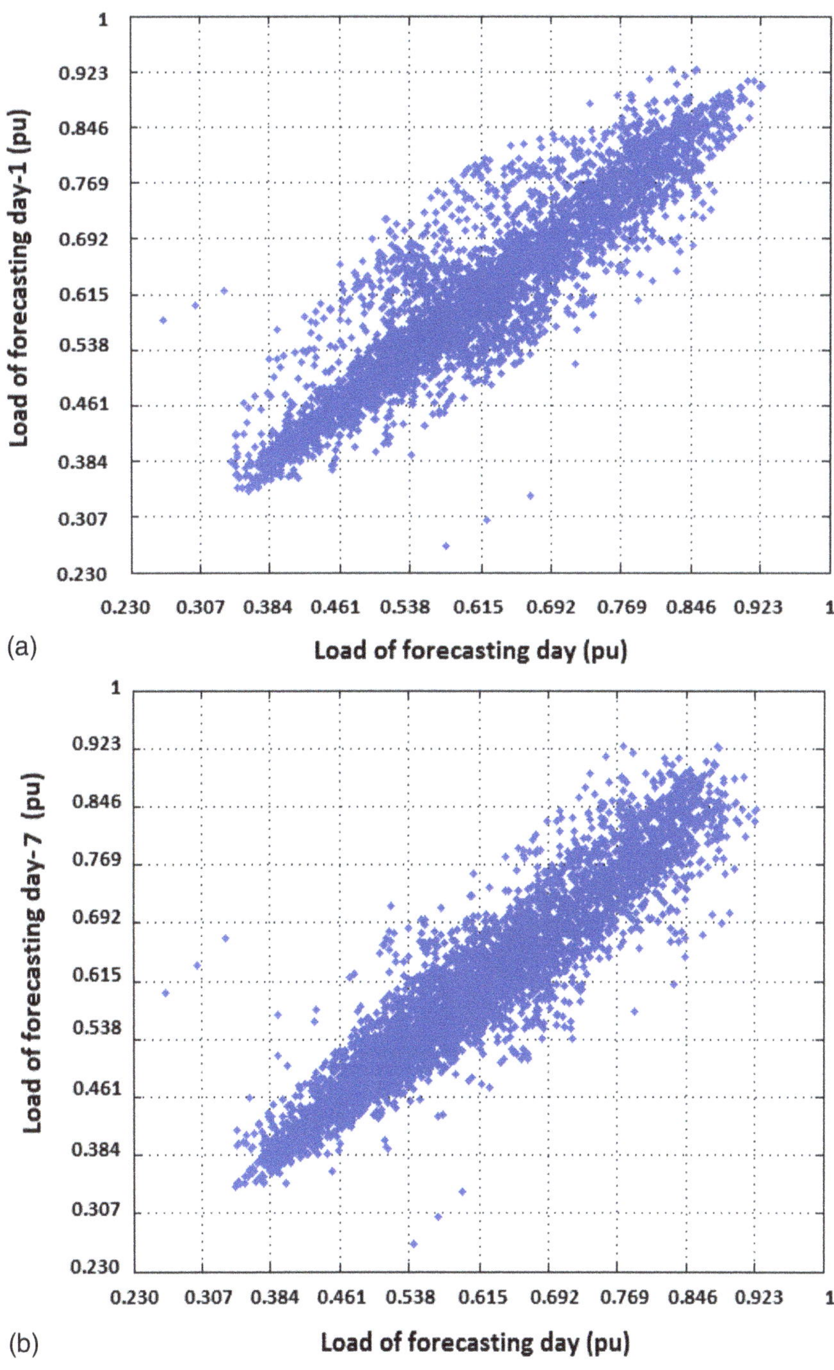

Fig. 15.8 (**a**) The correlation of load and the historical load database for $d - 1$ and day of forecasting, (**b**) the correlation of load and the historical load database for $d - 7$ and day of forecasting

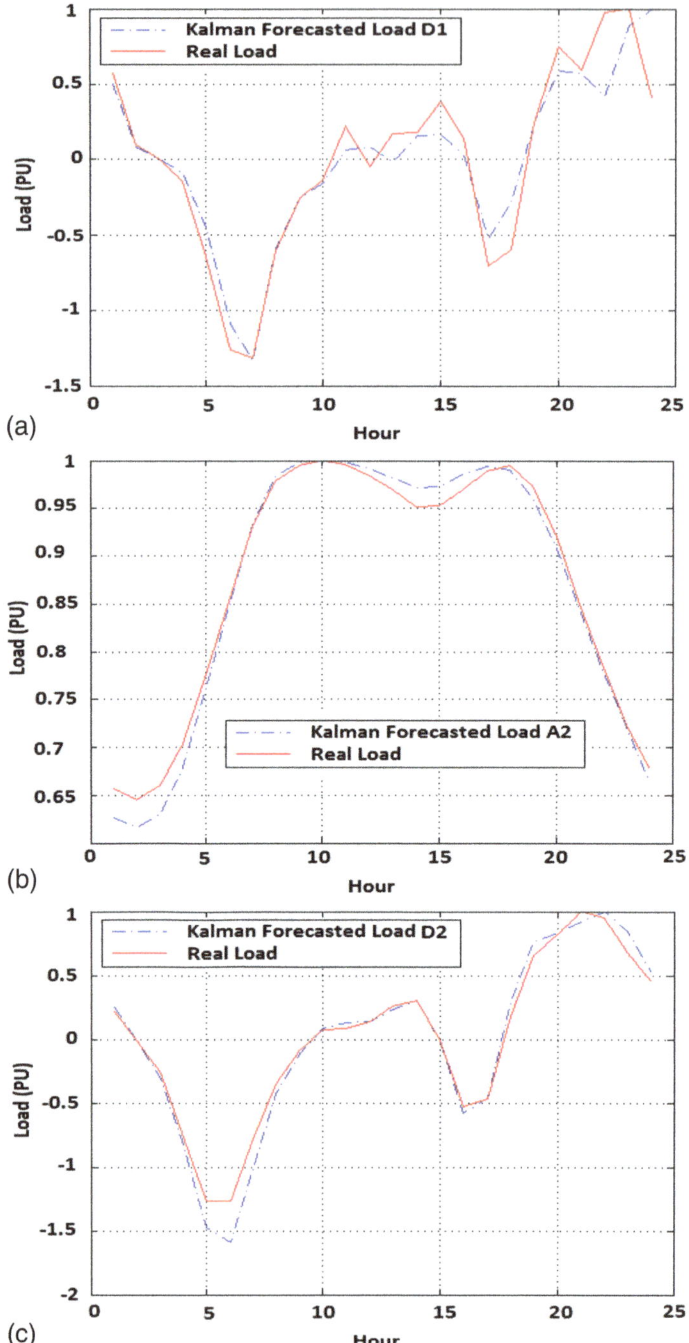

Fig. 15.9 (**a**) The Kalman filter forecasted load D_1 signal, (**b**) the Kalman filter forecasted load A_2 signal, (**c**) the Kalman filter forecasted load D_2 signal

The second stage of load forecasting was performed by the multilayer perceptron neural network. Figure 15.12 presents the second stage load forecasting of A_2 signal that was carried out by the neural network. Figure 15.13 depicts the maximum and minimum values of MAPE for neural network load-forecasting process. As shown in Fig. 15.13, the maximum and minimum values of MAPE took on a value of 3.88% and 1.31%, respectively.

The ANFIS forecasting engine forecasted LMPs in the second stage of LPF. Figure 15.14 presents the forecasted ALMP.

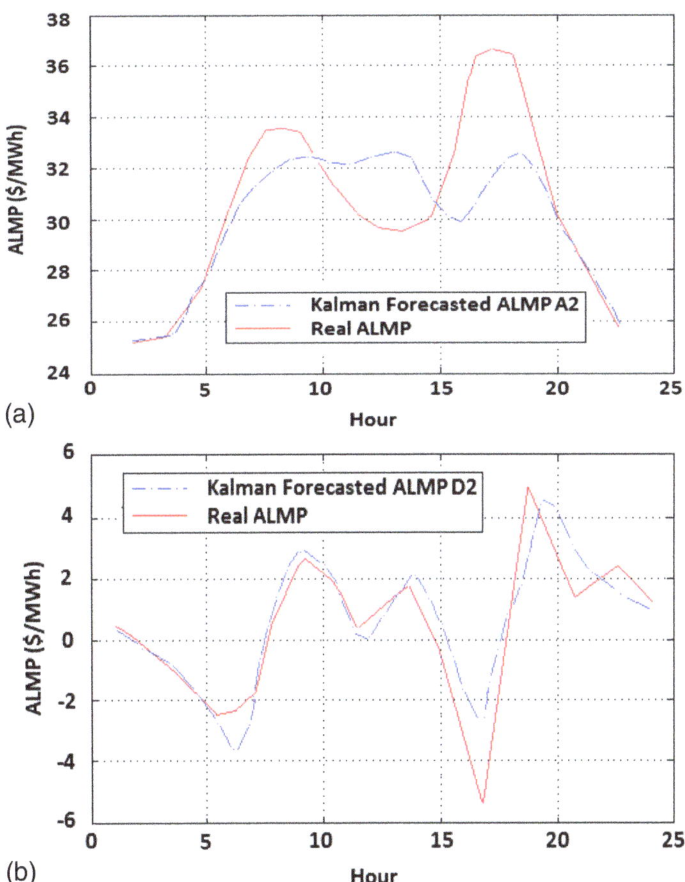

Fig. 15.10 (**a**) The Kalman filter forecasted ALMP D_1 signal, (**b**) the Kalman filter forecasted ALMP A_2 signal, (**c**) the Kalman filter forecasted ALMP D_2 signal, (**d**) the Kalman filter forecasted ALMP signal

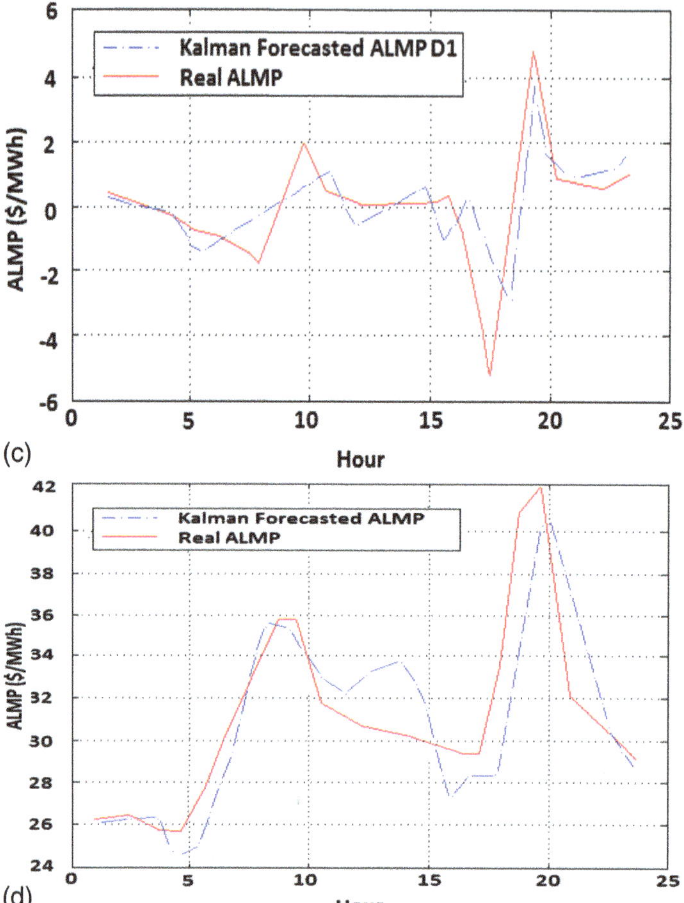

Fig. 15.10 (continued)

Figure 15.15 depicts the GA objective function values, which was calculated from Eq. (15.3). The maximum values of the GA objective function took on a value of 1242.432, 1358.784, and 1426.458 for the load factor 0.6, 0.8, and 1, respectively.

Tables 15.1 and 15.2 depict the CANFIS pattern recognition and GA optimization outputs for different load factor (LF) and ALMP values. The total number of the CANFIS machine pattern recognitions was 444 cases, and the number of recognition errors was six cases or 1.35%. The proposed forecasting machines were successfully determined the day-ahead load, price, and topology. More details of the justification process are presented in [12].

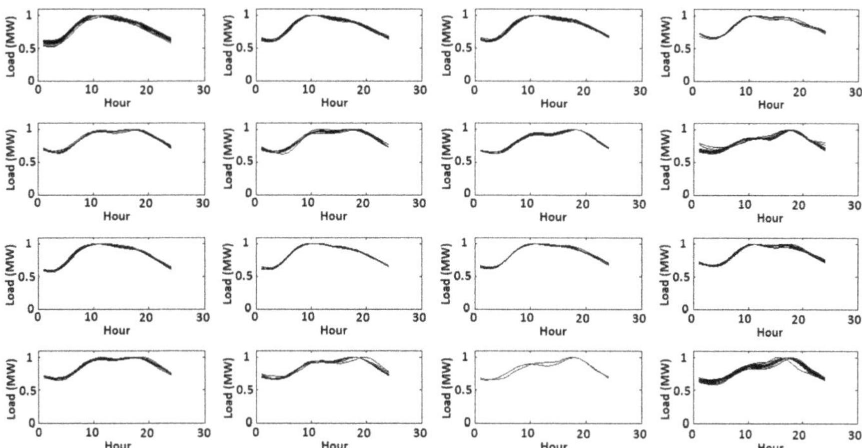

Fig. 15.11 The Kohonen machine clustered load data for the A_2 signal

15.4 Conclusion

A day-ahead load, price, and topology-forecasting process was reviewed in this chapter. The introduced framework employed feature selection and Kohonen clustering machines. Two-level load and locational marginal price-forecasting process were carried out. The wavelet machine decomposed the original signals into different frequencies, and the Kalman filter predicted the decomposed signals for the day-ahead horizon. Then, the Kohonen process clustered the first-level forecasted signals, and the multilayer perceptron neural network machine forecasted the decomposed frequencies of the load. The adaptive neuro-fuzzy inference system was utilized to forecast the decomposed signals of locational marginal prices. Finally, the load and price signals were restructured.

The topology-forecasting machine considered the contingency severity index, locational marginal price, and load of system as inputs, and a genetic algorithm optimized the topology of the system considering system constraints. Then, a co-active neuro-fuzzy inference system was used to predict the optimal topology of the system. The proposed method was successfully assessed for the 33-bus IEEE test system.

The authors are working on the new structure of the proposed algorithm that utilizes meta-heuristic optimization algorithm to speed up the process.

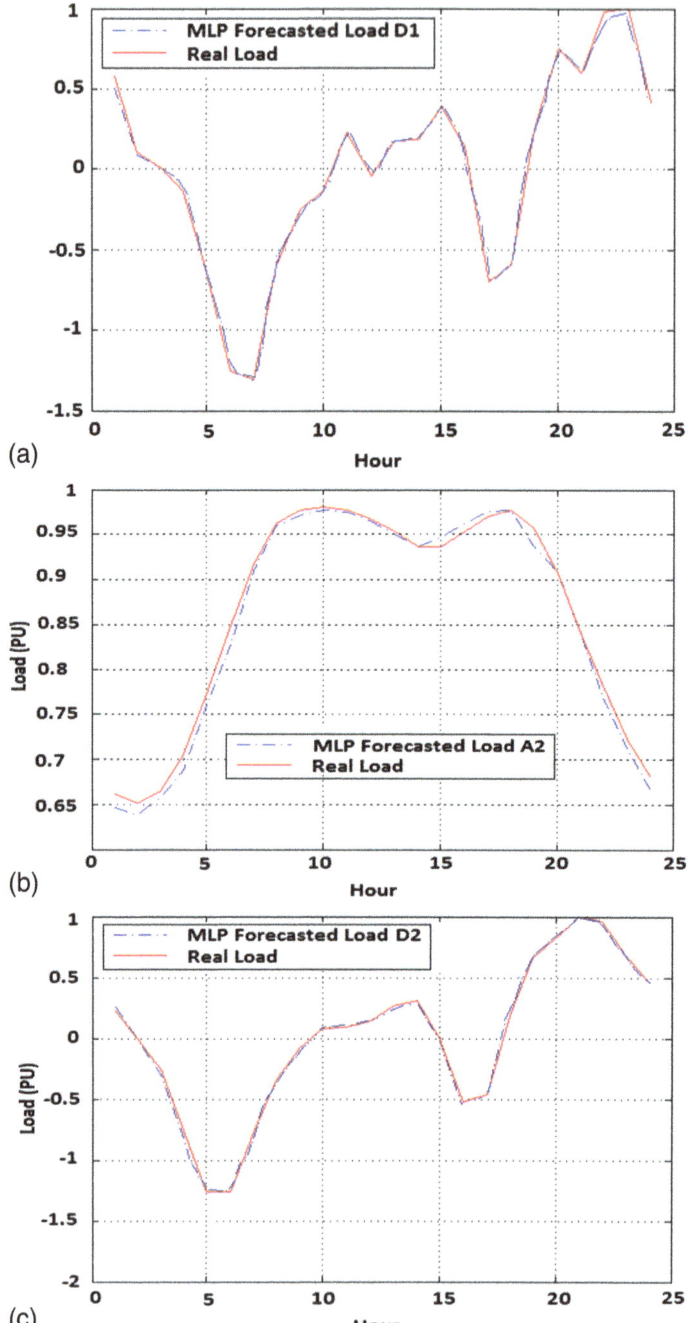

Fig. 15.12 (**a**) The MLP forecasted load D_1 signal, (**b**) the MLP forecasted load A_2 signal, (**c**) the MLP forecasted load D_2 signal

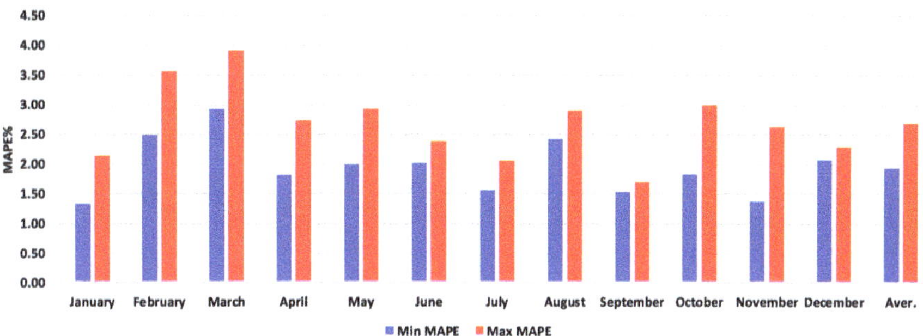

Fig. 15.13 The neural network MAPE for the forecasted load data

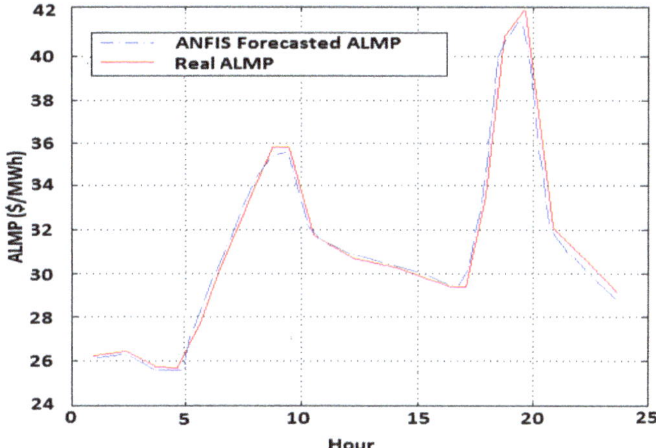

Fig. 15.14 The ANFIS machine forecasted ALMP

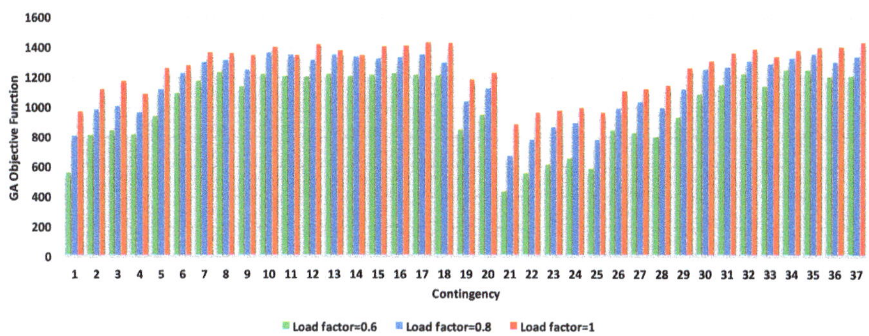

Fig. 15.15 The GA objective functions for different load levels

Table 15.1 Comparison of the GA outputs and the CANFIS machine for optimal topology recognition and different ALMP and LF values

	GA (ALMP = 1)	CANFIS (ALMP = 1)	GA (ALMP = 0.7)	CANFIS (ALMP = 0.7)	GA (ALMP = 0.5)	CANFIS (ALMP = 0.5)
Contingency	LF = 0.95	LF = 0.95	LF = 0.9	LF = 0.9	LF = 0.85	LF = 0.85
1	1	1	1	1	6	6
2	2	2	2	2	5	5
3	4	4	4	4	3	3
4	3	3	3	3	4	4
5	1	1	1	1	1	1
6	1	1	1	1	1	1
7	4	4	1	1	1	1
8	1	1	1	3	1	1
9	4	4	4	4	4	4
10	1	3	1	1	1	1
11	1	1	1	1	1	1
12	1	1	1	1	1	1
13	6	6	6	1	1	1
14	4	4	4	4	1	1
15	1	1	1	3	1	1
16	1	1	3	1	3	3
17	6	6	1	1	1	1
18	2	2	1	1	1	1
19	1	1	1	1	1	1
20	1	1	1	2	1	1
21	1	1	1	1	1	1
22	1	1	1	1	1	1
23	1	1	1	1	1	1
24	1	1	1	1	1	1
25	1	1	1	1	1	1

26	1	1	1	1	1	4
27	1	1	1	1	1	1
28	2	2	1	1	1	1
29	1	1	1	1	1	1
30	2	2	1	1	1	1
31	1	1	1	1	1	1
32	1	1	1	1	1	1
33	1	1	1	1	1	1
34	1	1	1	1	1	1
35	1	1	1	1	1	1
36	1	1	1	1	1	1
37	1	1	1	1	1	1

Table 15.2 Comparison of the GA outputs and the CANFIS machine for optimal topology recognition and different ALMP and LF values

	GA (ALMP = 1)	CANFIS (ALMP = 1)	GA (ALMP = 0.7)	CANFIS (ALMP = 0.7)	GA (ALMP = 0.5)	CANFIS (ALMP = 0.5)
Contingency	LF = 0.8	LF = 0.8	LF = 0.75	LF = 0.75	LF = 0.7	LF = 0.7
1	7	7	1	1	1	7
2	5	5	1	1	1	1
3	1	1	1	1	1	1
4	1	1	1	1	1	1
5	2	2	2	2	2	2
6	1	1	1	1	1	1
7	3	3	1	1	3	3
8	1	1	1	1	1	1
9	1	1	1	1	1	1
10	4	4	4	4	3	3
11	4	4	1	1	4	4
12	1	1	1	1	1	1
13	3	3	3	3	3	3
14	1	1	1	1	1	1
15	3	3	1	1	3	3
16	7	7	7	7	7	7
17	2	2	2	2	1	1
18	6	6	2	2	2	2
19	1	1	1	1	1	1
20	1	1	1	1	1	1
21	1	1	1	1	1	1
22	6	6	1	1	6	6
23	1	1	1	1	1	1
24	1	1	1	1	1	1
25	1	1	1	1	1	1

26	1	1	1	1	1	1
27	1	1	1	1	1	1
28	2	2	1	1	1	1
29	1	1	1	1	1	1
30	1	1	1	1	1	1
31	1	1	1	1	1	1
32	1	1	1	1	1	1
33	1	1	1	1	1	1
34	3	3	3	3	3	3
35	3	3	3	3	2	2
36	1	1	1	1	1	1
37	1	1	1	1	1	1

References

1. M.H. Amini, A. Kargarian, O. Karabasoglu, ARIMA-based decoupled time series forecasting of electric vehicle charging demand for stochastic power system operation. Electr. Power Syst. Res. **140**, 378–390 (2016)
2. P. Gupta, K. Yamada, Adaptive short-term load forecasting of hourly load using weather information. IEEE Trans. Power Syst. **91**, 2085–2094 (2007)
3. A.J. Amorim, T.A. Abreu, M.S. Tonelli-Neto, C.R. Minussi, A new formulation of multinodal short-term load forecasting based on adaptive resonance theory with reverse training. Electr. Power Syst. Res. **179**, 106096 (2020)
4. J. Xing Che, J. Zhou Wang, Y. Juan Tang, Optimal training subset in a support vector regression electric load forecasting model. Appl. Soft Comput. **12**, 1523–1531 (2012)
5. P.F. Pai, W. Chiang Hong, Support vector machines with simulated annealing algorithms in electricity load forecasting. Energy Convers. Manage. **46**, 2669–2688 (2005)
6. R. Mamlook, O. Badran, E. Abdulhadi, A fuzzy inference model for short-term load forecasting. Energy Policy **37**(4), 1239–1248 (2009)
7. S. Kouhi, F. Keynia, A new cascade NN based method to short-term load forecast in deregulated electricity market. Energy Convers. Manage. **71**, 76–83 (2013)
8. A. Ghanbari, S.M.R. Kazemi, F. Mehmanpazir, M.M. Nakhostin, A Cooperative Ant Colony Optimization-Genetic Algorithm approach for construction of energy demand forecasting knowledge-based expert systems. Knowl. Based Syst. **39**, 194–206 (2013)
9. J.P.S. Catalão, H.M.I. Pousinho, V.M.F. Mendes, Short-term electricity prices forecasting in a competitive market by a hybrid intelligent approach. Energy Convers. Manage. **52**, 1061–1065 (2011)
10. A.S. Pandey, D. Singh, S.K. Sinha, Intelligent hybrid models for short-term load forecasting. IEEE Trans. Power Syst. **25**, 1266–1273 (2010)
11. A. Heydari, M. Majidi Nezhad, E. Pirshayan, D. Astiaso Garcia, F. Keynia, L.D. Santoli, Short-term electricity price and load forecasting in isolated power grids based on composite neural network and gravitational search optimization algorithm. Appl. Energy **277**, 115503 (2020)
12. M.S. Nazar, A.E. Fard, A. Heidari, M. Shafie-khah, J.P.S. Catalão, Hybrid model using three-stage algorithm for simultaneous load and price forecasting. Electr. Power Syst. Res. **165**, 214–228 (2018)
13. Y. Zhang, C. Deng, R. Zhao, S. Leto, A novel integrated price and load forecasting method in smart grid environment based on multi-level structure. Eng. Appl. Artif. Intell. **95**, 103852 (2020)
14. W. Yang, J. Wang, T. Niu, P. Du, A novel system for multi-step electricity price forecasting for electricity market management. Appl. Soft Comput. J. **88**, 106029 (2020)
15. W. Yang, J. Wang, T. Niu, P. Du, A hybrid forecasting system based on a dual decomposition strategy and multi-objective optimization for electricity price forecasting. Appl. Energy **235**, 1205–1225 (2019)
16. S. Luo, Y. Weng, A two-stage supervised learning approach for electricity price forecasting by leveraging different data sources. Appl. Energy **242**, 1497–1512 (2019)
17. J. Ding, M. Wang, Z. Ping, D. Fu, V.S. Vassiliadis, An integrated method based on relevance vector machine for short-term load forecasting. Eur. J. Oper. Res. **287**, 497–510 (2020)
18. H. Manner, F. Alavi Fard, A. Pourkhanali, L. Tafakori, Forecasting the joint distribution of Australian electricity prices using dynamic vine copulae. Energy Econ. **78**, 143–164 (2019)
19. A. Yang, W. Li, X. Yang, Short-term electricity load forecasting based on feature selection and least squares support vector machines. Knowl.-Based Syst. **163**, 159–173 (2019)
20. Z. Guo, K. Zhou, X. Zhang, S. Yang, A deep learning model for short-term power load and probability density forecasting. Energy **160**, 1186–1200 (2018)
21. X. Kong, C. Li, C. Wang, Y. Zhang, J. Zhang, Short-term electrical load forecasting based on error correction using dynamic mode decomposition. Appl. Energy **261**, 114368 (2020)

22. M. Ghayekhloo, R. Azimi, R. Ghofrani, M.B. Menhaj, E. Shekari, A combination approach based on a novel data clustering method and Bayesian recurrent neural network for day-ahead price forecasting of electricity markets. Electr. Power Syst. Res. **168**, 184–199 (2019)
23. G. Hafeez, K. Saleem Alimgeer, I. Khan, Electric load forecasting based on deep learning and optimized by heuristic algorithm in smart grid. Appl. Energy **269**, 114915 (2020)
24. A. Ghasemi, H. Shayeghi, M. Moradzadeh, M. Nooshyar, A novel hybrid algorithm for electricity price and load forecasting in smart grids with demand-side management. Appl. Energy **177**, 40–59 (2016)
25. S. Salarkheili, M. Setayesh Nazar, Capacity withholding analysis in transmission constrained electricity markets. IET Gener. Transm. Distrib. **10**, 487–495 (2016)
26. S. Madadi, M. Nazari-Heris, B. Mohammadi-Ivatloo, S. Tohidi, Implementation of genetic-algorithm-based forecasting model to power system problems, in *Handbook of Research on Predictive Modeling and Optimization Methods in Science and Engineering*, (IGI Global, Hershey, 2018), pp. 140–155
27. H. Singh, Y.A. Lone, *Deep Neuro-Fuzzy Systems with Python: With Case Studies and Applications from the Industry* (Apress Book, Berkeley, 2020)
28. A. Bostan, M. Setayesh Nazar, M. Shafie-khah, J.P.S. Catalão, Optimal scheduling of distribution systems considering multiple downward energy hubs and demand response programs. Energy **190**, 116349 (2020)
29. S. Salarkheili, M. Setayesh Nazar, New indices of capacity withholding in power markets. Int. Trans. Electr. Energy Syst. **25**, 180–196 (2015)

Chapter 16
Application of Machine Learning for Predicting User Preferences in Optimal Scheduling of Smart Appliances

Milad Sadat-Mohammadi, Morteza Nazari-Heris, Alireza Ameli, Somayeh Asadi, Behnam Mohammadi-Ivatloo, and Houtan Jebelli

16.1 Introduction

The presence of innovative pricing strategies in the residential sector for electrical energy substantiates the necessity of home energy management systems [1]. Home energy management systems optimally operate the flexible appliances, based on predefined constraints by residents, to reduce the overall electricity bill [2, 3]. These systems consist of hardware and software layers [4, 5]. The hardware layer consists of smart meters and actuators to connect and establish the communication between appliances and the central home energy management unit. Designing the actuators and meters modules is a challenging issue since each appliance has its own control circuit, which should be modified to receive the control commands from the home

M. Sadat-Mohammadi
Department of Electrical Engineering and Computer Science, Pennsylvania State University, University Park, PA, USA
e-mail: miladsm@cpsu.edu

M. Nazari-Heris · S. Asadi (✉) · H. Jebelli
Department of Architectural Engineering, Pennsylvania State University, University Park, PA, USA
e-mail: mun369@psu.edu; m.nazariheris.2015@ieee.org; sxa51@psu.edu; asadi@engr.psu.edu; hkj5117@psu.edu; hjebelli@psu.edu

A. Ameli
Department of Civil Engineering, Malard Branch, Islamic Azad University, Tehran, Iran
e-mail: ameli@iaumalard.ac.ir

B. Mohammadi-Ivatloo
Department of Energy Technology, Aalborg University, Aalborg, Denmark

Faculty of Electrical and Computer Engineering, University of Tabriz, Tabriz, Iran
e-mail: mohammadi@ieee.org; bmo@et.aau.dk

M. Nazari-Heris et al. (eds.), *Application of Machine Learning and Deep Learning Methods to Power System Problems*, Power Systems,
https://doi.org/10.1007/978-3-030-77696-1_16

energy management unit and implement the control commands. Recently, smart appliances have been introduced which have embedded communication capability. The smart appliance can receive external signals for controlling its operation from the user, utility, or central home energy management unit without any need for modification of its control circuit. Consequently, the central control unit can send on/off commands to the appliance or receive information about the status of the appliance. The control commands are being determined by running and optimization algorithm embedded in the home energy management unit.

There are several heuristics algorithms proposed in the literature, such as particle swarm optimization, genetic algorithm, etc. to optimize the residential load scheduling models [6, 7]. These models consist of optimization parameters, constraints, decision variables, and an objective function. Constraints are defined to satisfy the technical requirement of operating the appliances. The objective function aims to reduce the monthly electricity bill, considering the constraints and hourly electricity rates. Moreover, some studies also consider the user's comfort level in the objective function [8]. The user comfort level is measured as the total waiting time to get the tasks done after running the scheduling unit. There are two types of parameters present in the scheduling models: fixed and dynamic parameters. Fixed parameters are total energy required by each appliance for completing the tasks, cycle times, etc. that are defined in the central home energy management unit once and are not required to be updated regularly. However, dynamic parameters such as start time and length of the desired scheduling window for each appliance should be updated at the beginning of the scheduling process. Updating the dynamic parameters is a challenging task and considered as one of the factors that makes the implementation of home energy management systems less attractive for residential customers. It is clear that users' lifestyle and activity patterns significantly affect the dynamic parameters. Consequently, analyzing the activity pattern of the residents can provide useful information in addressing the mentioned challenge in implementing the home energy management systems.

The activity prediction of the residents based on the sensor network data in the smart home has been studied in the literature. For instance, data motion sensors and door sensors, installed on the doors and cabinets, can be analyzed and used to predict the order of the activities, start times, as well as duration [9]. Studies have investigated the possibility of discovering the daily routines of the residents from the usage pattern of the appliances. The smart meters that are connected to the appliances record their energy consumption profile; then, the on/off status of the appliance is determined from the recorded profile. The results proved that human activity can be predicted with an acceptable accuracy using Bayesian network [10]. Not only human activities can be predicted using sensory data, but also several studies have reported the possibility of human preference detection. For instance, environmental information and historical data of the operation of appliances were used to predict the comfort preferences of the residents in operating the air conditioner and water heater [11]. Moreover, the GPS data from smart mobile phones and social application data, in addition to weather data, can be used to predict the arrival and living time of the residents and control the thermostatic loads in a way to not only reduce the electricity bill but also provide the required thermal comfort for the residents [12]. Inspiring from proposed approaches in the literature for the prediction of the residents' preferences in the operation of the thermostatic loads, this study aims to study the possibility of prediction of user preferences in scheduling the flexible loads, such as washing machine and dishwasher, based on the recorded energy consumption pattern by smart applicants. Our hypothesis is that users' preferences in operating the smart appliances can be learned from historical data collected by smart appliances; then, the user's preferences can be predicted for future observations. Smart appliances can provide a comprehensive database for this purpose without any need for installing additional hardware such as smart meters [13].

The rest of the study is organized as follows: Sect. 16.2 introduces the factors affecting the dynamic parameters of the residential load scheduling problem. Section 16.3 elaborates on supervised learning and data labeling. Section 16.4 investigates the accuracy of predicting the dynamic parameters with machine learning algorithms. Finally, the concluding remarks and some directions for future research are provided in Sect. 16.5.

16.2 Key Factors Affecting Dynamic Parameters

Dynamic parameters in scheduling problems are affected by several factors, such as the start time of the task, environmental factors, the weekly schedule of the residents, etc. [14]. However, it should be mentioned that there is some level of randomness in dynamic parameters that lie outside the scope of this study. Some of the mentioned factors can be detected by the energy consumption profile of the appliances, while others can be acquired from developed online databases, including environmental factors such as temperature. Investigating the relation between the factors and the dynamic parameters can provide insight into the feasibility of predicting the dynamic parameters.

16.2.1 The Start Time of the Task

The start time of the task is defined as the moment when an appliance becomes ready to start its operation cycle. This time can be detected from recorded data by smart appliances. By analyzing the energy consumption profile of the appliance, the beginning of its operation cycle can be detected whenever the energy consumption increases from the standby energy consumption limit. Recorded data of appliances of a case study for a 2-year period shows that the start time of the tasks can affect the desired length of scheduling interval. For instance, the end of the scheduling interval of the dishwasher is the typical time for serving the dinner or lunch, while the start time is variable and depends on the user to get the dishwasher ready to start. Consequently, whenever the user fills the dishwasher close to the meal serving time, the desired scheduling window is shorter in comparison to the scenario in which the dishwasher is filled earlier. In this study, the start time of the task is considered as a feature in training the machine learning algorithms.

16.2.2 *Day of the Week*

Analyzing the hourly electricity consumption pattern of 60 low-energy houses for 12 months showed that the weekend peak electricity consumption occurs at 19:00, while there is a small peak in electricity consumption at 8:00 in addition to the main peak during the weekdays [15]. The difference between weekday and weekend energy consumption profile is reasonable as people typically left the house in the morning during the weekdays, which causes a reduction in electricity consumption after 8:00. It can be concluded that weekdays and weekends significantly affect the activities of the residents. Consequently, the change in the activity at weekends can result in a change in the desired length of the scheduling window of the appliances. Moreover, each user has specific working hours and tasks, which should be done on a daily basis. Due to the possible importance of the day of the week in the length of the desired scheduling window of the appliances, this factor is considered as an input for training machine learning algorithms.

16.2.3 *Temperature*

Outdoor temperature is also an important factor in the daily activity of the residents. People spend more time doing outdoor activities whenever the temperature is pleasant while they prefer to stay at home whenever outside is snowy or rainy. Spending less time indoors limits the available time for getting daily tasks done; consequently, the desired scheduling window for scheduling the appliances may be preferred to be as short as possible. However, it should be noted that depending on the lifestyle of the residents, the impact of temperature on the desired scheduling window can be notably different. Therefore, the relationship between the outdoor temperature and the desired scheduling window for each appliance can be detected if this factor is considered as a feature for training the machine learning algorithm in the next steps.

16.3 Supervised Learning

Supervised learning refers to learning the relationship between input and output from labeled examples. The labeled examples are divided into two sets as the training and test sets. The training set is used during training the machine learning algorithm; the training process involves tuning the model parameters. Weights, support vectors, and coefficients are model parameters of artificial neural network, support vector machine, and linear regression machine learning algorithms, respectively [16]. Model parameters are often optimally determined using an optimization algorithm whose objective function is the training accuracy. Although optimization algorithms increase the accuracy of training, the trained model may not be generalized well on unseen dataset; this issue is referred to as overfitting. Overfitting occurs whenever the model learns the hidden pattern specific to the training dataset, which cannot completely represent the overall pattern. The test dataset is used to have an unbiased evaluation of the performance of the model and the accuracy of generalizing the model. The accuracy of the trained model in predicting future examples is investigated, comparing the predicted labels with the expected labels for the test set. One practical approach to avoid overfitting is to consider both training and test errors as the objective function of the optimization algorithm; by doing so, there will be a balance between the ability of the machine learning algorithm to learn from available data and its ability to predict the future observations.

Supervised machine learning algorithms such as support vector machine (SVM), artificial neural networks (ANN), and random forest (RF) have been used in the literature for a variety of prediction problems. Each algorithm has a specific basic principle, which makes it suitable for a particular problem. SVM algorithm seeks to determine a set of hyper plains in high dimensional space that can separate the nearest training classes with a higher margin. A higher margin can increase the accuracy of the generalization of the model to future observations. SVM has broad application in face detection, handwriting recognition, generalized predictive control, etc. ANN algorithm is a well-known machine learning algorithm that is inspired by the simplification of neurons in the human brain. The model consists of an interconnected artificial neuron. Each artificial neuron has weighted input, bias, transfer function, and one output [17]. The model parameters, weights, and biases are adjusted by a training algorithm during the learning process to find the optimum values which result in the higher accuracy of the prediction. Gradient descent algorithm, backpropagation algorithm, and heuristic algorithms such as particle swarm optimization (PSO) or genetic algorithms are some examples of training algorithms. There are several architectures for ANN, such as feed-forward neural networks, multilayer perceptron (MLP), and convolutional neural networks. RF consists of several decision trees that are trained on a bootstrapped sample [18]. Each decision tree is built in a way to reduce the correlation of the tress and achieve high accuracy in prediction than that of each individual tree. The random forest has been used widely in the literature due to being insensitive to high dimensional features, performing fast out of sample prediction, requiring less parameter tuning, and having feature ranking capability [19]. Each machine learning algorithm is the best fit for a specific problem; however, there is not any rule to determine which algorithm is the best choice for a specific problem without implementing and analyzing the accuracy of the results. In this study, we implemented ANN, SVM, and RF algorithms; then, the results were analyzed to determine the algorithm with the highest accuracy.

16.3.1 Data Labeling

Extracting the desired scheduling window from recorded data by smart appliances is the primary step for labeling the input data of machine learning algorithms. The desired scheduling window of appliances can be detected with the proceeding activity recorded in the dataset. For instance, the end of the preferred scheduling window for a washing machine can be determined using the recorded data by the dryer. The start time for the operation of the dryer represents the end of the scheduling window for the washing machine. Moreover, the end of the scheduling window of the dishwasher is typically the closest meal serving time. Instead of considering the continuous values for the desired scheduling window, each extracted value is assigned to a cluster with a specific length. For instance, if the desired scheduling window for the washing machine for an operation cycle is 2 h and 15 min, it is rounded to the nearest hour, which is 2 h. Then, the labeled data for each operation cycle and the corresponding value for the factors mentioned in the previous are used to train and test machine learning algorithms.

16.4 Simulation Results

The possibility of predicting the desired scheduling window based on the introduced factors using machine learning algorithms is elaborated in this section. A dataset including 1-min readings from meters connected to several flexible appliances such as dishwasher, dryer, washing machine, etc. is used. The dataset was collected for 2 years interval, from 2012 to 2014, and includes weather data [20]. There is not any dataset in the literature collected by smart appliances; consequently, we use this dataset to investigate the feasibility of the proposed approach. However, the result of this study can be generalized to any house with smart home appliances. In this study, we only investigate the accuracy of the predicting desired scheduling interval of the washing machine by collected data. Consequently, collected data is divided into two sets: training and test sets to learn the relation between input and labels; then, analyze the accuracy of prediction. Twelve labels are extracted where each label represents the length of the desired scheduling window of the washing machine. Figure 16.1 illustrates the relative frequency of each label.

Three machine learning algorithms, ANN, SVM, and RF are used to learn the hidden relations between the affecting factors and labels from the training set; then predict the respective labels of the test set. The results showed that the accuracy of prediction is significantly low for 12 labels, while including only three labels with the highest probability of occurrence results in higher prediction accuracy. The underlying reason for this outcome is the week relationship between some of the labels with the inputs. The labels with the scheduling window higher than 3 h do not represent the overall trend of using the washing machine by the residents. For instance, they may relate to operating the washing machine in some circumstances

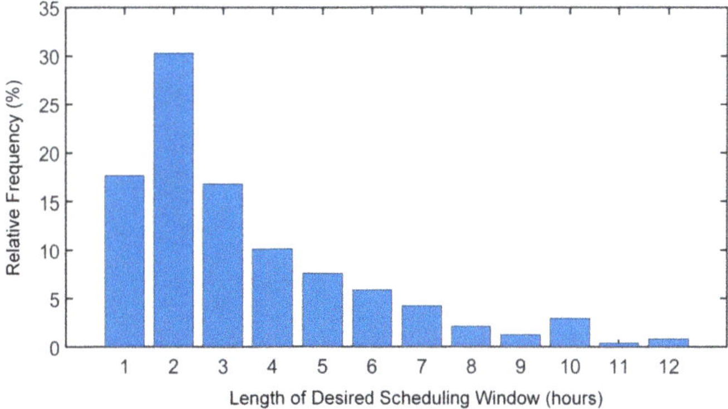

Fig. 16.1 Extracted length of scheduling window for washing machine

which are not part of daily routine such as special gatherings, events, parties, etc.; consequently, the pattern of these labels cannot be learned and predicted as they do not correlate with the daily life of the residents.

The results of predicting the three labels using ANN, SVM, and RF were analyzed to determine which of the mentioned algorithms is the best fit for this problem. The results showed that the KNN algorithm has higher accuracy in predicting the labels while being less prone to overfitting. The accuracy of prediction was 77.6% for the test set. In order to visualize the result of prediction, principal component analysis (PCA) is used to reduce the dimension of the input data [21]. PCA cuts down the dimension of the input data into two unitless dimensions, as shown in Fig. 16.2. Labels 1, 2, and 3 refer to the desired scheduling window with a length of 1, 2, and 3 h, respectively.

The background area in Fig. 16.2 illustrates the decision boundaries determined by the KNN algorithm in the training process, and the data points refer to the test set data. It can be noted that the KNN algorithm has better efficiency in predicting label 2 in comparison with the labels 1 and 3; one possible reason for this outcome is the frequency of occurrence for label 2. Label 2 has the highest frequency of the occurrence, which means the desired scheduling window with a length of 2 h is the dominant trend in using the washing machine; consequently, the algorithm can learn and predict this trend with higher accuracy.

Considering the results, it can be concluded that the proposed approach can be used for predicting the desired length of the scheduling window for residential load scheduling problems. The main contribution of this approach is in using machine learning algorithms for learning the hidden relationship between factors such as the start time of the task, day of the week and outdoor temperature, and the length of the desired scheduling window. Although this study only reports the result of applying the proposed approach for the washing machine as a highly used flexible home appliance, the proposed concept can be used for other flexible appliances such as dryer, dishwasher, etc. The proposed approach can be used in real-time applications where the smart applicants can report the start command by the user to the central

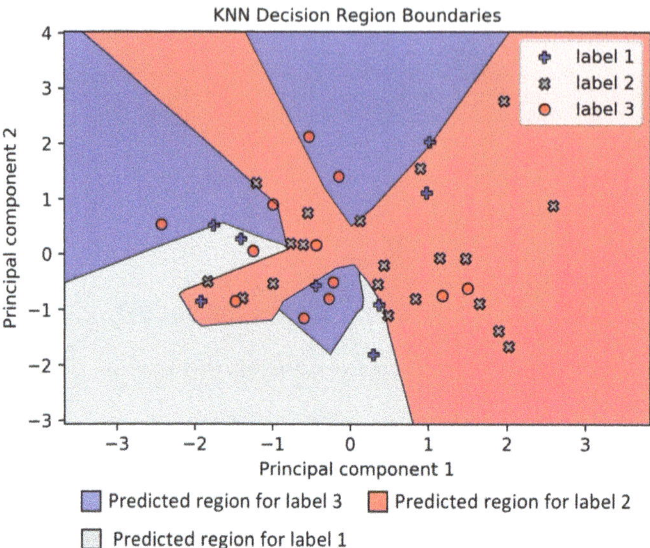

Fig. 16.2 Performance of the proposed approach

home energy management system. Then, the trained model with historical data can predict the desired length of the scheduling window. The predicted parameter can be fed into the optimal load schedule model to determine the optimal operation interval of the appliance, considering the price of the electricity and operational constraints. Finally, the on/off command can be sent to the smart appliance to finish its operation cycle within the defined time interval. Consequently, the proposed approach can make the implementation and operation of the home energy management systems less demanding and nonintrusive in residential sectors by reducing the amount of the data which should be provided to this system at the beginning of the scheduling process.

Despite the effectiveness of the proposed approach in predicting the length of the scheduling window, some challenges should be addressed properly to make this approach applicable in real systems. The training set should be dynamically updated after each operation of the appliances to record any change in the lifestyle of the residents. The model may be retrained in specific time intervals with the updated train set to keep the trained model updated. However, there may be another arising issue with including the new observation in the training set, such as a reduction in the accuracy of prediction. If there is a change in the lifestyle of the residents, some data of the training set will be related to the previous preferences of the residents, while the updated data will represent the current lifestyle. Consequently, the accuracy of the prediction will be decreased as there will be a weak relationship between input and output if the whole training set is fed to the model. Therefore, while adding new observations to the training set, some historical data may be removed to increase the accuracy of the prediction.

Moreover, future studies may identify the important factors which affect the length of the scheduling window of the other flexible appliances such as dishwasher, dryer, etc. Also, possible approaches in extracting the desired scheduling window of the flexible appliances may be investigated for labeling the input data. The application of wearable devices in providing a comprehensive dataset for implementing this approach may be studied in the future. Nowadays, wearable devices such as smartwatches are pervasively used by people. In addition to the health applications of these devices, they can provide data about the lifestyle of the residents and their location, which can be used for identifying the residents' preferences besides the data supplied by smart appliances and stationary sensors installed in houses for monitoring the temperature, humidity, etc. Considering more features and increasing the dimension of the input dataset can result in higher accuracy if the relevant features are selected and used for training the machine learning algorithms.

16.5 Conclusion

This study proposes an approach to predict the length of the desired scheduling window of flexible smart appliances. Smart appliances can collect the energy consumption profile and receive on/off command from external sources such as home energy management systems. The proposed approach was implanted on a dataset to predict the desired scheduling window of the washing machine in a residential building. The start time of the task, day of the week, and temperature are introduced as features in learning and predicting the length of the desired scheduling window. Three machine learning algorithms, ANN, RF, and SVM, were implemented on the labeled data to analyze the accuracy of the prediction. The results showed that KNN outperforms other algorithms by achieving 77.6% accuracy in predicting the desired scheduling windows with the length of 1, 2, and 3 h. Although there were scheduling windows with lengths of higher than 3 h, their frequency of occurrence was low. Consequently, they cannot represent the common trend in using the washing machine and cannot be predicted. The results demonstrated the possibility of predicting the desired scheduling window of the flexible appliance that reduces the difficulty of using home energy management systems in the residential sector. Including the proposed approach in the home energy management systems will eliminate the need for updating the scheduling window of each flexible appliance by the user at the beginning of the scheduling process. Future studies can be followed in three directions: (1) increasing the accuracy of the prediction by introducing and including other important factors besides what mentioned in this study; (2) implementing the proposed approach on a complete set of the flexible loads and charger of the plug-in hybrid electric vehicle and analyzing the results of prediction for all appliances; and (3) integrating the proposed approach in the residential load scheduling algorithms to make scheduling process less demanding for users.

References

1. S. Althaher, P. Mancarella, J. Mutale, Automated demand response from home energy management system under dynamic pricing and power and comfort constraints. IEEE Trans. Smart Grid **6**(4), 1874–1883 (2015)
2. B. Zhou et al., Smart home energy management systems: concept, configurations, and scheduling strategies. Renew. Sust. Energ. Rev. **61**, 30–40 (2016)
3. A. Anvari-Moghaddam, H. Monsef, A. Rahimi-Kian, Optimal smart home energy management considering energy saving and a comfortable lifestyle. IEEE Trans. Smart Grid **6**(1), 324–332 (2014)
4. F. Baig, A. Mahmood, N. Javaid, S. Razzaq, N. Khan, Z. Saleem, Smart home energy management system for monitoring and scheduling of home appliances using zigbee. J. Basic. Appl. Sci. Res **3**(5), 880–891 (2013)
5. J. Han, C.-S. Choi, W.-K. Park, I. Lee, S.-H. Kim, Smart home energy management system including renewable energy based on ZigBee and PLC. IEEE Trans. Consum. Electron. **60**(2), 198–202 (2014)
6. M. Daneshvar, M. Pesaran, B. Mohammadi-Ivatloo, 7 - Transactive energy in future smart homes, in *The Energy Internet*, ed. by W. Su, A. Q. Huang, (Woodhead Publishing, Sawston, 2019), pp. 153–179
7. M.S. Ahmed, A. Mohamed, T. Khatib, H. Shareef, R.Z. Homod, J. Abd Ali, Real time optimal schedule controller for home energy management system using new binary backtracking search algorithm. Energ. Buildings **138**, 215–227 (2017)
8. A.-H.H. Mohsenian-Rad, A. Leon-Garcia, Optimal residential load control with price prediction in real-time electricity pricing environments. IEEE Trans. Smart Grid **1**(2), 120–133 (2010)
9. E. Nazerfard, P. Rashidi, D.J. Cook, Discovering temporal features and relations of activity patterns, in *Proceedings - IEEE International Conference on Data Mining, ICDM*, (2010), pp. 1069–1075
10. A. Yassine, S. Singh, A. Alamri, Mining human activity patterns from smart home big data for healthcare applications. IEEE Access **99**, 1 (2017)
11. S. Chen, F. Gao, X. Guan, T. Liu, Y. Che, Y. Liu, A residential load scheduling approach based on load behavior analysis, in *2014 IEEE International Conference on Automation Science and Engineering (CASE)*, (2014), pp. 954–959
12. S. Chen et al., She: smart home energy management system based on social and motion behavior cognition, in *2015 IEEE International Conference on Smart Grid Communications (SmartGridComm)*, (2015), pp. 859–864
13. A. Moradzadeh, O. Sadeghian, K. Pourhossein, B. Mohammadi-Ivatloo, A. Anvari-Moghaddam, Improving residential load disaggregation for sustainable development of energy via principal component analysis. Sustainability **12**(8), 3158 (2020)
14. M. Sadat-Mohammadi, M. Nazari-Heris, E. Nazerfard, M. Abedi, S. Asadi, H. Jebelli, Intelligent approach for residential load scheduling. IET Gener. Transm. Distrib. **14**(21), 4738 (2020)
15. S. Lee, D. Whaley, W. Saman, Electricity demand profile of Australian low energy houses. Energy Procedia **62**(4), 91–100 (2014)
16. M.A. Nielsen, *Neural Networks and Deep Learning*, vol 2018 (Determination Press, San Francisco, 2015)
17. S. Agatonovic-Kustrin, R. Beresford, Basic concepts of artificial neural network (ANN) modeling and its application in pharmaceutical research. J. Pharm. Biomed. Anal. **22**(5), 717–727 (2000)

18. G. Biau, E. Scornet, A random forest guided tour. TEST **25**(2), 197–227 (2016)
19. J. Xia, P. Ghamisi, N. Yokoya, A. Iwasaki, Random forest ensembles and extended multiextinction profiles for hyperspectral image classification. IEEE Trans. Geosci. Remote Sens. **56**(1), 202–216 (2018)
20. S. Makonin, B. Ellert, I.V. Bajić, F. Popowich, Electricity, water, and natural gas consumption of a residential house in Canada from 2012 to 2014. Sci. Data **3**(1), 1–12 (2016)
21. S. Karamizadeh, S.M. Abdullah, A.A. Manaf, M. Zamani, A. Hooman, An overview of principal component analysis. J. Signal Inf. Process. **4**(3B), 173 (2013)

Chapter 17
Machine Learning Approaches in a Real Power System and Power Markets

Peyman Razmi and Mahdi Ghaemi Asl

17.1 Introduction

An electric power grid is a network of electrical elements employed to supply, transfer, and utilize electric power that are often referred to as the largest and the most complex industry in the world. A model of a power system is the network which feeds a region's considerable areas such as houses and industry. The power system is widely divided into the generators (feeds the power), the transmission system (transfers the power from the generating stations to the load hubs), and the distribution system (supplies the power to close houses and industries). Thus, the mixture of these parts that work systematically and seamlessly together is known as power grid. Small power grids are also existing in industry, commercial buildings, hospitals, and houses. Electrical energy is not able to be kept in large quantities and therefore should be utilized the moment it is generated, so there should always be an electrical connection between the generator and the load. This means that generation, transmission, and distribution parts of power system should carry out their work with complete consistency and safety. To maintain the safety and continuity of the system, these parts need to be utilized and operated in an optimal fashion. Here, operating means monitoring, controlling, and coordinating all components of the power system, and it is a task that is heavily influenced by quality of human resources (operator) and equipment.

But the growing demand for electricity and the advent of concepts such as microgrids have added to complexity and nonlinearity of power systems, which in

P. Razmi (✉)
Faculty of Electrical Engineering, University of Ferdowsi, Mashhad, Iran

M. Ghaemi Asl
Faculty of Economics, Kharazmi University, Tehran, Iran
e-mail: m.ghaemi@khu.ac.ir

M. Nazari-Heris et al. (eds.), *Application of Machine Learning and Deep Learning Methods to Power System Problems*, Power Systems,
https://doi.org/10.1007/978-3-030-77696-1_17

turn has caused the conventional and inflexible method of system safety to become ineffective in a broad array of performing points. Power system is a dynamic system with a wide range of different working conditions and uncertainties, which make it susceptible to a wide range of problems that could emerge at any moment. The quality of the stability is one of the main issues in the power system. Each power system that after an interruption fails to return to its pre-fault operating condition (or a close proximate of that condition) is called an unstable power system. In such circumstances, the interconnected power system either loses its integration and starts to operate as a segmented system or fails to supply the electrical energy from the generator to the point of consumption. It is clear that operating such a complex system needs advanced studies and extensively developed computer programs and algorithms to be used across planning, control, and protection cycles for technical investigation of safety as well as economic viability of electricity generation and supply mechanism. In addition to stability, power systems also have a number of other uncertainties which include:

1. Change in the system's functional conditions: change in the magnitude of load or output of power system.
2. Faults in the power system: type, location, and time of the faults can never be predicted.
3. Uncertainties originating from the present-day problems of the power system, e.g., electricity marketing and deregulation issues.
4. Inaccurate information of operator about control, function, management, and planning of power systems.

Many articles have investigated the applications of ML in power systems, and some have achieved very desirable results. The aim of this paper is to review the works accomplished in the context of ML within the framework of the power systems and express the importance and possible roles of ML in power networks.

17.2 Control-Themed Studies

The majority of studies that devoted to control-themed applications of ML in power systems are emphasized on model of power system stabilizers (PSSs), control of synchronous machine (SM), power system frequency, and flexible alternating current transmission system (FACTS) components. On this basis, the concepts and applications of ML in dealing with the uncertainties of power systems can be classified into four major categories as shown in Table 17.1. With automatic voltage regulators (AVRs) approaching in late 1950s, establishment of AVRs on power generation units turns into a usual procedure [1]. Regrettably, the high performance of AVRs brought about instability in the power system. The majority of the challenges are related to the low-frequency fluctuation in interconnected power systems, particularly in the deregulated models. Small magnitude and low-frequency fluctuation usually remained a problem for a prolonged time. To

Table 17.1 Four major categories of machine learning application in power system

Control	Detection	Function	Optimization
Design of PSS	Detection of fault's exact location	Evaluation and stability	Distribution system
Control of synchronous machine Control of AGG Control of FACTS elements	Detection type of fault in power system location System monitoring Protection of transmission lines	Load prediction State estimation Load distribution	Production planning Reliability System protection

enhance the dynamic performance of the system by providing a quick damping, an extra control sign in excitation system or/and the governor system of a generation block could be utilized. Being the most economical damping controller, PSS has been extensively used to restrain the low-frequency fluctuation and improve the system dynamic stability. PSSs provide a reliable performance to retain the power system's stability by adding a supplementary signal to the excitation system [1].

Over the recent years, artificial neural network controllers (ANNCs) [2] and fuzzy logic controllers (FLCs) [3] have been applied and examined as PSSs. In contrast to other conventional control approaches, ANNCs and FLCs are model-free controllers, i.e., they do not need an exact mathematical design of the controlled system. In addition, their rapidity and strength are the most important features compared with the other conventional methods. Since these controllers applied to the Fuzzy PSS design, the laws and the participation function of the controller are tuned independently, causing the design process to be arduous and time-consuming. Regarding ANNCs, they have the ability to learn and adapt, but they perform like a "black box," and it is complicated to comprehend the action of the network. The examination has showed that the adequate operation during the system upsets is connected with the proper selection of the conventional power system stabilizer (CPSS) parameters. The PSS parameters' tuning difficulty is one of the optimization challenges in power system stability. The artificial neural network (ANN), when adequately trained, could be utilized as a controller in place of the CPSS.

Most of the ANN functions in the power systems utilize a multilayer feed-forward network. ANNs are practical tools exerted for numerous years for recognition and control of intricate systems because of their nonlinear mapping features. To attain the finest operation, the ANN should be prepared for various operating circumstances to adapt the CPSS parameters [1]. The learning strategies lead to interference by the typical back propagation network under diverse circumstances. To solve this hindrance, a modular ANN was introduced instead of a back propagation network [4]. This design comprises three local proficient networks and one alone gate network which has three layers for each one of them. The ANN was disciplined straightly through the output and input of the CPSS. The model outcomes validated that the modular PSS was more capable of damping oscillations and bringing about proper performances. To advance a neural adaptive PSS, a feed-forward neural network with a single hidden layer was examined in [5–7]. The recommended neural adaptive PSS consist of two sub-networks: adaptive neuro-identifier and adaptive neuro-controller which tracked the dynamic features of the plant and damp the low-frequency oscillations, respectively. These two sub-networks were worked out in an online mode using the back propagation approach.

An innovative method for online adaptive tuning of PSS parameters employing radial basis function neural network (RBFNN), was cited by Abido and Abdel-Magid [8]. The suggested RBFN was operated over a broad range of circumstances and system parameter alterations for readjusting PSS parameters online on the basis of real-time calibrations of machine loading situations. The mentioned RBFN-based PSS was studied for diverse operating circumstances and system parameter alternations. To enhance the ephemeral stability of power systems, a recurrent neural network (RNN) stabilization controller was offered in [9]. The suggested technique was employed for both the AVR and governor. The weights of introduced controller were modified online. For excitation control, the signal output of the first RNN was added to the PSS signal output. In addition, to have a stabilization signal for the governor unit, the signal output of the second RNN was utilized. To enhance the damping performance, a probabilistic PSS was presented by Ping et al. employing a single-neuron model [10]. The alteration of system performing conditions could be recorded by adding ancillary self-adjusted gains to PSSs. In an analogous paper, adaptive critics and ANNs were offered to plan a real-time digital signal processor (DSP) to accomplish an optimal wide-area control system (WACS) for a power system [11].

In 1964, fuzzy logic (FL) was developed by Lotfi Zadeh to give an attention to inaccuracy and uncertainty that usually occur in engineering problems [12]. In the mentioned method, FL is able to be utilized as a general technique to integrate knowledge, heuristics, or theory into controllers and decision-makers [1]. The pros are as follows: (1) precise description of the functional restrictions of the power systems; and (2) fuzzified restrictions will be softer than traditional limitations [13]. In 1979, for the first time FL was presented to solve power system conundrums. To make classic controllers, it is prerequisite to linearize nonlinear systems. As a result, control regulations are determined on the basis of the new design. These kinds of controllers are used to monitor the system. Fuzzy logic controllers (FLCs) are nonlinear, do not require controlled device models, and are not delicate to changes of the device parameters. Reference [14] proposed a self-constructing fuzzy neural network (SCFNN)-based static synchronous series control (SSSC) to soften the inter-area fluctuations in interconnected power grids that consist of an online equipped fuzzy neural network (FNN) controller with adaptive learning rates (ALRs) and self-constructing structure. The Lyapunov paradigm is used to gain the adaptive learning rates. Reference [15] offers fuzzy particle swarm optimization of PID controller (PSO-FPIDC) utilized as a CPSS to enhance the dynamic stability efficiency of generating unit during low-frequency fluctuations. Reference [16] presents an adaptive fuzzy sliding-mode controller (AFSMC) with a PI switching surface to damp power system fluctuations. A wavelet neural network (WNN) and sliding mode as a monittor unit can be used to surmount the complications of designing a sliding-mode controller, which are the conjecture of known uncertainty restricts and the chattering incident in the control effort. In [17] the author proposed an adaptive fuzzy logic PSS (AFLPSS) model to damp electromechanical forms and improving the first-swing synchronized stability margins. The plan was to arrange a multi-zonal PID formation and fuzzy logic variable gain to make the damping operation more optimal. Ramirez and Malik introduced a self-tuned FLPSS to improve the damping of power grid fluctuations [18]. The model of self-tuned FLPSS was consisted of a simple FL controller which has a considerably simple architecture and few tuning parameters which was implemented through a basic control algorithm. The fuzzy tuner was utilized to nonlinearly and online adjust the reactivity of the simple FL controller to its input variable, which obliquely alters the relative sensitivity of areas of related input membership functions.

The tuning of traditional power system stabilizer for single-machine infinite bus power system is presented in [19]. With applying the bat algorithm (BA) an eigenvalue-based objective function, the author optimized the time and gain constants (pole and zero) of CPSS. The speed feedback of the generators for the single-machine infinite bus system (SMIB) excluding PSS was presented in [20]. With PSO-CPSS and employing the BA-based advanced CPSS, power grid schemes are examined in various operation circumstances. It is revealed the feedback with PSO-CPSS is able to stabilize the grid during unwanted situations. But by using continued settling time over decreased settling time and applying BA-CPSS can stabilize the system for whole plants situations. The admirable work of BA-CPSS is more demonstrated with utilizing performance indexes as the value is least in comparison to the grid with PSO-CPSS and excluding PSS. The little signal stability is assured with applying BA-CPSS since all electromechanical mode eigenvalue (for 133 plants) is shifted to left-hand side (LHS) of s-plane under the D-shape sector. On the other hand, by using PSO-CPSS, only nine plant situations are confirmed to be stable.

In [21], an intelligent power system stabilizer on the basis of a new system centric control device is presented. The presented approach utilized hybrid structure of two algorithms: the first one is a neural network (NN)-based control device with specific neuro-identifier, and the second one is an adaptive control device employed as a model reference adaptive controller (MRAC). An identifier is utilized to estimate the nonlinear functional dynamics of the electrical grid. The MRAC control device modifies when electrical grid (plant) parametric set varies. The notable uniqueness and benefit of the presented model are the controller's capability to supplement one another during functional and parametric uncertainty and develop in the case of unstable system dynamics. Work [22] demonstrates an improvement in a smart neighborhood signs on the basis of a control device in order to damp low-frequency fluctuations in electrical grids. The control device is developed off-line for functioning adequately during a broad range of electrical grids functioning sites, permitting it to manage the intricate, haphazard, and time-varying characteristic of electrical grids. NN-based unite recognition erases the demand for developing precise designs through primary rules for control model, culminating a methodology which is thoroughly drive by data.

A unified control device was replaced by the traditional PSS, and AVR for adjusting the generators' terminal voltage and lessening the power system fluctuations was presented in [23]. This brilliant control device is termed as online trained self-recurrent wavelet neural network controller (OTSRWNNC). For attaining the previous purposes, two control faults were decreased at the same time with modifying the characteristic of OTSRWNNC. Since the mentioned control device has great capability to learn, it doesn't need an identifier for estimating the dynamic of monitored electrical grid. In [24], a model for a grid with a real-time closed-loop wide-area decentralized power system stabilizer (WD-PSSs) is examined. For this purpose, real-time wide-area measurement information are refined and used to plan a series of stability factors on the basis of a reinforcement learning (RL) approach. Current technical developments in wide-area measurement system (WAMS) cause the ability of using the system-wide signals in constructing electrical grid control devices. The primary purpose of such control devices is to maintain the grid after serious disruptions and mitigate the fluctuations afterward.

An adaptive neuro-fuzzy inference system (ANFIS) approach on the basis of ANN for planning a SSSC-based control device to enhance the transient stability was demonstrated in [25]. The offered ANFIS control device merges the benefits of a fuzzy control device along with the fast feedback and flexibility characteristic of ANN. ANFIS design was trained by utilizing the created database from fuzzy control device in SSSC. It's illustrated that the introduced SSSC control device enhances the system's voltage profile markedly during serious disruptions.

ML algorithms have been lately deemed as a method for monitoring SMs. Authors in [26] introduce a function of fuzzy logic for controlling a SM's speed. In accordance with the examination of the SM transient feedback and fuzzy logic, the fuzzy control device was advanced. The fuzzy control device produces the alterations of the reference current vector of the SM speed control on the basis of the speed error and its variation. Authors in [27] examine a wide practical range of the synchronous generator (SG), its intricate dynamics, transient operation, non-linearities, and an altering unit formation. Hence this generator is not able to exactly be designed as a linear device; the mentioned work offers a new optimal neuro-controller which substitutes the conventional controller (CONVC). A combination of the AVR and turbine governer can be used, for controlling a SG in an electrical grid by utilizing a multilayer perceptron neural network (MLPN) and a RBFNN.

In the last years, developments in magnetic materials, semiconductor power devices, and control theories caused permanent magnet synchronous motor (PMSM) drives to perform significant role in motion-control applications [28]. They own admirable characteristics, namely, small size, immense torque to weight ratio, as well as lack of rotor losses. Moreover, higher power of the developed magnetic materials causes broader applications of PMSM [29–32]. In Reference [28] a strong speed control approach of a PMSM drive employing an adaptive neural network model following speed controller (ANNMFC) is presented. The robust speed control device is a compound of a feedback neural network controller (NNC) along with an online trained NN model following controller (NNMFC) to enhance the dynamic operation of the drive unit. Outputs of NNMFC are added to the NN speed control device output to offset the difference among the reference design and the PMSM drive unit output during load disruptions as well as parameters

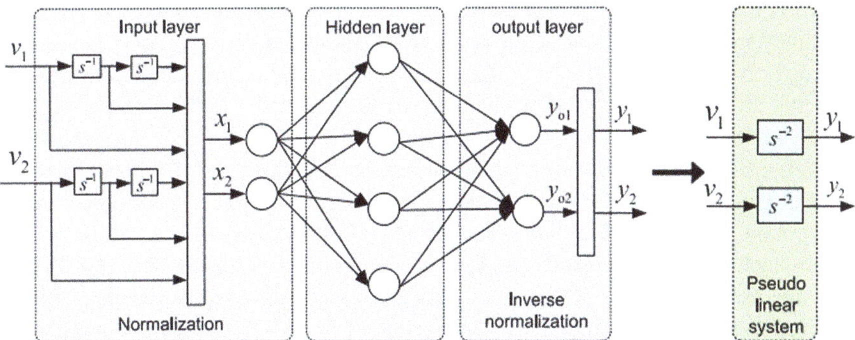

Fig. 17.1 The topological structure of the RBFNN

uncertainties. A speed control device for PMSM during the field-oriented control (FOC) approach is explained in [33]. A new adaptive neuro-control design, single artificial neuron goal representation heuristic dynamic programming (SANGRHDP) for speed adjusting of PMSMs, is discussed. Paper [34] proposes a smart speed control device for the interior PMSM, in accordance with a single artificial neuron. Reference [35] presents a new decoupling paradigm for a bearing-less PMSM (BPMSM) to obtain quick response and high accuracy performances. The mentioned control design includes the NN inverse (NNI) technique and two-degree-of-freedom (DOF) internal design control devices. With presenting the NNI schemes into the original BPMSM structure, a decoupled pseudo-linear arrangement could be formed. In [36] a new internal model control (IMC) method on the basis of the back propagation neural network inverse (BPNNI) control plan was suggested for the purpose of effectively decoupling the PMIWM. Reference [37] proposes a new nonlinear decoupling monitor plan for a permanent magnet in-wheel motor (PMIWM), in which the radial basis function neural network inverse (RBFNNI) and the state feedback robust pole placement (RPP) are both utilized. The offered NNI is settled on the left side of the PMIWM shown in Fig. 17.1, which gives rise to a pseudo-linear scheme for the system decoupling. The NNI control method is able to enhance the robustness and reject the disruption.

17.3 Detection and Protection-Themed Studies

To sustain the efficiency and reliability of the power systems, it is undoubtedly essential for the TL faults to be identified and settled in a reliable and precise approach [38]. Furthermore, by noticeable growing of smart grids, the significance of designing a smart fault controlling and diagnosis unit is to classify and locate various kinds of errors is indisputable [39]. Procedures for error detecting, categorizing, and locating in TLs and DSs have been extensively researched during the recent years. A novel method on the basis of generalized neural network (GNN) with wavelet transform (WT) is proposed in [38] for fault locating evaluation. Acquired qualities are employed as an input to the GNN design for approximating the zone of the error in a provided TL. The differences between the results attained from GNN scheme are examined with ANN which leads to establish mathematical paradigms with more accuracy [40].

Advance TL fault location methods consists of stationary WT (SWT), determinant function feature (DFF), support vector machine (SVM), and support vector regression (SVR) which are discussed in [41]. Different kinds of errors at various zones such as fault impedance and fault inception angles on a 400 kV, 361.297 km TL are examined [41]. The system only uses single-end measurements. DFF is utilized to separate specific fault characteristics from 1/4 cycle of post-fault signals after noise, and the declining DC offset has been wiped out by filtering method on the basis of SWT. Afterward, a classifier (SVM) and a regression (SVR) approaches are prepared with characteristics attained from DFF, which is employed at the fault zone on the TL. The outcomes of the discussed paper illustrate that fault zone on TLs is able to be detected quickly and accurately regardless of fault impedance. Reference [42] presents k-nearest neighbor (k-NN)-based approach for fault zone estimation of all kinds of errors in parallel lines with utilizing one-terminal measurement. Discrete Fourier transform (DFT) is applied in the mentioned work for pre-analyzing the signals. After that, the standard alteration of one cycle of pre-fault and one cycle of post-fault examples is utilized as inputs to k-NN algorithm. Reference [43] mentions the significance of fault location within the context of smart grids and microgrids that can be investigated with all the capabilities of this new paradigm. This reference proposes a multi-stage technique that consists of decision trees (DT) and ANNs for fault location aims. Authors in [44] discuss the significance of ANNs for fault location and compare it with other soft computing methods such as fuzzy logic technique, wavelet approach, SVM, and other traditional procedures.

In [45], a unified framework with combination of error categorization and locating is presented by employing a novel ML algorithm: the summation wavelet extreme learning machine (SW-ELM) which incorporates quality derivation during the learning procedure. Another essential subject requires further analysis in the transmission and DS in categorizing and detecting the fault's type. Fault designation is important for dependable and quick response protective relays followed by digital distance protection. Categorization of power system errors is the initial step for enhancing power quality and guaranteeing the system protection. For this reason, equipping a unique method in identifying the class of the faults is needed. Reference [46] offers a unique plain and efficient approach for faulty feeder detecting in resonant grounding DSs in accordance with the continuous WT (CWT) and convolutional NN (CNN). At the same time, the characteristics derivation for error signals and the defective feeder detection are performed by the instructed CNN.

In [47], categorizing the electrical grid errors by utilizing the empirical mode decomposition (EMD) and SVMs is discussed. EMD is utilized to decompose the TL voltages into intrinsic mode functions (IMFs). A multiple SVM design is presented to classify the fault status in a group of ten power system faults. Reference [48] introduces an approach employing discrete WT (DWT) and SVM to identify and categorize faults through a TL. DWT is applied to find the high-frequency components within a fault signal range. In this paper, various faults (short circuits) on disparate positions of a transmission line are formed and simulated. In [49], a brilliant method for fault categorization in a TL is presented. Ten various types of faults (LAG, LBG, LCG, LABG, LBCG, LCAG, LAB, LBC, LCA, and LABC) have been studied as one normal status on a simulated TL system. Post-fault current signs have been utilized for characteristic derivation for additional research. EMD approach is employed for decomposing post-fault current signals into IMFs. These IMFs are utilized as input variables to an ANN on the basis of intelligent fault classification method. Reference [50] presents a novel fault identifier algorithm for photovoltaic (PV) systems in accordance with ANN and fuzzy logic system interface. Considering the small number of examples of ML methods used in fault detecting algorithms in PV systems, the major focus of the mentioned paper is to design a system with an ability to find possible faults in PV systems by utilizing radial basis function (RBF), ANN, and both Mamdani and Sugeno fuzzy logic systems interface.

A new fault detection and categorization method for high-voltage DC transmission lines employing K-nearest neighboring was offered in [51]. In [52], problems of monitored learning, namely, necessity for historical information and incapability of classification of further errors precisely, are solved by a novel method utilizing unmonitored learning for quick execution of predictive maintaining activity. This method consists of error kind and forecasting for familiar and unfamiliar errors by employing density approximation through Gaussian method. Combination of design clustering and K-means algorithm is employed and compared their outcomes with an actual case vibration information. A new approach to trace the ephemeral errors and to categorize the error kind in DSs employing WTs and ANFIS has been evolved in [53]. It implements on developed methods of signal processing in accordance with WTs, utilizing data sampled from the main feeder current to collect major features and dynamic qualities of the fault signal. In [40] a fault categorization method on the basis of Haar-WT (HWT) and ANN is introduced for six-phase TL in opposition to phase-to-phase faults. Reference [54] attempts to describe four multiwavelet packet entropies to elicit the qualities of various TL faults and employs RBFNN to detect and categorize ten fault types of transmission lines. In [55] DWT is utilized to elicit transient data from the measured voltages. Then SVM classifiers are used to detect the faulty section and faulty half. Moreover, writers in [56, 57] employed characteristics elicited by DWT as inputs to SVMs. Authors in [58, 59] applied principal component analysis (PCA) to diminish the dimensionality of the wavelet coefficients in advance of sending the coefficients to the SVMs for fault-type categorization.

In [60] a novel categorization approach on the basis of NN is presented to decrease the training time and dimensions of an ANN. Employing the mentioned technique, high precision of fault categorization is obtained. In the mentioned reference, primary elements of pre-fault and post-fault as well as positive sequence elements of currents and voltages of three phases have been applied as inputs to presented ANN. The output of the ANN is the approximated fault zone. Reference [61] proposes a novel fault categorization method for high-speed relaying utilizing minimal RBFNN. This novel method reduces the training time noticeably and presents a systematic framework for choosing the number of neurons in the hidden layer. In addition, the minimal radial basis function network provides an exact error kind categorization of a TL even in the existence of high fault impedance in the fault line. The suggested method in [62] is a compound of a preprocessing block in accordance with DWT and probabilistic neural network (PNN). The DWT operates as extractor of distinct qualities in the input current signal, which are accumulated at source end. The data is then fed into PNN to categorize the faults. Analysis of the possible and obscure faults within the electric transformers is the means of guaranteeing a steady power supply for costumers [63].

SVM is a modern ML approach on the basis of statistical learning theory, which is a robust mechanism to overcome the challenges of nonlinearity, small sampling, and high dimension. Selecting the SVM characteristics has a notable effect on the categorization exactness of SVM. In [63] SVM with genetic algorithm (SVMG) is employed to fault analysis of a power transformer, in which genetic algorithm (GA) is applied for selecting the proper free characteristics of SVM. Empirical outcomes illustrate that the SVMG approach is able to attain higher diagnostic precision than International Electrotechnical Commission (IEC) three ratios, normal SVM arranger, and ANN. Paper [64] proposes a brilliant fault categorization method to power transformer dissolved gas analysis (DGA), handling highly adaptable or noise-corrupted information. Bootstrap and genetic programming (GP) are applied to enhance the interpretation precision for DGA of power transformers. Bootstrap preprocessing is used to roughly adjust the sample numbers for various fault classes to enhance subsequent fault categorization with GP aspect extraction. GP is utilized to set up categorization qualities for each class in accordance with the collected gas information. The qualities elicited from GP are then applied to the inputs of ANN, SVM, and K-nearest neighbor (KNN) arrangers for error categorization. The categorization exactness of the mixed GP-ANN, GP-SVM, and GP-KNN arrangers is compared sequentially with the ones extracted from ANN, SVM, and KNN arrangers [65]. presents a novel differential protection design in accordance with the SVM, which yields efficient distinction between internal faults in a power transformer with the other disruptions, namely, different types of over excitation and inrush currents states. The characteristic derivation is accomplished employing WT, which will be given as input to the SVM classifier.

A novel approach to diagnose the transformer fault on the basis of relevance vector machine (RVM) is presented in [66]. A Bayesian estimator is implemented to SVM in the unique algorithm, which made the performance of the fault analysis system more efficient. In paper [66], a diagnosis paradigm is offered in which the resolutions of RVM have the characteristics of sparseness and RVM is able to achieve global resolutions under confined samples. The outcomes proved that this approach has discernible benefits of diagnosis time and precision in comparison with back propagation (BP) NNs and general SVM models. Reference [67] presents a fault direction separator that utilizes an ANN for protecting TLs. The separator employs different features to achieve a decision and tends to imitate the traditional pattern categorizatio*n* challenge. In reference [67] an equation of the boundary explaining the categorization is embedded in the multilayer feed-forward neural network (MFNN) by training and employing a proper learning algorithm. The separator utilizes immediate values of the line voltages and line currents to make decisions. Consequences of the ANN-based separator operation in [67] illustrates that it is quick, powerful, and precise. Reference [68] offers an anomaly detection method utilizing ANNs and the WT for the condition monitoring of wind turbines. In this method, nonlinear autoregressive NNs are applied to evaluate the temperature signals of the gearbox.

Author in [69] proposes an ANN-based method control the voltage stability margin (VSM) in electrical grids online. In Reference [70] ANN-based approach is modified for rapidly approximating the long-term VSM. The applied examinations in the mentioned work demonstrated that phase angles and node voltage magnitudes are the most excellent forecasting systems of VSM. Moreover, the mentioned article illustrates that the presented ANN-based approach is able to effectively evaluate the VSM under normal situations and under N-1 contingency conditions as well. The distance relays which are used to protect the TLs are commonly planned based on set settings [71]. The operation of these relays will be affected by any alterations to network situation. Appling a pattern discriminator for the power system diagnosis is able to make major improvements in the protection field. Paper [71] illustrates the utilization of an ANN as a model discriminator for a distance relay operation. The method uses the magnitudes of three phase voltage and current phasors as inputs. In Reference [72] authors discuss the potential ways to improve the digital power transformer preservation. Establishing the inrush in power transformers is turning into unreliability in existent numerical preservation. An ANN is implemented to inrush sensing. Unless with an appropriate dimensioning, the saturation of protective current transformers (CT) could not be completely eradicated. ANN has been utilized to reconstruct the deformed secondary CT currents as a result of saturation. In each case, an ANN has been provided in the preservation scheme as a development of the existent designs, which enhanced the preservation system's trustworthiness.

17.4 Function and Protection-Themed Studies

Electricity carries a significant value for national economies to grow. As a result, electric services attempt to provide a balance between power generation and consumption with intention of granting a satisfying service at a reasonable price [72]. Consequently, these services require electric load prediction at the most precise level. Since electric load is influenced by numerous factors (such as day of the week, month of the year, etc.), load predicting becomes a very involved process which demands something greater than statistical approaches. Authors in [72] offer a load prediction structural paradigm on the basis of an ANN that operates short-term load forecasting (STLF). Reference [72] illustrates significant outcomes as well as a plain, efficient design. In [73] authors offer a hybrid scheme that consists of SVR, RBFNN,

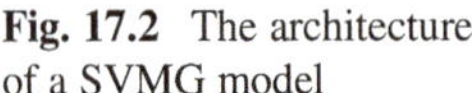

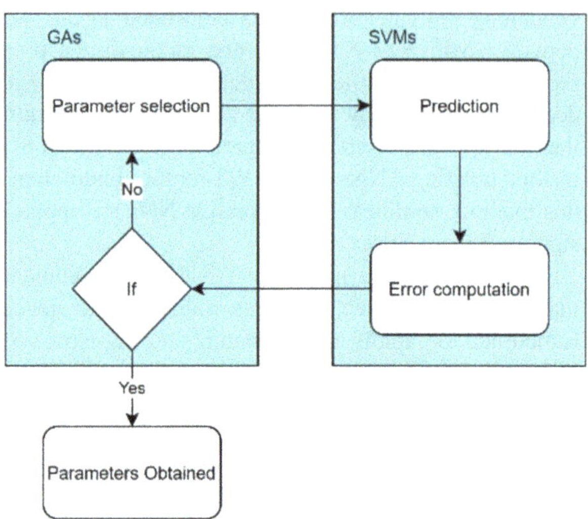

Fig. 17.2 The architecture of a SVMG model

and dual extended Kalman filter (DEKF) to create a forecast method (SVR-DEKF-RBFNN) for short-term load prediction. In the presented design, at first the SVR method is applied to ascertain the form and primary characteristics of the RBFNN. After initializing, the DEKF is utilized as the learning method to advance the features of the RBFNN. At last, the optimized RBFNN method is applied to forecast short-term load. In [74] authors tackle the trouble of forecasting hourly load demand by employing adaptive ANNs. A PSO algorithm is used to adapt the network's weights in the training phase of the ANNs.

In Reference [75], recurrent support vector machines with genetic algorithms (RSVMG) are presented to predict electricity load. Moreover, GAs are applied to identify free characteristics of SVMs. The experimental outcomes from Reference [75] illustrate that RSVMG paradigm offers an auspicious alternative for predicting electricity load in power system. Figure 17.2 illustrates the structure of the presented SVMG design. GAs are applied to provide a smaller mean absolute percentage error (MAPE) by probing for better mixture of three parameters in SVMs. Authors in Reference [76] proposed a model which combines WT, adaptive GA, and fuzzy system with GNN and employed to predict the short-term weekday electrical load. In [77], a new hybrid developmental fuzzy design with characteristic advancement is presented. Considering the fact that determining optimal values for the fuzzy laws and weights is a very involved work, the process of optimization is undertook by a bio-inspired optimizer, named group evolutionary strategy (GES), which develops from a mixture between two heuristic methods, namely, the evolution strategies and the greedy randomized adaptive search procedure (GRASP). Valid information from electric services obtained from the literature is utilized to justify the presented approach. In Reference [78], V. L. Paucar utilized multilayer perceptron NNs trained with the second order Levenberg/Marquardt design to compute voltage magnitudes and angles of the power flow (PF) problem on account of the admirable speed of

ANN over traditional PF approaches. The discussed ANN procedure has been triumphantly examined by applying the IEEE-30 bus system.

Arnagiriet in [79] has suggested an application of ANNs to identify bus voltages of a radial DS for any allotted load without performing the load-flow algorithm. Reference [80, 81] presented a novel, satisfying approach to perform stochastic load-flow analysis by employing the standard back propagation method for training the NN. In [82] Karami and Mohammadi suggested an approach for dealing with the load-flow issue of the electric power systems by utilizing RBFNN with a rapid hybrid training design. Accordingly, they tackled the function approximation problem by using RBFNN in order to solve the load-flow. State estimation (SE) is an indispensable part in overall monitoring and controlling of the transmission networks. It is chiefly employed to provide a reliable approximation of the system voltages. The data from the state estimator streams into control centers and database servers within the network [83]. Authors in [84] propose a novel method in accordance with ANN for power system network topology identification and static SE. In the mentioned reference, the state estimator design uses the dynamic alteration in network topology and bad data processing. The performance of RBFNN for SE is examined in [85] by analyzing its pertinence on IEEE 14 bus system, and the suggested estimator is compared with traditional weighted least squares (WLS) state estimator based on time, exactness, and robustness. Reference [86] proposed a fuzzy clustering and pattern matching for topology determination, bad data identification, and SE. In this approach, a fuzzy pattern vector (FPV) is produced on the basis of the obtained analog measurement vector (telemetry data). The topology determination and gross errors are identified from the difference between the analog measurement data and FPV. The gross errors and topology errors in the measurement data are determined and modified by utilizing the FPV which can be utilized directly as an efficient data measurement for SE.

Voltage stability is a crucial factor that should be considered during the planning and controlling the power systems for the purpose of avoiding voltage collapse and partial or full system blackout afterward. Studying the voltage collapse incident may yield an approach to avert this event from occurring [87]. Authors in [88] propose an approach to monitor the voltage uncertainty in a power system with a hybrid artificial NN which is a compound of a multilayer perceptron and the Kohonen NN. In the presented approach, the Kohonen network is applied to categorize the system operating status; the Kohonen produced patterns that are utilized as inputs to train a multilayer perceptron for determination of alarm states which are threatening for the system security [88]. In [69] authors studied the performance of real-time system monitoring models which can ensure a timely warning ahead of the voltage collapse happened in the electrical grid. In the mentioned paper, various kinds of line voltage stability indices (LVSI) are distinguished to determine their capability to find the weakest lines in the electrical grid. The suggested real-time voltage stability monitoring is implemented by applying ANN. Authors in [89] modified the ANN-based approach to rapidly evaluate the long-term VSM. The discussed examination in [89] illustrates that the phase angles and node voltage magnitudes are the most effective estimators for determining the VSM. Moreover, the paper indicates that the presented ANN-based approach is able to satisfactorily evaluate the VSM under normal operation as well as under N-1 contingency conditions. In [90] authors offer a SVM regression network to assess voltage stability for both normal situation and contingency cases. SVM receives real and reactive power from all buses in the system and provides the loading margin. Eventually the outcomes of the presented approach are compared with NN, ELM, online sequential ELM, and extreme SVM regression models.

17.5 Optimization and Protection-Themed Studies

Power system optimization has developed with advancements in computation and optimizing theories [91]. The optimal power flow (OPF) is an optimization method in power system which is used for monitoring the generation-consumption to optimize definite aims, namely, reducing the generation cost or the system power losses. Much research has dealt with OPF problem by applying AI and ML algorithms [92]. In paper [93] two kinds of neuron transfer functions are utilized to examine the robustness and exactness of the presented method by considering the OPF. Nguyen in [93] modified the NN which executes the Newton-Raphson-based OPF calculations to reduce the real power losses. Hartati and El-Hawary proposed an application of applying an adapted Hopfield NN to resolve the OPF problem with the aim being the incremental fuel cost [94]. Reference [95] has offered an artificial NN method on the basis of Lagrangian multiplier to deal with the economic load-flow problem as well as increasing the convergence speed by applying the momentum technique in the power system.

A novel approach to resolve the economic power dispatch problem (EPDP) with piecewise quadratic cost function by employing the Hopfield NN was proposed in [96]. Authors in [97] propose an Improved Hopfield Neural Network (IHNN) to solve the economic dispatch problem (EDP). The presented IHNN method has quick convergence and moves effectively toward possible equilibrium points. Paper [98] presents an approach on the basis of quadratic programming (QP) and augmented Lagrange Hopfield network (ALHN) for solving EDP through piecewise quadratic cost functions and prohibited locations. In the mentioned approach, the QP model is firstly utilized to identify the fuel cost curve for each of the units and set initial parameters for the ALHN model; after that a heuristic search is applied to repair prohibited location violations, and the ALHN approach is finally used to solve the problem if any violations identified.

An automatic learning structure in order to control the dynamic security of an electrical grid was presented in [99]. The presented approach uses RBFNN, which serves to determine the dynamic security conditions of the electrical grid and to evaluate the impact of a corrective control action employed in the event where disruption happens. PSO is used to detect the optimum control action, where the objective function is provided by the RBFNN for optimization. In Reference [100], a novel dynamic security evaluation and generation rescheduling approach applying

GAs which are unified with PNNs and ANFISs are presented for the precautionary control of vast electrical grids in the event of fleeting disruptions. In the mentioned method, PNNs are applied in a practical manner to measure the security zones precisely during the evaluation and control. The security confined generation rescheduling is applied using a GA which improves the generation shifting or amount of the fuel cost during the precautionary control.

Authors in Reference [100] presented DT-based preventive/corrective control (PCC) approaches to advance the dynamic security of electrical grids against the possible and unforeseen circumstances leading to fleeting disruptions. PCCs, namely, generation rescheduling as well as load shedding models, are advanced on the basis of security zones and boundaries, respectively. These two PCCs could be computed in the space of proper decision variables. The security zones and boundaries are decided in accordance with laws of DTs which are modified by the generated knowledge principles. A novel approach is proposed in [101] for security-constrained corrective rescheduling of real power by utilizing the Hopfield NN. The minimum aberrations in real power generations and loads at power system buses are merged to form the objective function for optimization.

In order to decrease the energy losses in DSs, one of the approaches is demonstrated in [102]. It is in accordance with heuristic methods, namely, branch exchange method and fuzzy logic method. This algorithm is developed by taking the loads as fuzzy numbers, which deals better with instabilities in the power demand in DSs. Reference [103] presents a computational framework for the voltage adjustment of DSs supplied by dispersed generation systems (DGS). The purpose is to determine an efficient solution for the optimal regulation problem by unifying a traditional nonlinear programming algorithm with an adaptive local learning algorithm (LLA). The explanation for the method is that the LLA is able to quickly learn based on a confined number of historical observations.

Reference [104] proposes a method to combine the feed-forward NN and the simulated annealing approach for solving the problem of unit commitment, which is a mixed integer combinational optimization issue in power system. ANN is applied to identify the discrete variables in accordance with the status of each unit at its time interval. The simulated annealing approach is utilized to produce the continuous variables in accordance with the output power of each unit and the production cost. The outcomes in [104] proved that the presented method is able to deal with unit commitment in a less computational time with an optimal generation plan. H. Sasaki [105] examines the practicability of using the Hopfield NN for combinational optimization issues in power systems, specifically for unit commitment. Lots of inequality constraints contained in unit commitment could be dealt with dedicated NNs.

The monitoring exactness and speed are major parameters for the fixed step perturb-and-observe (P&O) maximum power point tracking (MPPT) approach. Paper [106] presents a new method to balance the trade-off between operation and cost based on a PV system. The disturbance step size is identified off-line for a certain zone on the basis of local irradiance information. The SVM is utilized to automatically categorize the desert or coastal zones by applying historical irradiance information. The disturbance step size is modified for better system operation without enhancing the control intricacies. Assessing the solar energy and thermal transmission mechanism, efficiency needs a great deal of time, financial charge, and human resources. Concurrently, straight forecasting the efficiency of these mechanisms is an onerous task since they have involved internal designs. Luckily, a knowledge-based ML approach is able to yield an auspicious forecasting and improvement procedure for the efficiency of energy networks. In Reference [107], authors illustrate how they apply the ML methods to train an immense experimental database to achieve exact forecast for an enhancement on a solar water heater (SWH) model.

A novel energy system improvement approach on the basis of a high-throughput screening (HTS) procedure is suggested in [107]. The mentioned paper's structure is a compound of (1) comparative research on various ML methods: ANNs, SVM, and ELM to forecast the efficiencies of SWHs; (2) advancement of an ANN-based program to support the fast forecasting; and (3) an explanation for a computational HTS approach for planning a SWH model with a high efficiency. In paper [108], authors researched the advantages of applying deep reinforcement learning (DRL) in the smart grid context, a hybrid type of approaches which unites RL with deep learning (DL), in order to execute an online optimization for building energy management systems. The learning process was examined by applying two approaches, deep Q-learning and deep policy gradient; both of these methods were being expanded to complete multiple actions at the same time. In Reference [109], the improved chicken swarm optimization algorithm support vector machine (ICSO-SVM) design is presented for forecasting the wind power. Regarding the conventional chicken swarm optimization method (CSO) features, it usually falls into a local optimal while dealing with high-dimensional matters. As a result, the CSO approach is modified and the ICSO technique is suggested. Paper [110] proposes a developed non-dominated sorting genetic algorithm II (NSGA-II) method including a parameter-free self-tuning RL approach named NSGA-RL in order to deal with environmental and EDPs with several purposes.

17.6 Conclusion

In this chapter a broad range of ML methods in order to solve timely matters in the fields of generation, transmission, and distribution of recent electrical grids have been discussed. Scope of ANNs, kernel regression, decision trees, fuzzy networks, and GAs have been presented for security assessment, control, EDP, PF computations, prediction, rehabilitation, plant supervising, transformer construction, faulted TL issues, and inductive inference matters. At every instance, the applications of ML methods have demonstrated their effectiveness for tackling complicated electrical grid operations, design, and construction problems. The discussed applications showed the thriving activity in electrical grid applied study, as illustrated through the considerable number of relevant publications. To sum up, ML approaches along with developed systematic methods guarantee a great supportive effect for future modern electrical grids.

References

1. M.S. Eslami, Application of artificial intelligent techniques in PSS design: A survey of the state-of-the-art methods. Przegld Elektrotechniczny (Electr. Rev.) **87**(4), 188–197 (2011)
2. R.S. Segal, A self-tuning power system stabilizer based on artificial neural network. Int. J. Electr. Power Energy Syst. **26**(6), 423–430 (2004)
3. S. Pillutla, Power system stabilization based on modular neural network architecture. Int. J. Electr. Power Energy Syst. **19**(6), 411–418 (1997)
4. N. Hosseinzadeh, A. Kalam, A hierarchical neural network adaptive power system stabilizer. Int. J. Electr. Power Energy **19**, 28–33 (1999)
5. P. Shamsollahi, Design of a neural adaptive power system stabilizer using dynamic back-propagation method. Int. J. Electr. Power Energy Syst **22**, 29–34 (2000)
6. P. Shamsollahi, An adaptive power system stabilizer using online trained neural networks. IEEE Trans Energy Convers. **12**, 382–387 (1997)
7. M.A.-M. Abido, Adaptive tuning of power system stabilizers using radial basis function networks. Electr. Power Syst. Res. **49**, 21–29 (1999)
8. T. Senjyu, Recurrent neural network supplementary stabilization controller for automatic voltage regulator and governor. Electr. Power Components Syst. **31**, 693–707 (2003)
9. H.K. Ping, Studies of the improvement of probabilistic PSSs by using the single neuron model. Int. J. Electr. Power Energy Syst. **29**(3), 217–221 (2007)
10. S. Ray, A wide area measurement based neuro control for generation excitation systems. Eng. Appl. Artif. Intell. **22**(3), 473–481 (2009)
11. L.A. Zadeh, Fuzzy sets. Inf. Control. **8**(3), 338–353 (1956)
12. S. Pal, D. Mandal, Fuzzy logic and approximate reasoning: an overview. IETE J. Res. **37**, 548–559 (1991)
13. A.R. Tavakoli, A.R. Seifi, M.M. Arefi, Designing a selfconstructing fuzzy neural network controller for damping power system oscillations. Fuzzy Sets Syst. **356**, 63–76 (2019)
14. K. Eltag, M.S. Aslamx, R. Ullah, Dynamic stability enhancement using fuzzy PID control technology for power system. Int. J. Control. Autom. Syst. **17**(1), 234–242 (2019)
15. M. Farahani, S. Ganjefar, Intelligent power system stabilizer design using adaptive fuzzy sliding mode controller. Neurocomputing **226**, 135–144 (2017)
16. T.T. Lie, An adaptive fuzzy logic power system stabilizer. Electr. Power Syst. Res. **38**(1), 75–81 (1996)
17. M. Ramirez-Gonzalez, Self-tuned power system stabilizer based on a simple fuzzy logic controller. Electr. Power Components Syst. **38**(4), 407–423 (2010)
18. D.K. Sambariya, R. Prasad, Robust tuning of power system stabilizer for small signal stability enhancement using metaheuristic bat algorithm. Int. J. Electr. Power Energy Syst. **61**, 229–238 (2014)
19. H.M. Soliman, E.H.E. Bayoumi, M.F. Hassan, Power system stabilizer design for minimal overshoot and control constraint using swarm optimization. Electr. Power Components Syst. **37**(1), 111–126 (2008)
20. S. Kamalasadan, G.D. Swann, R. Yousefian, A novel systemcentric intelligent adaptive control architecture for power system stabilizer based on adaptive neural networks. IEEE Syst. J. **8**(4), 1074–1085 (2014)
21. D. Molina, G.K. Venayagamoorthy, J. Liang, R.G. Harley, Intelligent local area signals based damping of power system oscillations using virtual generators and approximate dynamic programming. IEEE Trans. Smart Grid **4**(1), 498–508 (2013)

22. M. Farahani, A multi-objective power system stabilizer. IEEE Trans. Power Syst. **28**(3), 2700–2707 (2013)
23. R. Hadidi, B. Jeyasurya, Reinforcement learning based real-time wide-area stabilizing control agents to enhance power system stability. IEEE Trans. Smart Grid **4**(1), 489–497 (2013)
24. S.R. Khuntia, S. Panda, ANFIS approach for SSSC controller design for the improvement of transient stability performance. Math. Comput. Model. **57**(1), 289–300 (2013)
25. A.G. Aissaoui, M. Abid, H. Abid, A. Tahour, A.K. Zeblah, A fuzzy logic controller for synchronous machine. J. Electr. Eng. Bratislava **58**(5), 285 (2007)
26. J.W. Park, R.G. Harley, G.K. Venayagamoorthy, Adaptivecritic-based optimal neurocontrol for synchronous generators in a power system using MLP/RBF neural networks. IEEE Trans. Ind. Appl. **39**(5), 1529–1540 (2003)
27. F.M.E.S. Fayez, Robust adaptive wavelet-neural-network sliding-mode speed control for a DSP-based PMSM drive system. J Power Electr (JPE) **10**(5), 50517 (2010)
28. K.C. Yu, S.P. Hsu, Y.H. Hung, Optimization of fuzzy controller of permanent magne synchronous motor. J. Appl. Sci. **7**(19), 272535 (2007)
29. E. Cetin, U. Oguz, H.S. Hasan, A neuro-fuzzy controller for speed control of a permanent magnet synchronous motor drive. Expert Syst. Appl. **34**(1), 65764 (2008)
30. J. Faiz, A. Azami, A. Keyhani, A. Proca, Closed-loop control stability for permanent magnet synchronous motor. Int. J. Electr. Power Energy Syst. **19**(5), 3317 (1997)
31. K. Murat, I.E. Hasan, Speed and current regulation of a permanent magnet synchronous motor via nonlinear and adaptive backstepping control. Math. Comput. Model. **53**(910), 201530 (2011)
32. Q. Wang, H. Yu, M. Wang, X. Qi, A novel adaptive NeuroControl approach for permanent magnet synchronous motor speed control. Energies **11**(9), 2355 (2018)
33. C.B. Butt, M.A. Rahman, Untrained artificial neuron-based speed control of interior permanent-magnet motor drives over extended operating speed range. IEEE Trans. Ind. Appl. **49**(3), 1146–1153 (2013)
34. X. Sun, L. Chen, H. Jiang, Z. Yang, J. Chen, W. Zhang, High performance control for a bearingless permanent-magnet synchronous motor using neural network inverse scheme plus internal model controllers. IEEE Trans. Ind. Electron. **63**(6), 3479–3488 (2016)
35. Y. Li, B. Zhang, X. Xu, Decoupling control for permanent magnet in-wheel motor using internal model control based on back-propagation neural network inverse system. Bull. Pol. Acad. Sci. Tech. Sci. **66**(6), 961 (2018)
36. Y. Li, B. Li, X. Xu, X. Sun, A nonlinear decoupling control approach using RBFNNI-based robust pole placement for a permanent magnet in-wheel motor. IEEE Access **6**, 1844–1854 (2018)
37. Z. Frijet, A. Zribi, M. Chtourou, Adaptive neural network internal model control for PMSM speed regulation. J. Electr. Syst. **14**(2), 118–126 (2018)
38. K. Chen, C. Huang, J. He, Fault detection, classification and location for transmission lines and distribution systems: a review on the methods. High Volt. **1**(1), 25–33 (2016)
39. A.A. Yusuff, A.A. Jimoh, J.L. Munda, Fault location in transmission lines based on stationary wavelet transform, determinant function feature and support vector regression. Electr. Power Syst. Res. **110**, 73–83 (2014)
40. Z. Liu, Z. Han, Y. Zhang, Q. Zhang, Multiwavelet packet entropy and its application in transmission line fault recognition and classification. IEEE Trans. Neural Netw. Learn. Syst. **25**(11), 2043–2052 (2014)
41. A. Swetapadma, A. Yadav, A novel single-ended fault location scheme for parallel transmission lines using k-nearest neighbor algorithm. Comput. Electr. Eng. **69**, 41–53 (2018)
42. A.L. da Silva Pessoa, M. Oleskovicz, P.E.T. Martins, A multi-stage methodology for fault location in radial distribution systems, in *2018 18th International Conference on Harmonics and Quality of Power (ICHQP)* (IEEE, 2018), pp. 1–6

43. A. Prasad, J.B. Edward, Importance of artificial neural networks for location of faults in transmission systems: a survey, in *2017 11th International Conference on Intelligent Systems and Control (ISCO)* (IEEE, 2017), pp. 357–362
44. Y.Q. Chen, O. Fink, G. Sansavini, Combined fault location and classification for power transmission lines fault diagnosis with integrated feature extraction. IEEE Trans. Ind. Electron. **65**(1), 561–569 (2018)
45. M.F. Guo, X.D. Zeng, D.Y. Chen, N.C. Yang, Deep learning-based earth fault detection using continuous wavelet transform and convolutional neural network in resonant grounding distribution systems. IEEE Sens. J. **18**(3), 1291–1300 (2018)
46. N.R. Babu, B.J. Mohan, Fault classification in power systems using EMD and SVM. Ain Shams Eng. J. **8**(2), 103–111 (2017)
47. F.I. Lozada, F.L. Quilumba, F.E. Prez, Fault detection and classification in transmission lines using wavelet transform and support vector machines. Revista Tecnica Energia **14** (2018)
48. H. Malik, R. Sharma, EMD and ANN based intelligent fault diagnosis model for transmission line. J. Intell. Fuzzy Syst. **32**(4), 3043–3050 (2017)
49. M. Dhimish, V. Holmes, B. Mehrdadi, M. Dales, Comparing Mamdani Sugeno fuzzy logic and RBF ANN network for PV fault detection. Renew. Energy **117**, 257–274 (2018)
50. J.M. Johnson, A. Yadav, Fault detection and classification technique for HVDC transmission lines using KNN, in *Information and Communication Technology for Sustainable Development*, (Springer, Singapore, 2018), pp. 245–253
51. N. Amruthnath, T. Gupta, Fault class prediction in unsupervised learning using model-based clustering approach, in *2018 International Conference on Information and Computer Technologies (ICICT)* (IEEE, 2018), pp. 5–12
52. A. Khaleghi, M.O. Sadegh, M. Ghazizadeh-Ahsaee, A.M. Rabori, Transient fault area location and fault classification for distribution systems based on wavelet transform and adaptive NeuroFuzzy inference system (ANFIS). Adv. Electr. Electron. Eng. **16**(2), 155–166 (2018)
53. R. Kumar, E. Koley, A. Yadav, A.S. Thoke, Fault classification of phase to phase fault in six phase transmission line using Haar wavelet and ANN, in *2014 International Conference on Signal Processing and Integrated Networks (SPIN)*, (IEEE, 2014), pp. 5–8
54. H. Livani, C.Y. Evrenosoglu, A machine learning and waveletbased fault location method for hybrid transmission lines. IEEE Trans. Smart Grid **5**(1), 51–59 (2014)
55. V. Malathi, N.S. Marimuthu, S. Baskar, Intelligent approaches using support vector machine and extreme learning machine for transmission line protection. Neurocomputing **73**(10), 2160–2167 (2010)
56. S.R. Samantaray, P.K. Dash, G. Panda, Distance relaying for transmission line using support vector machine and radial basis function neural network. Int. J. Electr. Power Energy Syst. **29**, 551556 (2007)
57. J.A. Jiang, C.L. Chuang, Y.C. Wang, C.H. Hung, J.Y. Wang, C.H. Lee, Y.T. Hsiao, A hybrid framework for fault detection, classification, and location. Part I: concept, structure, and methodology. IEEE Trans. Power Deliv. **26**(3), 1988–1998 (2011)
58. J.-A. Jiang, C.-L. Chuang, Y.-C. Wang, C.-H. Hung, J.-Y. Wang, C.-H. Lee, Y.-T. Hsiao, A hybrid framework for fault detection, classification, and locationpart II: implementation and test results. IEEE Trans. Power Deliv. **26**(3), 1999–2008 (2011)
59. M.T. Hagh, K. Razi, H. Taghizadeh, Fault classification and location of power transmission lines using artificial neural network, in *IPEC 2007. International Power Engineering Conference* (IEEE, 2007), pp. 1109–1114
60. P.K. Dash, S.R. Samantaray, An accurate fault classification algorithm using a minimal radial basis function neural network. Eng. Intell. Syst. **4**, 205–210 (2004)
61. J. Upendar, C.P. Gupta, G.K. Singh, Discrete wavelet transform and probabilistic neural network based algorithm for classification of fault on transmission systems, in *INDICON 2008. Annual IEEE India Conference*, vol. 1 (IEEE, 2008), pp. 206–211
62. S.W. Fei, X.B. Zhang, Fault diagnosis of power transformer based on support vector machine with genetic algorithm. Expert Syst. Appl. **36**(8), 11352–11357 (2009)

63. A. Shintemirov, W. Tang, Q.H. Wu, Power transformer fault classification based on dissolved gas analysis by implementing bootstrap and genetic programming. IEEE Trans. Syst. Man Cybernet. C Appl. Rev. **39**(1), 69–79 (2009)
64. A.M. Shah, B.R. Bhalja, Discrimination between internal faults and other disturbances in transformer using the support vector machine based protection scheme. IEEE Trans. Power Deliv. **28**(3), 1508–1515 (2013)
65. L. Liu, Z. Ding, Modeling analysis of power transformer fault diagnosis based on improved relevance vector machine. Math. Prob. Eng. **2013**, 636374 (2013)
66. T.S. Sidhu, H. Singh, M.S. Sachdev, Design, implementation and testing of an artificial neural network based fault direction discriminator for protecting transmission lines. IEEE Trans. Power Deliv. **10**(2), 697–706 (1995)
67. Y. Cui, P. Bangalore, L.B. Tjernberg, An anomaly detection approach using wavelet transform and artificial neural networks for condition monitoring of wind turbines' gearboxes, in *2018 Power Systems Computation Conference (PSCC)* (IEEE, 2018), pp. 1–7
68. A.R. Bahmanyar, A. Karami, Power system voltage stability monitoring using artificial neural networks with a reduced set of inputs. Int. J. Electr. Power Energy Syst. **58**, 246–256 (2014)
69. D.Q. Zhou, U.D. Annakkage, A.D. Rajapakse, Online monitoring of voltage stability margin using an artificial neural network. IEEE Trans. Power Syst. **25**(3), 1566–1574 (2010)
70. D.V. Coury, D.C. Jorge, Artificial neural network approach to distance protection of transmission lines. IEEE Trans. Power Deliv. **13**(1), 102–108 (1998)
71. J. Pihler, B. Grar, D. Dolinar, Improved operation of power transformer protection using artificial neural network. IEEE Trans. Power Deliv. **12**(3), 1128–1136 (1997)
72. L. Hernandez, C. Baladrn, J.M. Aguiar, B. Carro, A.J. SanchezEsguevillas, J. Lloret, Short-term load forecasting for microgrids based on artificial neural networks. Energies **6**(3), 1385–1408 (2013)
73. C.N. Ko, C.M. Lee, Short-term load forecasting using SVR (support vector regression)-based radial basis function neural network with dual extended Kalman filter. Energy **49**, 413–422 (2013)
74. Z.A. Bashir, M.E. El-Hawary, Applying wavelets to short-term load forecasting using PSO-based neural networks. IEEE Trans. Power Syst. **24**(1), 20–27 (2009)
75. P.F. Pai, W.C. Hong, Forecasting regional electricity load based on recurrent support vector machines with genetic algorithms. Electr. Power Syst. Res. **74**(3), 417–425 (2005)
76. D.K. Chaturvedi, A.P. Sinha, O.P. Malik, Short term load forecast using fuzzy logic and wavelet transform integrated generalized neural network. Int. J. Electr. Power Energy Syst. **67**, 230–237 (2015)
77. V.N. Coelho, I.M. Coelho, B.N. Coelho, A.J. Reis, R. Enayatifar, M.J. Souza, F.G. Guimares, A self-adaptive evolutionary fuzzy model for load forecasting problems on smart grid environment. Appl. Energy **169**, 567–584 (2016)
78. V.L. Paucar, M.J. Rider, Artificial neural networks for solving the power flow problem in electric power systems. Electr. Power Syst. Res. **62**(2), 139–144 (2002)
79. A. Arunagiri, B. Venkatesh, K. Ramasamy, Artificial neural network approach-an application to radial loadflow algorithm. IEICE Electron. Exp. **3**(14), 353–360 (2006)
80. A. Jain, S.C. Tripathy, R. Balasubramanian, K. Grag, Y. Kawazoe, Neural network based stochastic load flow analysis, in *PowerCon 2004. 2004 International Conference on Power System Technology, 2004*, vol. 2 (IEEE, 2004), pp. 1845–1850
81. A. Karami, M.S. Mohammadi, Radial basis function neural network for power system load-flow. Int. J. Electr. Power Energy Syst. **30**(1), 60–66 (2008)
82. Y.F. Huang, S. Werner, J. Huang, N. Kashyap, V. Gupta, State estimation in electric power grids: meeting new challenges presented by the requirements of the future grid. IEEE Sign. Process. Mag. **29**(5), 33–43 (2012)
83. D.V. Kumar, S.C. Srivastava, S. Shah, S. Mathur, Topology processing and static state estimation using artificial neural networks, in *IEE Proceedings-Generation, Transmission and Distribution*, vol. 143, no. 1 (IET, 1996), pp. 99–105

84. D. Singh, J.P. Pandey, D.S. Chauhan, Radial basis neural network state estimation of electric power networks, in *Proceedings of the 2004 IEEE International Conference on Electric Utility Deregulation, Restructuring and Power Technologies, 2004*. (DRPT 2004), vol. 1 (IEEE. 2004), pp. 90–95
85. D. Singh, J.P. Pandey, D.S. Chauhan, Topology identification, bad data processing, and state estimation using fuzzy pattern matching. IEEE Trans. Power Syst. **20**(3), 1570–1579 (2005)
86. F. Larki, M. Joorabian, H.M. Kelk, M. Pishvaei, Voltage stability evaluation of the Khouzestan power system in Iran using CPF method and modal analysis, In *2010 Asia-Pacific Power and Energy Engineering Conference (APPEEC)* (IEEE, 2010), pp. 1–5
87. A. Zhukov, N. Tomin, D. Sidorov, D. Panasetsky, V. Spirayev, A hybrid artificial neural network for voltage security evaluation in a power system, in *2015 5th International Youth Conference on Energy (IYCE)* (IEEE. 2015), pp. 1–8
88. H.H. Goh, Q.S. Chua, S.W. Lee, B.C. Kok, K.C. Goh, K.T.K. Teo, Evaluation for voltage stability indices in power system using artificial neural network. Proc. Eng. **118**, 1127–1136 (2015)
89. M.V. Suganyadevi, C.K. Babulal, S. Kalyani, Assessment of voltage stability margin by comparing various support vector regression models. Soft. Comput. **20**(2), 807–818 (2016)
90. M.B. Cain, R.P. Oneill, A. Castillo, *History of Optimal Power Flow and Formulations* (Federal Energy Regulatory Commission, Washington, DC, 2012)
91. L. Gan, N. Li, U. Topcu, S.H. Low, Optimal power flow in distribution networks, in *Proceeding of 52nd IEEE conference on decision and control*, December 2013
92. T.T. Nguyen, Neural network optimal-power-flow, in *Fourth International Conference on Advances in Power System Control, Operation and Management*, 1997, pp. 266–271
93. R.S. Hartati, M.E. El-Hawary, Optimal active power flow solutions using a modified Hopfield neural network, in *2001 Canadian Conference on Electrical and Computer Engineering*, vol. 1 (IEEE. 2001), pp. 189–194
94. M. Mohatram, P. Tewari, N. Latanath, 2011, April Economic load flow using Lagrange neural network, in *2011 Saudi International Electronics, Communications and Photonics Conference (SIECPC)*, (IEEE, 2011), pp. 1–7
95. J.H. Park, Y.S. Kim, I.K. Eom, K.Y. Lee, Economic load dispatch for piecewise quadratic cost function using Hopfield neural network. IEEE Trans. Power Syst. **8**(3), 1030–1038 (1993)
96. S.S. Reddy, J.A. Momoh, Economic dispatch using improved Hopfield neural network, in *2015 North American Power Symposium (NAPS)*, (IEEE, Chicago, 2015), pp. 1–5
97. V.N. Dieu, P. Schegner, Augmented Lagrange Hopfield network initialized by quadratic programming for economic dispatch with piecewise quadratic cost functions and prohibited zones. Appl. Soft Comput. **13**(1), 292–301 (2013)
98. E.M. Voumvoulakis, N.D. Hatziargyriou, A particle swarm optimization method for power system dynamic security control. IEEE Trans. Power Syst. **25**(2), 1032–1041 (2010)
99. C.F. Kucuktezcan, V.I. Genc, A new dynamic security enhancement method via genetic algorithms integrated with neural network based tools. Electr. Power Syst. Res. **83**(1), 1–8 (2012)
100. I. Genc, R. Diao, V. Vittal, S. Kolluri, S. Mandal, Decision tree-based preventive and corrective control applications for dynamic security enhancement in power systems. IEEE Trans. Power Syst. **25**(3), 1611–1619 (2010)
101. S. Ghosh, B.H. Chowdhury, Security-constrained optimal rescheduling of real power using Hopfield neural network. IEEE Trans. Power Syst. **11**(4), 1743–1748 (1996)
102. G. Crtina, C. Bonciu, M. Musat, Z. Zisman, Application of fuzzy logic for energy loss reduction in distribution networks, in *9th Mediterranean Electrotechnical Conference, 1998. MELECON 98*, vol. 2 (IEEE, 1998), pp. 974–977
103. D. Villacci, G. Bontempi, A. Vaccaro, An adaptive local learning-based methodology for voltage regulation in distribution networks with dispersed generation. IEEE Trans. Power Syst. **21**(3), 1131–1140 (2006)

104. R. Nayak, J.D. Sharma, A hybrid neural network and simulated annealing approach to the unit commitment problem. Comput. Electr. Eng. **26**(6), 461–477 (2000)
105. H. Sasaki, M. Watanabe, D. Kubokawa, N. Yorino, R. Yokoyama, A solution method of unit commitment by artificial neural networks. IEEE Trans. Power Syst. **7**(3), 974–981 (1992)
106. K. Yan, Y. Du, Z. Ren, MPPT perturbation optimization of photovoltaic power systems based on solar irradiance data classification. IEEE Trans. Sustain. Energy **10**, 514–521 (2018)
107. H. Li, Z. Liu, Performance prediction and optimization of solar water heater via a knowledge-based machine learning method, in *Handbook of Research on Power and Energy System Optimization*, (IGI Global, Hershey, 2018), pp. 55–74
108. E. Mocanu, D.C. Mocanu, P.H. Nguyen, A. Liotta, M.E. Webber, M. Gibescu, J.G. Slootweg, On-line building energy optimization using deep reinforcement learning. IEEE Trans. Smart Grid **10**(4), 3698–3708 (2018)
109. C. Fu, G.Q. Li, K.P. Lin, H.J. Zhang, Short-term wind power prediction based on improved chicken algorithm optimization support vector machine. Sustainability **11**(2), 512 (2019)
110. T.C. Bora, V.C. Mariani, L. dos Santos Coelho, Multiobjective optimization of the environmental-economic dispatch with reinforcement learning based on non-dominate sorting genetic algorithm. Appl. Therm. Eng. **146**, 688–700 (2019)

Index

M. Nazari-Heris et al. (eds.), *Application of Machine Learning and Deep Learning Methods to Power System Problems*, Power Systems,
https://doi.org/10.1007/978-3-030-77696-1

E

I

K

L

GPSR Compliance
The European Union's (EU) General Product Safety Regulation (GPSR) is a set of rules that requires consumer products to be safe and our obligations to ensure this.

If you have any concerns about our products, you can contact us on

ProductSafety@springernature.com

In case Publisher is established outside the EU, the EU authorized representative is:

Springer Nature Customer Service Center GmbH
Europaplatz 3
69115 Heidelberg, Germany

www.ingramcontent.com/pod-product-compliance
Ingram Content Group UK Ltd.
Pitfield, Milton Keynes, MK11 3LW, UK
UKHW021009290726
14059UKWH00001BA/31